Aeglidae

Life History and Conservation Status of Unique Freshwater Anomuran Decapods

Advances in Crustacean Research

Ingo S. Wehrtmann
University of Costa Rica, San Jose

The *Advances in Crustacean Research* series publishes internationally significant contributions to the biology of Crustacea. The thematic focus of individual volumes includes particular aspects from various fields of research, such as molecular biology, comparative morphology, developmental biology, systematics, phylogenetics, natural history, evolution, palaeontology, zoogeography conservation biology, (eco-) physiology, ecology, extreme environments, behavioural biology, and fisheries and aquaculture.

Crayfish in Europe as Alien Species
edited by Francesca Gherardi and David M. Holdich

The Biodiversity Crisis and Crustacea - Proceedings of the Fourth International Crustacean Congress
edited by J. Carel von Vaupel Klein

Isopod Systematics and Evolution
edited by Richard C. Brusca

Evolutionary Developmental Biology of Crustacea
edited by Gerhard Scholtz

Crustacea and Arthropod Relationships
edited by Stefan Koenemann and Ronald Jenner

The Biology and Fisheries of the Slipper Lobster
edited by Kari L. Lavalli, Ehud Spanier

Decapod Crustacean Phylogenetics
edited by Joel W. Martin, Keith A. Crandall, and Darryl L. Felder

Phylogeography and Population Genetics in Crustacea
edited by Christoph Held, Stefan Koenemann, and Christoph D. Schubart

The Biology of Squat Lobsters
edited by Gary C.B. Poore, Shane T. Ahyong, and Joanne Taylor

Aeglidae: *Life History and Conservation Status of Unique Freshwater Anomuran Decapods*
edited by Sandro Santos and Sergio Luiz de Siqueira Bueno

For more information about this series, please visit: https://www.crcpress.com/Advances-in-Crustacean-Research/book-series/CRCCRUSTARES

Aeglidae

Life History and Conservation Status of Unique Freshwater Anomuran Decapods

Edited by

Sandro Santos and Sergio Luiz de Siqueria Bueno

CRC Press
Taylor & Francis Group
Boca Raton London New York

CRC Press is an imprint of the
Taylor & Francis Group, an **informa** business

CRC Press
Taylor & Francis Group
6000 Broken Sound Parkway NW, Suite 300
Boca Raton, FL 33487-2742

First issued in paperback 2022

ISBN-13: 978-1-138-29472-1 (hbk)
ISBN-13: 978-1-03-233792-0 (pbk)
DOI: 10.1201/b22343

Library of Congress Cataloging-in-Publication Data

Names: Santos, Sandro, editor.
Title: Aeglidae : life history and conservation status of unique freshwater anomuran decapods / edited by Sandro Santos and Sergio Luiz de Siqueira Bueno.
Description: Boca Raton : Taylor & Francis, [2020] | Includes bibliographical references and index. | Summary: "Aeglidae focuses on these unique crustaceans who are endemic to South America. The book is the first to summarize the diverse aspects of the Aeglidae, whose taxonomic features and phylogenetic relationships, evolutionary history and biogeographical background, biological characteristics, and current conservation awareness make them stand out among all other decapods. This will be an important reference not only for carcinologists working with this family of decapods, but also readers interested in the evolution, biogeography, taxonomy, phylogenetics, physiology, and reproductive ecology"-- Provided by publisher.
Identifiers: LCCN 2019025182 (print) | LCCN 2019025183 (ebook) | ISBN 9781138294721 (hardback) | ISBN 9781315100937 (ebook)
Subjects: LCSH: Aeglidae.
Classification: LCC QL444.M33 A34 2020 (print) | LCC QL444.M33 (ebook) | DDC 595.3--dc23
LC record available at https://lccn.loc.gov/2019025182
LC ebook record available at https://lccn.loc.gov/2019025183

Visit the Taylor & Francis Web site at
http://www.taylorandfrancis.com

and the CRC Press Web site at
http://www.crcpress.com

Contents

John Campbell McNamara and Samuel Coelho Faria

Harry Boos, Paula Guimarães Salge, and Marcelo A. A. Pinheiro

Roberto Munehisa Shimizu and Sergio Luiz de Siqueira Bueno

Preface

A preface to a unique taxon: *Aegla* Leach, 1820, a crown jewel among South American freshwater decapods.

This book is about one single taxon: *Aegla* Leach, 1820. And what a remarkable taxon it is! Those who had—and those who are having—the experience of studying these unique freshwater decapods could not agree more with Schmitt's remarks written down on the first page (Schmitt, 1942; p. 431) in his seminal monography on aeglids: "There are no freshwater Crustacea at all like *Aegla* anywhere else in the world."

The production of this book comes in a special moment because we find ourselves at the brink of celebrating 200 years since the first taxonomic entry of an extant aeglid, as *Galathea laevis*, in the scientific literature. Over these two centuries, hundreds of investigations on *Aegla* have been published. A brief search for Aeglidae on Google Scholar, for example, retrieves more than 1,600 entries.

Aegla is the only taxon within the Anomura whose representatives are entirely adapted to the freshwater environment. As of 2018, there are now 87 known valid species, all endemic to subtropical and temperate South America. This figure makes *Aegla* the most species-rich genus of all true freshwater decapods in this subcontinent. The tally is certainly bound to go up considerably as putative new species are being recognized and still waiting for the necessary formal description (Chapter 1), and as unexplored or poorly explored areas within the known range of distribution continues to be systematically investigated. It is only reasonable to expect that the number of valid species may soon surpass the barrier of 100 species within the next few years ahead.

This book is also about perhaps the most endangered freshwater decapod in the Neotropical Region (Chapter 9). About 70% of the 87 known species are currently threatened with extinction, having been assessed as critically endangered, endangered, or vulnerable threatened categories, as defined by the International Union for Nature Conservation. The main threats to aeglids include the removal of riparian forest, habitat fragmentation and destruction, industrial, agricultural, livestock, and domestic pollution of the water bodies.

One unique feature about *Aegla* is the fact that its evolutionary history can be told based on sound scientific evidence, starting from marine fossil representatives to the successful adaptation of *Aegla* to freshwater habitats and the subsequent dispersal routes through paleobasins of continental South America that neatly explain the distributional pattern we see today (Chapter 1). The successful adaptation to freshwater environments demanded the acquisition of adaptive life history strategies, most importantly those regarding physiological ecology (Chapter 8), postembryonic development and parental care (Chapter 6).

Morphological studies have been a strong line of investigation starting right from the beginning. Schmitt's monography (1942) may still be the most revered landmark publication on the taxonomy of *Aegla*, but other South American leading investigators have published several equally important papers on this topic since the 1980s (see Chapters 1 and 2 for references therein). Together, this bulk of publications on aeglid taxonomy has provided a great contribution to the knowledge of *Aegla* distribution and diversity. More recently, molecular analyses have made a huge impact in systematic studies of aeglids, providing valuable insights and hypotheses regarding the phylogenetic relationships among *Aegla* species as well as the phylogenetic position of the family Aeglidae within the Anomura (Chapter 1).

Throughout the pages of this book, the reader will also have the opportunity to check out fine compilations on topics such as population structure and maturity (Chapter 3), trophic ecology (Chapter 4) as well as reproduction and gonadal development (Chapter 5) and behavior (Chapter 7). Finally, Chapter 10 deals with sampling techniques, handling procedures, and provides a discussion on analytical treatments of data obtained under field working conditions.

For us, the editing experience involved in the production of this book has been a quite extraordinary one. We are really grateful to our colleagues Dr. Ingo Wehrtmann and Dr. Célio Magalhães for having invited us to carry out this task, which we humbly accepted without hesitation. Also, we wish to demonstrate our gratitude to all who have directly or indirectly contributed to this book. We thank all authors of the chapters: Alexandre V. Palaoro, Bianca Laís Zimmermann, Carlos G. Jara, Carolina Sokolowicz, Georgina Bond-Buckup, Harry Boos, John Campbell McNamara, Juliana Cristina Bertacini Moraes, Keith A. Crandall, Marcelo A. A. Pinheiro, Marcelo M. Dalosto, Marcos Pérez-Losada, Marlise Ladvocat Bartholomei-Santos, Pablo Collins, Paula Guimarães Salge, Roberto Munehisa Shimizu, Samuel Coelho Faria, and Setuko Masunari. We are also especially grateful to the researchers who kindly collaborated with us reviewing the chapters: Antônio Leão Castilho, Christopher Tudge, Ingo Wehrtmann, Marlise L. Bartholomei-Santos, Marcos Tavares, Neil Cumberlidge, Roberto Shimizu, and Rodney Feldmann.

Sandro Santos and Sérgio Bueno
May 2019

REFERENCE

Schmitt, W. 1942. The species of *Aegla*, endemic South American freshwater crustaceans. *Proceedings of the United States National Museum* 91:431–520.

Editors

Sandro Santos is a full professor of the Department of Ecology and Evolution at the Universidade Federal de Santa Maria, Brazil. He holds a Ph.D. in zoology from Universidade Estadual Paulista (UNESP), a post-doctorate in systematics of Aeglidae from Universidade Federal do Rio Grande do Sul (UFRGS) and phylogeography of Aeglidae from George Washington University (GWU). For twenty years he has been studying aeglid crustaceans, focusing on population structure, reproductive biology, behavioral ecology, systematics, and conservation.

Sérgio Luiz de Siqueira Bueno is associate professor at the Department of Zoology, Institute of Bioscience, University of São Paulo, Brazil. He has dedicated the past 15 years of his research career to the study of aeglids. These studies are focused on the relative growth, reproduction, life cycle, postembryonic development, taxonomy, estimation of population size, and conservation of these remarkable freshwater decapods.

List of Contributors

Marlise Ladvocat Bartholomei-Santos
Departamento de Ecologia e Evolução
Universidade Federal de Santa Maria
Santa Maria, Brazil

Georgina Bond-Buckup
Departamento de Zoologia
Universidade Federal do Rio Grande
    do Sul
Porto Alegre, Brazil

Harry Boos
Instituto Chico Mendes de Conservação
    da Biodiversidade
Itajaí, Brazil

Pablo Collins
Facultad de Humanidades y Ciencias
Universidad Nacional del Litoral
Santa Fé, Argentina

Keith A. Crandall
Computational Biology Institute
George Washington University
Ashburn, Virginia

Marcelo Dalosto
Departamento de Ecologia e Evolução
Universidade Federal de Santa Maria
Santa Maria, Brazil

Samuel Faria
Departamento de Fisiologia Geral
Instituto de Biociências
Universidade de São Paulo
São Paulo, Brazil

Carlos G. Jara
Instituto de Zoología
Universidad Austral de Chile
Valdivia, Chile

John Campbell MacNamara
Departamento de Biologia - Faculdade
    de Filosofia, Ciências e Letras
Universidade de São Paulo
Ribeirão Preto, Brazil

Setuko Masunari
Departamento de Zoologia
Universidade Federal do Paraná
Curitiba, Brazil

Juliana Cristina Bertacini Moraes
Departamento de Zoologia
Universidade de São Paulo
São Paulo, Brazil

Alexandre Palaoro
Departamento de Ecologia
Universidade de São Paulo
São Paulo, Brazil

Marcos Pérez-Losada
Computational Biology Institute
George Washington University
Ashburn, Virginia

Marcelo A. A. Pinheiro
Instituto de Biociências
Universidade Estadual Paulista
São Vicente, Brazil

Paula Guimarães Salge
Instituto Chico Mendes de Conservação
    da Biodiversidade
Itajaí, Brazil

Sandro Santos
Departamento de Ecologia e Evolução
Universidade Federal de Santa Maria
Santa Maria, Brazil

Roberto Munehisa Shimizu
Departamento de Ecologia
Universidade de São Paulo
São Paulo, Brazil

Sérgio Luiz de Siqueira Bueno
Departamento de Zoology
Universidade de São Paulo
São Paulo, Brazil

Carolina C. Sokolowicz
Departamento de Zoologia
Universidade Federal do Rio Grande
    do Sul
Porto Alegre, Brazil

Bianca Laís Zimmermann
Instituto Federal de Educação Ciência e
    Tecnologia
Ibirubá, Brazil

CHAPTER 1

Evolutionary History and Phylogenetic Relationships of Aeglidae

Marlise Ladvocat Bartholomei-Santos, Sandro Santos, Bianca Laís Zimmermann, Marcos Pérez-Losada, and Keith A. Crandall

CONTENTS

1.1 EVOLUTIONARY HISTORY: MARINE ANCESTORS AND CONTINENTAL INVASION

The family Aeglidae Dana, 1852 has had a fairly successful history of diversification within South American fresh waters. The family comprises only one extant genus, *Aegla*, with 87 known species (Bueno et al. 2017; Santos et al. 2017; Jara et al. 2018; Páez et al. 2018) and many others to be described (Crivellaro et al. 2017; Zimmermann et al. 2018). But how and when did this history start? Why has this anomuran group succeeded in fresh water? Although we still do not have complete answers for many aspects of Aeglidae evolution, and numerous points remain to be elucidated, several studies have shed light on many of these issues.

The family Aeglidae belongs to the infraorder Anomura, which has succeeded in colonizing a variety of ecosystems, including marine, brackish, terrestrial, freshwater, and hydrothermal vent habitats (Bracken-Grissom et al. 2013). *Aegla* has its life cycle entirely restricted to freshwater environments. Beyond aeglids, only a single species of Diogenidae, the hermit crab *Clibanarius fonticola*, is known to permanently inhabit freshwater habitats (McLaughlin and Murray 1990).

Currently, the family Aeglidae is distributed across rivers, streams, and lakes in multiple basins flowing to both the Atlantic and Pacific coasts of southern South America, between the latitudes of 20°18′S in Brazil (Bueno et al. 2007) and 50°34′S in Chile (Oyanedel et al. 2011). Although the extant aeglids are all living in fresh water, the family originated in the marine environment, as revealed by two fossil species of Aeglidae (Feldmann 1984; Feldmann et al. 1998).

Before the description of the first fossil member of the family Aeglidae (see Feldmann 1984), some naturalists had already speculated about its origin and dispersion within the continent. Schmitt (1942, p. 443) stated that "the marine origin of *Aegla* appears indisputable." Mentioning that less ornamented species could be more primitive, Schmitt (1942) hypothesized that *Aegla jujuyana* would be the closest taxon to the *Aegla* ancestor. This would place the center of distribution of the genus in the Province of Jujuy, in northwestern Argentina. From there, variants with the "Pacific type of rostrum" (flatter and troughed) would have spread out westward to the Andes and Chile and eastward to the Serra do Mar in Brazil, while the forms with the "Atlantic type of rostrum" (spine-like and ridge-roofed) dispersed throughout the Paraná River and Uruguay River basins. Based on information provided by an article about the geology of South America (Berry 1922), Schmitt (1942) also speculated that since Jujuy had a marine history, with marine deposits antedating the Devonian, up to the Carboniferous, the gradual elevation of the land above the sea level allowed the *Aegla* ancestors to adapt progressively to less and less salty and increasingly fresh water; from the Cretaceous, the Jujuy region would have been totally continental, and its waters would be no longer marine. *Aegla franca* (found in São Paulo state, Brazil) would not fit into this scenario due to its intermediate type of rostrum, more similar to that of *A. jujuyana*. Thus, the first species could be a northeastern offshoot of the ancestral or original *jujuyana* stock (Schmitt 1942). As the opening of the South Atlantic Ocean is accepted to have occurred progressively from south to north starting in the early Jurassic (reviewed in Seton et al. 2012), one can infer that if Schmitt (1942) was correct in his speculations, the aeglid ancestor would have come from the Pacific Ocean occupying the Jujuy region in Argentina in the past; however, he did not mention either a Pacific or an Atlantic origin for Aeglidae in his monograph. Ringuelet (1948) disagreed with Schmitt's view and pointed out that accepting Jujuy as the dispersion center would require descending to the Paleozoic to find marine sediments, and that would be much too old to find the marine ancestors of the extant aeglids. In fact, the oldest known anomuran fossil dates from the Triassic (Chablais et al. 2011). It is worthy to note that the Atlantic and Pacific types of rostra identified by Schmitt (1942) occur in both South American coasts among the currently described species. Hence, Schmitt's classification of aeglids based on rostra does not hold. Actually, Ringuelet (1948, 1949a) subsequently disagreed with Schmitt observing that there are different intermediate types of rostra. Moreover, he hypothesized that the most primitive species of *Aegla* would have a prominent rostrum, with wide extra-orbital sinus; from this type, there would be two possible evolutionary scenarios, one leading to species with an elevated (carinated), but short rostrum, with somewhat obtuse, somewhat excavated carina and a narrow extra-orbital sinus; or another leading to species with short, depressed rostra (non carinated) with a narrow extra-orbital sinus. Both scenarios would lead to the complete disappearance of the orbital spines and extra-orbital sinuses. Based on this idea, Ringuelet (1948, 1949a) also rejected the primitive status of *Aegla jujuyana* proposed by Schmitt (1942).

Forty years before Schmitt (1942), Ortmann (1902) had already hypothesized a possible Pacific origin for aeglids by comparing the distributions of *Aegla* and *Parastacus* Huxley, 1879, in South America. He speculated that as the family Parastacidae was present in Australia in the Upper Cretaceous, it could have spread into Antarctica and southern Chile, and in the early Tertiary into Northern Argentina and Southern Brazil. Since the genus *Aegla* had a close distribution to *Parastacus*, the pathway could have been similar for aeglids, although Ortmann (1902) believed that an Antarctic origin was improbable for the latter group, and did not discard the possibility of the inverse path, that is, from the Atlantic to the Pacific side. Moreover, he pointed out that the presence of the genus *Aegla* (as well as *Parastacus*) on both sides of the Andean Cordillera indicates that this distribution predates the complete uplift of the Andes, as the mountain chain would act as a barrier for their dispersion. Schmitt (1942) partially disagreed with this idea, considering that *Aegla* might not have had a wide distribution before the Andes reached their present height because passages in the Chilean and Argentinean lake regions could have allowed dispersion. Later, by using a panbiogeographic analysis, Morrone and Lopretto (1994) suggested a single generalized track oriented from northeast to southwest, indicating the pre-existence of ancestral biotas, based on the congruence of individual tracks for three freshwater decapod groups (Aeglidae, Parastacidae, and Trichodactylidae). Their analysis identified *Aegla uruguayana* from the Atlantic side as the most primitive species.

The discovery of a fossil member of the family Aeglidae, *Haumuriaegla glaessneri*, from marine sediments in New Zealand corresponding to the Haumurian stage in the Late Cretaceous (Feldmann 1984), has definitively supported the marine origin for the group. Feldmann (1984) suggested that the family could have originated in the Indo-Pacific region and dispersed eastward, reaching South America. The dispersion possibly occurred before the end of the Oligocene, preceding the separation of Australia and New Zealand from Antarctica and the development of the circum-Antarctic current system (Feldmann 1986). The circumpolar current isolated Antarctica, blocking the heat transfer from the low latitudes and allowing glaciation to develop (reviewed in Martin 2006), contributing to the isolation of New Zealand from South America in post-Oligocene time (Feldmann 1986). Although the extant aeglids do not present a larval phase, a possible adaptation to a freshwater life cycle (Pérez-Losada et al. 2002a; McLaughlin et al. 2007), their marine ancestors may have spread through larval dispersal in the main counter-clockwise gyre of ocean circulation in the southern Pacific Ocean (Feldmann 1986). An even older fossil from a member of the family Aeglidae, dating from the Albian stage in the Early Cretaceous, was found in Mexico (Feldmann et al. 1998). The description of *Protaegla miniscula* (Feldmann et al. 1998) not only supported the hypothesis of a Pacific origin for the group but also added a third genus to the family Aeglidae. Therefore, we currently accept two marine fossil genera (*Haumuriaegla* and *Protaegla*) and one extant freshwater genus (*Aegla*) for the aeglids. The oldest known anomuran fossil is from the Late Triassic (Chablais et al. 2011), and most anomuran superfamilies were already present in the Jurassic fossil record, so the origin of the family Aeglidae could be earlier than its oldest known fossil (Pérez-Losada et al. 2004).

At the transition between the Cretaceous and the Paleogene, many taxa experienced a mass extinction; a phenomenon commonly referred to as Cretaceous-Paleogene (K/Pg) event (Renne et al. 2013). The K/Pg event does not seem to have affected all taxa similarly, since some decapod families show high survival rates across the K/Pg boundary (Schweitzer and Feldmann 2005). The geographic distribution was an essential factor to cross the boundary, and decapod genera inhabiting temperate and high-latitude areas had higher survival rates than lower latitude genera (Schweitzer and Feldmann 2005). Although the two aeglid fossil genera apparently did not cross the boundary, the family survived into the present. Schweitzer and Feldmann (2005) hypothesized that the family Aeglidae could have survived because it was either a refugium taxon or inhabited a buffered habitat. Refugia taxa could migrate to secondary habitats not as impacted by the event causing the mass extinction; most of them represent species or their descendants that have been forwarded to more restrained habitats by competitive displacement from their marine environments (Harries et al. 1993). On the other hand, buffered habitats were not greatly disturbed by the mechanisms causing the mass extinction, as could be the case in temperate and high-latitude regions (Harries et al. 1993). The marine ancestor of the extant aeglids could have originated in the high southern latitudes during the Cretaceous (Feldmann and Schweitzer 2006) or earlier.

An alternative way to investigate the question of a Pacific or Atlantic origin of Aeglidae is by means of a phylogeny of the extant species. The basal taxa would be the first ones to diverge, and their area of occurrence (if in the Atlantic or the Pacific side) would indicate the possible path by which a marine ancestral might have invaded the continental waters. However, the conservative general morphotype and the low number of shared apomorphic characters relative to the large number of extant species (Bond-Buckup and Buckup 1994), coupled with the presence of homoplasic characters, present challenges to building a morphology-based phylogeny for Aeglidae. In situations like this, molecular approaches are very useful to elucidate the evolutionary history of a group of organisms by increasing the number of characters available for analysis. Initially, a molecular phylogeny was built using four mitochondrial genes from 17 Chilean aeglid species, two trans-Andean *Aegla* species collected in Argentina, with one galatheid and one porcellanid species used as outgroups (Pérez-Losada et al. 2002a). *Aegla papudo* from Chile stood in a basal position, as the sister group of the other *Aegla* species in the phylogeny, supporting a Pacific origin for Aeglidae. The basal position of *A. papudo* within the aeglids was also confirmed by constructing phylogenies using four mitochondrial and two nuclear genes and five anomuran members as outgroups (Lomisidae, Porcellanidae, Chirostylidae, Galatheidae, and Paguroidea), with three different approaches: Maximum Likelihood (ML), Maximum Parsimony, and Bayesian (BA) analyses (Pérez-Losada et al. 2004).

Due to computational limitations, Pérez-Losada et al. (2004) used *A. papudo* as a functional outgroup to investigate the phylogenetic relationships among 58 out of 63 known *Aegla* species at the time and six undescribed species. Their ML and BA consensus trees placed *A. ringueleti* as the most basal ingroup taxon and *A. scamosa* in the next most basal position. The remaining taxa, except *A. marginata* and *A. spinipalma*, were clustered into five major clades, named as clades A, B, C, D, and E. Clade A was the first to diverge, followed by clade B, both encompassing species from Chile and southern Argentina. The tree splitting pattern suggested an eastward radiation for Aeglidae. An alternative westward radiation hypothesis or two independent radiations in both directions were also tested and rejected by the authors. Thus, Pérez-Losada et al. (2004) supported the Pacific origin hypothesis, in agreement with fossil evidence (Feldmann 1984; Feldmann et al. 1998). Moreover, the age of the most basal node in the ML tree was estimated as 74 million years ago (Mya), also in accordance with the fossil data, which indicated that the *Aegla* relatives were at least 75 My old (Pérez-Losada et al. 2004).

By integrating the history of South American drainages and phylogenetics, Pérez-Losada et al. (2004) proposed a phylogeographic hypothesis to explain the Aeglidae radiation. The species arrangements in the phylogenetic trees for the genus *Aegla* do not perfectly match the current bifurcating pattern of river systems where these species occur, which was not surprising since the present drainages were established approximately 8 Mya (Potter 1997; Lundberg et al. 1998). The uplift of the Andean Cordillera and the opening of the Atlantic Ocean significantly shaped the Paleodrainage history in southern South America (Potter 1997; Lundberg et al. 1998; Ribeiro et al. 2006). Most drainages from the western continental shields flowed westward into the Pacific Ocean during the Jurassic and Early Cretaceous (Coney and Evenchick 1994; Potter 1997). The drainage pattern for the South America platform and adjacent Africa for that time is not well known, but large main streams may not have been present since the region was probably arid or semi-arid (Potter 1997). The rifting in the South Atlantic was associated with substantial intracontinental deformation within South America (Seton et al. 2012), as broad uplifts along the southeastern Brazilian coast, causing an inward and coast-parallel flow of Gondwanic drainages into the current La Plata River Estuary (Potter 1997; Lundberg et al. 1998). In the far southern paleodrainages of Patagonia (southwestern Gondwana), where the rifting began in the Mesozoic (~190 Mya, Jurassic period; Seton et al. 2012), the Magallanes and Neuquén basins were connected to the Paleo-Pacific, but starting in the Late Cretaceous, the paleo-rivers Colorado and Negro inverted the flow into the Atlantic (Potter 1997). Also in that period, the uplift of the Andean proto-cordillera caused the inversion of the flow in western drainages from westward to eastward (Coney and Evenchick 1994; Potter 1997). From the Late Cretaceous to the Early Paleocene, two large marine transgressions from the Pacific occurred, from the North through the underfilled foreland basin and parallel to the thrust front, reaching a river flowing northward in northwestern Argentina, between the Sierras Pampeanas Massif and the uplifting Andes (Potter 1997; Sempere et al.

1997; Lundberg et al. 1998; Bloom and Lovejoy 2011). In addition, two less extended marine incursions from the South Atlantic penetrated the lower paleo-Paraná basin, overlapping the Sierras Pampeanas Massif (Gayet et al. 1993) (Figure 1.1).

Pérez-Losada et al. (2004) suggested, by overlaying their phylogenetic hypothesis onto the paleodrainage scenario, that the marine ancestor of the extant Aeglidae radiated from the Pacific Ocean to the South American continent with one of the two marine transgressions, that is, at least 60 Mya, when the second marine transgression occurred. It is worthy to note that the introduction of aeglids into South America took place only once, and the subsequent reproductive and physiological adaptations to fresh water occurred in descendants of the ancestral invader population. The clustering of the Argentinean species *A. ringueleti* and *A. scamosa* to the Chilean species could be explained since the northward river flowing along the foreland basin

Figure 1.1 Distribution of *Aegla* species in southern South America, with an approximated delimitation of the phylogenetic clades A to E from Pérez-Losada et al. (2004). Named species are not included in any of the previous clades. (Modified from Pérez-Losada et al. [2004] to encompass the new records of occurrence and the species recently described.)

was separated from the paleo-Paraná drainage by the Sierras Pampeanas Massif, from the Late Cretaceous to the middle Eocene (Pérez-Losada et al. 2004). Several modifications occurring in the western paleodrainages could have favored multiple vicariance and migration events, producing the mixed pattern of present-day species locations between clades A and B (Figure 1.1). The eastward propagation of the Andean thrust front helped the Sierras Pampeanas to lose their influence as a barrier from the Middle Eocene, and the western drainage between ~20° and ~35° S was captured by the enlarging paleo-Paraná River (Lundberg et al. 1998), making possible an eastward radiation of the Chilean aeglids (Pérez-Losada et al. 2004). The significant uplift of the Andean Cordillera from the Late Oligocene (Sempere et al. 1997) might have isolated the Chilean species from the eastern aeglids (Pérez-Losada et al. 2004).

Clades C, D, and E radiated eastward following, to some extent, the pattern of the current drainages primarily established in the Eocene (Potter 1997), where the species occur: clade C over the Paraná River Basin, clade D over the western tributaries of River Paraná and the Uruguay River Basin, and clade E over the Guaíba River Basin (Pérez-Losada et al. 2004). The incongruences between some phylogenetic clusters and the present-day distribution in drainage systems might be connected to the paleodrainage changes (Pérez-Losada et al. 2004) occurring over the last Tertiary periods, mainly the Paranan Sea and the uplifting of Serra do Mar (Lundberg et al. 1998).

The update of the phylogenetic tree of the family Aeglidae, including the species described since Pérez-Losada et al. (2004), integrated with new studies on South America geological and hydrological history, will help to better clarify how this unique group radiated within southern South America.

1.2 PHYLOGENETIC RELATIONSHIPS

The phylogenetic relationships of the family Aeglidae with their marine anomuran relatives as well as among its species have long interested researchers. Initially, the family Aeglidae was classified within the superfamily Galatheoidea. Latreille (1818) drew for the first time an aeglid, previously unknown, without describing it, which he named *Galathea laevis*. Schmitt (1942) speculated that Latreille might have been unaware of the freshwater habitat of the species since he placed it in an exclusively marine genus. Leach (1820) noticed that the specimen drawn by Latreille (1818) represented, in fact, a new genus, naming it *Aegla*. Latreille (1829) highlighted the similarities between the genus "*Aeglea*"* and the galatheids. Dana (1852) separated the subtribe "Aegleidea" from the Galatheidea within the "Anomoura inferiora," and Girard (1855) graphed for the first time the name "Aegleidae" corresponding to the current taxonomic level of the family. Schmitt (1942) believed that the closest

---

* According to Schmitt (1942), the misspelling *Aeglea* was introduced by Desmarest in 1825 and followed by several authors (including Latreille) until Rathbun (1910) called attention to the first orthography.

relatives of the aeglids were marine and were to be found within the galatheids. From that time on, aeglids have been included in the superfamily Galatheoidea by most naturalists. However, the uniqueness of some characteristics of the aeglids led to question this traditional view (Martin and Abele 1986; 1988). The Aeglidae is the only anomuran family entirely restricted to fresh waters, it is endemic to southern South America, and its gills and carapace sutures are different from those of galatheids (Martin and Abele 1986, 1988). These and other features make the aeglids unique ecologically, biogeographically, and morphologically (Martin and Abele 1986, 1988; Bond-Buckup and Buckup 1994).

Martin and Abele (1988) hypothesized a relationship between aeglids and hermit crabs (Paguroidea) due to similarities in some morphological characteristics, but when the hypothesis was tested Aeglidae grouped with Galatheoidea instead of Paguroidea in their analyses (Martin and Abele 1986). They proposed a phylogeny in which aeglids would be the most primitive among Galatheoidea, with a sister-group relationship between them (Martin and Abele 1986). The unique spermatozoal structure of *Aegla* provided some support to the elevation of the Aeglidae to the superfamily rank, besides suggesting a close affinity to Lomoidea (Tudge and Scheltinga 2002). A relationship between *Aegla* and *Lomis* was also suggested based on mitochondrial gene rearrangements (Morrison et al. 2002), along with morphological and molecular data (Ahyong and O'Meally 2004; Porter et al. 2005). A Bayesian tree constructed using an amino acid dataset from 13 protein coding mitochondrial genes from 22 anomurans showed an unresolved relationship between *Aegla* and *Lomis*, placing both in a basal position in relation to Chirostyloidea (Chirostylidae and Kiwaidae), while the mitochondrial gene order analysis supported Aegloidea as sister group to the clade formed by Lomisoidea and Chirostyloidea (Tan et al. 2018).

Based on a phylogenetic analysis of the nuclear 18S gene, Pérez-Losada et al. (2002b) suggested the elevation of the family Aeglidae to the superfamily rank due to its explicit separation from the galatheoid families. In their phylogenies, Galatheoidea, excluding Aeglidae, presented a sister-relationship to Paguroidea with Aeglidae being sister to the cluster Galatheoidea + Paguroidea. McLaughlin et al. (2007) eventually proposed the superfamily Aegloidea based on a morphological phylogeny. While the phylogenetic position of Aegloidea within the monophyletic Anomura (Pérez-Losada et al. 2002b; Porter et al. 2005) has been controversial, a recent study by Bracken-Grissom et al. (2013) clearly placed the Aeglidae as the basal taxon in a strongly supported clade with Lomisidae, Eumunididae, Kiwaidae, and then Chirostylidae branching off respectively.

The phylogenetic relationships within the Aeglidae have been speculated upon since Schmitt (1942) suggested that *A. jujuyana* would be the most primitive species, spreading out from the center of dispersion in Jujuy, Argentina, and giving origin to species with the "Atlantic" and "Pacific" types of rostrum—except for *A. franca* with an intermediate type of rostrum. Ringuelet (1948, 1949a) recognized the difficulties in separating some species based on morphological characters due to both the uniformity of the genus and intraspecific variation and even observing that some specimens seemed to be hybrids. He reevaluated the taxonomic status of *Aegla affinis*, allocating it as a subspecies of *A. neuquensis* (see Ringuelet 1948).

It is interesting to observe that these two species represent a sister relationship in the phylogeny of Pérez-Losada et al. (2004).

Ringuelet (1949a,b) considered *A. parana, A. platensis, A. singularis*, and *A. uruguayana* the most primitive species or the closest to the ancestral form. He split the aeglids into five groups and presented a phylogenetic scheme for these groups (Ringuelet 1949b). For matter of comparison, we present Ringuelet's assemblages in their corresponding clades (between parentheses) in the phylogenetic study of Pérez-Losada et al. (2004) and in the present study in the case of *A. franca* and *A. paulensis* (formerly *A. odebrechtii paulensis*). Group I encompassed *A. singularis* (D) as the primitive species, originating* *A. prado* (D) and *A. denticulata* (A). Group II had an unknown primitive species, but *A. castro* (C) as a stem species, originating *A. franca* (C) and, by a collateral branch, *A. odebrechtii paulensis* (C) and finally *A. odebrechtii* (C). Group III also had an unknown primitive species, with *A. scamosa* (no clade) as the stem species, originating *A. neuquensis* (A) and at last *A. neuquensis affinis* (A); through another branch, *A. scamosa* would have originated "*A. spec.*" from El Sosneado, in Mendoza, Argentina [later described as *Aegla montana* Ringuelet, 1960 and further considered junior synonymy with *A. affinis* (Bond-Buckup and Buckup 1994)]. Group IV presented as the primitive species possibly *A. uruguayana* (D), *A. abtao* (B) as the stem species, from which gave rise to *A. abtao abtao* (B) and *A. abtao riolimayana* (B), subspecies that Ringuelet (1949b) proposed based on its similarities. On another branch, *A. abtao* (B) would have originated *A. laevis* (B), and this one would originate *A. papudo* (most basal species) and *A. concepcionensis* (not evaluated). Group V encompassed *A. parana* (C) as the primitive species, originating in a straight-line *A. sanlorenzo* (D), *A. jujuyana* (D), and *A. humahuaca* (D). Some but not all of the affinities found by Ringuelet (1949b) are in accordance with the current phylogeny of the aeglids.

Lopretto (1978, 1979, 1980, 1981) recognized four groups based on the fifth pair of male pereopods: *platensis* group—*A. platensis, A. singularis, A. uruguayana*, and *A. neuquensis affinis* (*A. affinis*); patagónico group—*A. neuquensis neuquensis* (*A. neuquensis*) and *A. abtao riolimayana* (*A. riolimayana*); northwestern group—*A. humahuaca, A. franca, A. jujuyana*, and *A. sanlorenzo*; cuyano group—*A. montana* (*A. affinis*) and *A. scamosa*. The groups "northwestern" and "cuyano" present some similarities to the groups V and III of Ringuelet (1949b), respectively, but Lopretto (1979) noticed large differences between *A. neuquensis neuquensis* and *A. neuquensis affinis* concerning the studied appendix and questioned the validity of the subspecies status.

Schuldt et al. (1988) proposed a preliminary cladogram based on morphological characters for the species *A. abtao abtao, A. abtao riolimayana, A. montana, A. neuquensis affinis, A. neuquensis neuquensis, A. scamosa*, and *A. uruguayana* from central-western Argentina, in which *A. uruguayana* would be the most basal taxon. Pérez-Losada et al. (2002a) presented a molecular phylogeny for 16 Chilean species,

---

* Ringuelet (1949b) used the verb "originar" (Spanish) in his study and we opted to keep his idea using the English word "originate."

using four mitochondrial genes, placing *A. papudo* as the most basal aeglid species and separating the remaining species in two major clades.

Pérez-Losada et al. (2004) performed the most complete study to date on the phylogenetic relationships within the Aeglidae. By using mitochondrial and nuclear genes, they obtained robust trees for 58 species and six undescribed new species, in which *Aegla papudo* from Chile was used as a functional outgroup after confirming its basal position within the family. *Aegla ringueleti* was the most basal ingroup taxon, followed by *A. scamosa*, both from western Argentina. The other species, except *A. marginata* and *A. spinipalma*, were grouped into five major clades. The species clustered in each clade and the countries where they occur are shown in Table 1.1. Clade A (the first to radiate) and clade B (the second to diverge) included the Chilean and southern Argentinean species. Relationships among clades C, D, and E were not strongly supported, and alternative arrangements between these clades could not be rejected; they encompassed the northern Argentinean, Uruguayan, and Brazilian species.

The study of Pérez-Losada et al. (2004) did not include five species known at the time: *Aegla concepcionensis* and *A. expansa*, from Chile, and *A. franca*, *A. lata*, and *A. microphthalma* from Brazil. The latter one is a stygobiotic species inhabiting a single cave with difficult access (Bond-Buckup and Buckup 1994) and the other four species could not be found at the time of the study. On the other hand, the authors included six putative new species (named n. sp. 1 to n. sp. 6 in the tree), which were described later under the names *A. muelleri* Bond-Buckup and Buckup, 2010 (*Aegla* sp. n. 1 and 5); *A. pomerana* Bond-Buckup and Buckup, 2010 (*Aegla* n. sp. 2); *A. brevipalma* Bond-Buckup and Santos, 2012 (*Aegla* n. sp. 3); *A. renana* Bond-Buckup and Santos, 2010 (*Aegla* n. sp. 4); and *A. saltensis* Bond-Buckup and Jara, 2010 (*Aegla* n. sp. 6).

Moreover, 18 new species not included in the study of Pérez-Losada et al. (2004) were also described after 2004: *Aegla manuinflata* Bond-Buckup and Santos, 2009; *Aegla leachi* Bond-Buckup and Santos, 2012; *Aegla oblata* Bond-Buckup and Santos 2012; *Aegla georginae* Santos and Jara, 2012; *Aegla ludwigi* Santos and Jara, 2013; *Aegla leachi* Bond-Buckup and Santos, 2013; *Aegla carinata* Bond-Buckup and Gonçalves, 2014; *Aegla lancinhas* Bond-Buckup and Buckup, 2015; *Aegla loyolai* Bond-Buckup and Santos, 2015; *Aegla meloi* Bond-Buckup and Santos, 2015; *Aegla japi* Moraes, Tavares, and Bueno, 2016; *Aegla jaragua* Moraes, Tavares, and Bueno, 2016; *Aegla jundiai* Moraes, Tavares, and Bueno, 2016; *Aegla vanini* Moraes, Tavares, and Bueno, 2016; *Aegla charon* Bueno and Moraes, 2017; *Aegla quilombola* Moraes et al. 2017; *Aegla okora* Páez and Teixeira, 2018; and *Aegla chilota* Jara, Pérez-Losada, and Crandall, 2018. Also, the species *Aegla rosanae*, Campos Jr., 1998, which was synonymized to *Aegla paulensis* (Bond-Buckup and Buckup 2000), was revalidated by Moraes et al. (2016). Thus, the number of *Aegla* species described to date is 87, although at least another 15 new species will be described in the near future. Crivellaro et al. (2017) demonstrated that *A. longirostri* encompasses a complex of 14 cryptic species; *A. longirostri* sensu stricto (from the type-locality) and nine undescribed species grouped with species from clade E; the remaining three species clustered with species belonging to clade D (Figures 1.2 and 1.3; Table 1.1). The widely distributed *A. platensis* encompasses three distinct species (Zimmermann et al. 2018), all clustering with species from clade D (Figures 1.2

Table 1.1 Species Clustering in Major Clades (A to E) in the ML/BA Consensus Phylogeny of the Genus *Aegla*, According to Pérez-Losada et al. (2004), and Placement of the Species Not Studied Before or Described After 2004 (in Bold) within These Clades

No Clade	Clade A	Clade B	Clade C	Clade D	Clade E
Aegla papudo	*Aegla bahamondei*	*Aegla riolimayana*	*Aegla camargoi*	*Aegla singularis*	***Aegla muelleri***
Aegla ringueleti	*Aegla occidentalis*	*Aegla abtao*	*Aegla paulensis*	*Aegla platensis sensu stricto*	*Aegla leptochela*
Aegla scamosa	*Aegla neuquensis*	*Aegla spectabilis*	*Aegla perobae*	*Aegla rossiana*	*Aegla inconspicua*
Aegla marginata	*Aegla affinis*	*Aegla araucaniensis*	*Aegla parva*	*Aegla uruguayana*	*Aegla serrana*
Aegla spinipalma	*Aegla alacalufi*	*Aegla pewenchae*	*Aegla parana*	*Aegla intercalata*	*Aegla franciscna*
	Aegla manni	*Aegla laevis*	*Aegla castro*	*Aegla prado*	*Aegla ligulata*
	Aegla hueicollensis	*Aegla talcahuano*	*Aegla schmitti*	*Aegla violacea*	*Aegla obstipa*
	Aegla denticulata denticulata	*Aegla cholchol*	*Aegla cavernicola*	*Aegla humahuaca*	***Aegla renana***
	Aegla denticulata lacustris	*Aegla rostrata*	*Aegla strinatii*	***Aegla saltensis***	*Aegla itacolomiensis*
		Aegla leachi(?)	***Aegla pomerana***	*Aegla septentrionalis*	*Aegla plana*
		Aegla oblata(?)	*Aegla leptodactyla*	*Aegla jujuyana*	*Aegla grisella*
			Aegla jarai	*Aegla sanlorenzo*	*Aegla inermis*
			Aegla brevipalma	***Aegla manuinflata***	*Aegla longirostri sensustricto*
			Aegla spinosa	***Aegla carinata***	***Aegla georginae***
			Aegla odebrechtii	***Aegla platensis species complex 1***	***Aegla ludwigi***
			Aegla lancinhas	***Aegla platensis species complex 2***	***Aegla longirostri species complex 1***
			Aegla loyolai	***Aegla longirostri species complex 10***	***Aegla longirostri species complex 2***
			Aegla meloi	***Aegla longirostri species complex 11***	***Aegla longirostri species complex 3***

(Continued)

Table 1.1 (Continued) Species Clustering in Major Clades (A to E) in the ML/BA Consensus Phylogeny of the Genus *Aegla*, According to Pérez-Losada et al. (2004), and Placement of the Species Not Studied Before or Described After 2004 (in Bold) within These Clades

No Clade	Clade A	Clade B	Clade C	Clade D	Clade E
			Aegla franca	*Aegla longirostri* species complex 12	*Aegla longirostri* species complex 4
			Aegla japi	*Aegla longirostri* species complex 13	*Aegla longirostri* species complex 5
			Aegla jaragua		*Aegla longirostri* species complex 6
			Aegla jundiai		*Aegla longirostri* species complex 7
			Aegla vanini		*Aegla longirostri* species complex 8
			Aegla rosanae		*Aegla longirostri* species complex 9

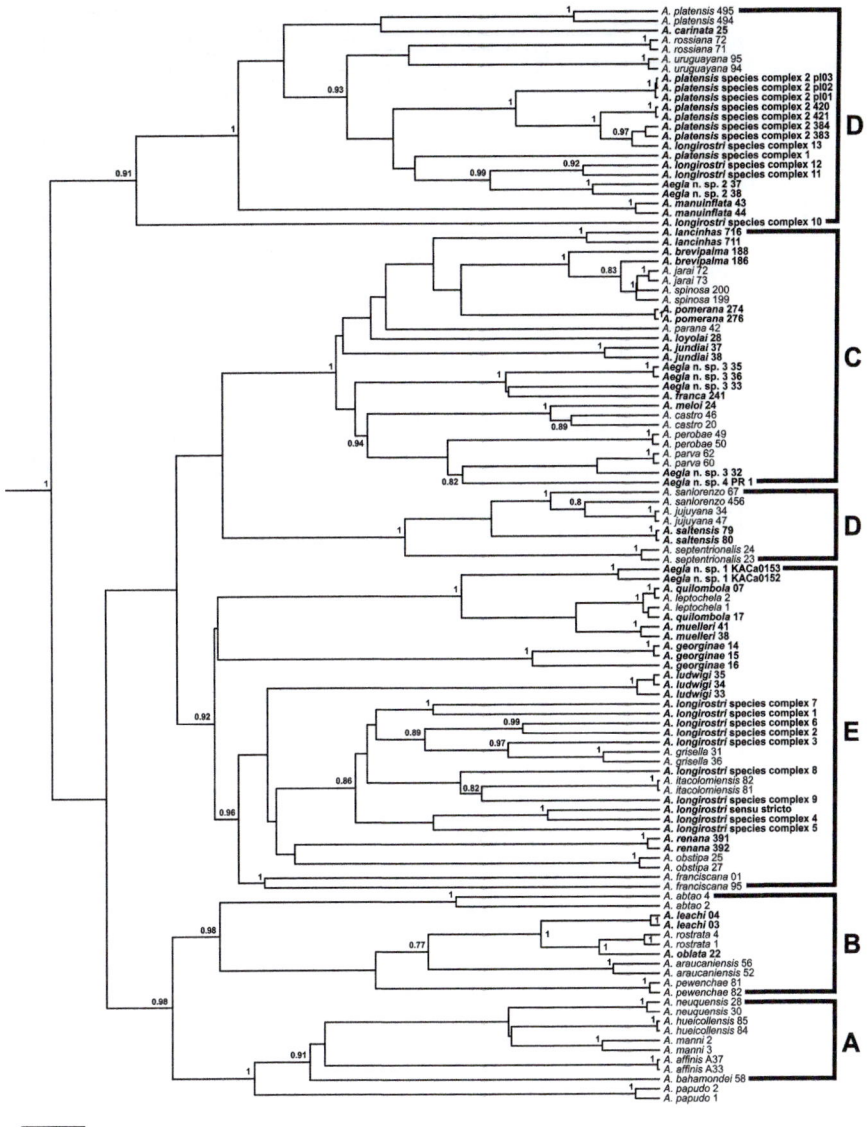

Figure 1.2 Bayesian tree based on 16S-COI haplotypes from a subset of species representing each phylogenetic clade of Pérez-Losada et al. (2004) and including species recently described (in bold). Clades A to E are highlighted according to Pérez-Losada et al. (2004). Bayesian posterior probabilities > 0.75 are shown above the branches.

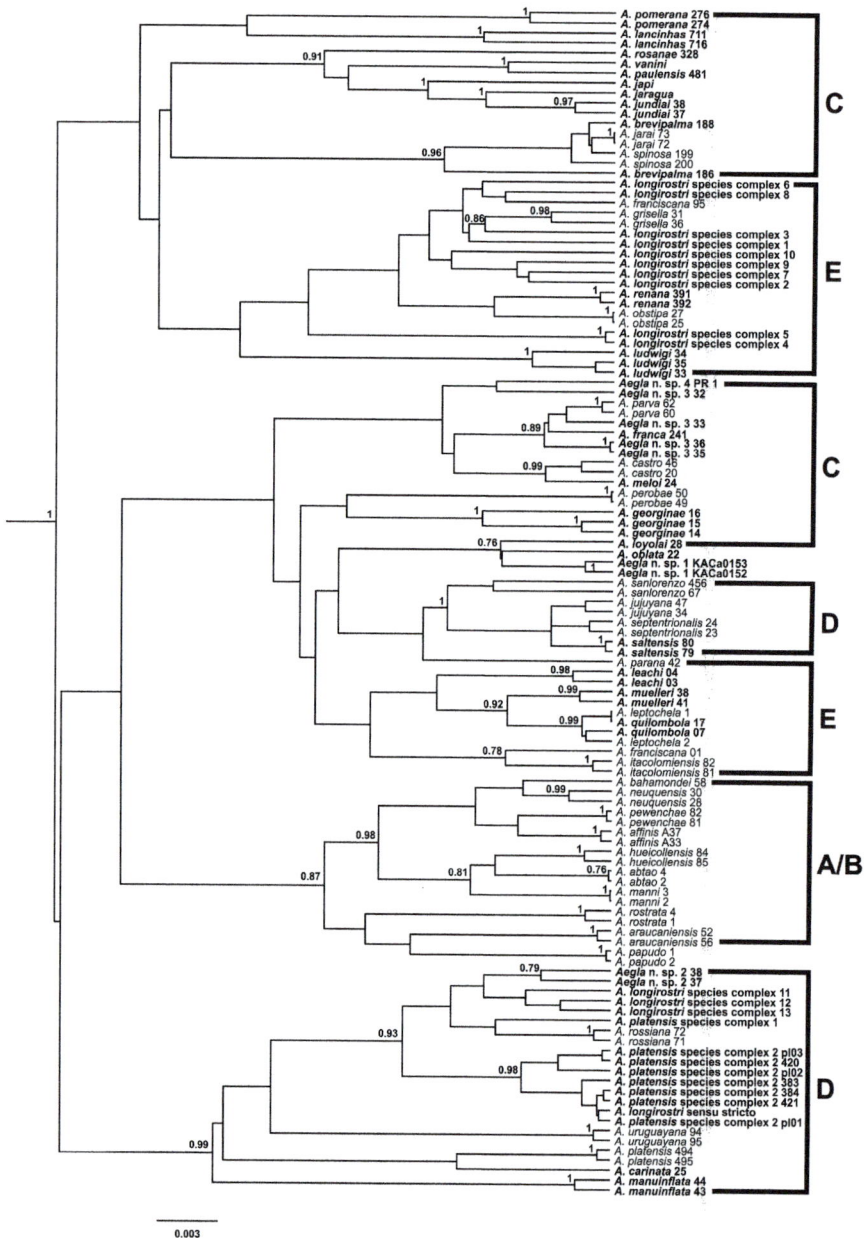

Figure 1.3 Bayesian tree based on 16S sequences from a subset of species representing each phylogenetic clade of Pérez-Losada et al. (2004) and including species recently described (in bold). Clades A to E are highlighted according to Pérez-Losada et al. (2004). The asterisk denotes species "misplaced" in relation to the 16S-COI tree. Bayesian posterior probabilities > 0.75 are shown above the branches.

and 1.3; Table 1.1). We also present evidence for six new species (see following paragraph). After the description of these species, the total number of species of the genus *Aegla* will exceed 100.

The species *A. japi*, *A. jaragua*, *A. jundiai*, *A. lancinhas*, *A. paulensis*, *A. rosanae*, and *A. vanini* belong to the *A. paulensis* complex, all of them occurring in southeastern Brazil (Moraes et al. 2016). The specimens of *A. paulensis* used by Pérez-Losada et al. (2004) were from the recently described *A. jundiai*, belonging to clade C. In addition, Moraes et al. (2017) revised the taxonomic status of *A. marginata*, splitting it into two species, one redescribed from the type locality and another described as a new species, *A. quilombola*, from the Ribeira de Iguape Basin (São Paulo state, Brazil). *Aegla marginata* was paraphyletic in the phylogenetic analysis of Pérez-Losada et al. (2004), with specimens from the type locality not included in any major clade, and specimens from the Ribeira de Iguape Basin clustering in clade E (current *A. quilombola*), with a very close sister-relationship to *A. leptochela*, and these two species being sympatric in one cave.

We obtained mitochondrial gene sequences (16S rRNA and COI) for the recently described *A. carinata*, *A. georginae*, *A. lancinhas*, *A. leachi*, *A. loyolai*, *A. ludwigi*, *A. manuinflata*, *A. meloi*, *A. oblata*, and also for *A. franca*, not previously included in the aeglid phylogeny (GenBank accessions FJ360714-15, FJ360706-07, KT319222, KT319210, KT319218, KT319206, MH998634-63). Primer pairs already described in the literature were used to amplify both genes (Pérez-Losada et al. 2002a; Xu et al. 2009). Standard Polymerase Chain Reaction (PCR) was conducted and PCR products were sequenced in both directions. Sequences were aligned with Muscle (Edgar 2004). We performed phylogenetic analyses for single and concatenated genes, using sequences both from all the species analyzed by Pérez-Losada et al. (2004) and for a subset of species representing each clade (AY050031-2; AY050035-6; AY0500042-4; AY050054-7; AY050065-6; AY050071-4; AY050077-8; AY050081-2; AY050088-90; AY050100-3; AY050111-2; AY50117-20; AY595549-51; AY595561; AY595565-7; AY595576-7; AY595581-2; AY595584-8; AY595591-2; AY595594-9; AY595603; AY595605; AY595611-2; AY595623-4; AY595627-32; AY595637-8; AY595641-6; AY595650-5; AY595658-9; AY595662-3; AY595667-70; AY595803-5; AY595815; AY595819-21; AY595830-1; AY595835-6; AY595839-42; AY595845-6; AY595848-53; AY595857; AY595859; AY595865-6; AY595877-8; AY595881-6; AY595891-2; AY595895-900; AY595904-9; AY595912-3; AY595916-7; AY595921-4; JQ844885-6; JQ844889; JQ844891-2; MF442421-2; MF442424-5), and including sequences from undescribed new species of the cryptic species complex *A. longirostri* (see Crivellaro et al. 2017) and *A. platensis* (see Zimmermann et al. 2018), along with sequences from six putative new species. We also conducted phylogenetic analyses including 16S rRNA sequences from *A. japi*, *A. jaragua*, *A. paulensis*, *A. rosanae*, *A. vanini*, the only gene available in GenBank for these species (GenBank accessions KU948368-73). The best-fit model of sequence evolution selected by jModelTest 2.1.10 (Darriba et al. 2012) was GTR+Gamma+I for all gene regions. Bayesian Inference was carried out using the Monte Carlo Markov Chain method implemented in Beast 1.8.0 (Drummond et al. 2012). Analyses were run for 30 million chains and sampled every 1000 generations. Posterior probabilities were calculated after a burn-in of three million states and checked for convergence using Tracer 1.6 (Rambaut et al. 2014). Results were visualized using FigTree 1.4.2 (Rambaut 2014).

For better data visualization, we are showing the results for the analyses using some representatives of each clade (Figures 1.2 and 1.3), instead of trees with all the species, since the results did not differ between analyses. The clades within each new species are inserted according to our analyses (Figures 1.2 and 1.3) and are also shown in Table 1.1. Our concatenated 16S-COI tree recovered the same clades as in Pérez-Losada et al. (2004), except that clade D was split into two different clades: one clustering only northwestern Argentinean and a southern Bolivian species from the Paraná River Basin, and the other including species occurring in southern Brazil (Rio Grande do Sul state), Uruguay, southeastern Paraguay, and northeastern Argentina. It is interesting to note that in the phylogenetic tree of Pérez-Losada et al. (2004), clade D was split into two subclades and the species within each of them were the same as in our tree (Figure 1.1). Another interesting finding was that two recently described species from Rio Grande do Sul state, Brazil, *A. leachi* and *A. oblata*, were clustered with species from clade B, from Chile, in a well-supported clade. A phylogenetic tree constructed using 16S and COII mitochondrial genes (Santos et al. 2012) placed *A. leachi* within clade E and *A. oblata* within clade C.

The 16S tree produced mostly weakly supported clades (Figure 1.2) and did not recover exactly the same clades as those in Pérez-Losada et al. (2004). Species from clades A and B were mixed, and clades C, D, and E were split into two groups. Species from the *A. paulensis* complex as well as *A. franca* clustered with some species from clade C. *Aegla leachi* and *A. oblata* were grouped with species from clade D. Although these clustering patterns may be easier to explain than those seen in the concatenated tree, the support was low. By sequencing more genes for these species, we will have a better understanding of their phylogenetic relationships. In the 16S tree, *A. georginae* clustered with species from clade C while in our 16S-COI tree as well as in a COII mitochondrial gene tree (Santos et al. 2013), it was placed within clade E. *Aegla parana*, which was grouped within clade C in the multi-locus tree of Pérez-Losada et al. (2004) and also in our 16S-COI concatenated tree and in a concatenated 16S-COII tree (Santos et al. 2012), clustered with the northwestern Argentinean species from clade D in the 16S tree. These results suggest that the mitochondrial gene 16S alone may not be suitable to investigate phylogenetic relationships across all the Aeglidae.

1.3 *AEGLA* SPECIATION AND DISPERSION IN FRESHWATER ENVIRONMENTS

The evolutionary history of *Aegla* is a formidable example of successful colonization of freshwater habitats. The genus is represented by 87 described species, which are found across Argentina, Bolivia, Brazil, Chile, Paraguay, and Uruguay. The distribution of this impressive number of species includes all main rivers of southern South America, except the southernmost Patagonian drainages (Bond-Buckup et al. 2008;

Moraes et al. 2017). However, although the genus is widely distributed, most species of *Aegla* have narrow distributional ranges; several of them have few records of occurrence and are restricted to small creeks located on hillsides or high mountainous regions (Bond-Buckup and Buckup 1994; Santos et al. 2015; Santos et al. 2017). The available evidence allows us to infer that the dispersion and speciation of *Aegla* species were shaped by a series of factors, both biological and historical, which were initiated when the ancestors of this group invaded the mainland waters of South America.

Robust phylogenetic analyses indicate that the genus *Aegla* evolved from Pacific relatives about 75 Mya having subsequently dispersed toward the Atlantic coast (Pérez-Losada et al. 2004). Consequently, species located near the Atlantic should be the most recent ones (Pérez-Losada et al. 2004; Crivellaro et al. 2017). The western *Aegla* species radiated, approximately, 40–45 Mya (Figure 1.1: clades A and B), but the speciation of the central and eastern taxa took place later, around 23–35 Mya (clades C, D, and E). Thus, the radiation and dispersion of *Aegla* species predates current river drainage patterns of South America (for review see Pérez-Losada et al. 2004). Basically, after the middle Eocene, the Sierras Pampeanas Massif lost their influence as a barrier and added to drainage directions that changed from westward to eastward as a consequence of the uplift of the early Andes, which created the possibility for the Chilean aeglids to radiate over the vast eastern territories. Subsequently, the uplift of the Andean Cordillera (Pliocene) probably isolated the Chilean species from their eastern relatives. During the middle Tertiary periods, the present Paraná-Uruguay drainage pattern was largely established and the eastern radiation of the group seemed to have grossly followed the present river pattern.

Surprisingly, although aeglids have colonized a large part of the watersheds of southern South America, they are not considered good dispersers (Baumart et al. 2015). On the other hand, biological characteristics of these animals may have driven the high degree of endemism and speciation observed in the group. Due to their poor dispersal ability, small population sizes, and habitat fragmentation, high population differentiation seems to be common in *Aegla* (Xu et al. 2009). First of all, as well as other crustaceans, *Aegla* species had to face challenges and develop key adaptations for the invasion of fresh water, such as the reduction of egg numbers, direct postembryonic development (see Chapter 6), extension of brood care, and lecithotrophy of the first post-hatching life stages (Anger 2016). Direct development and extension of brood care are associated with the reduction of dispersal and gene flow among populations (Vogt 2013). In fact, characteristics such as the presence of direct development and parental care, habitat specialization, preference for headwaters, and low dispersion capabilities (López Greco et al. 2004; Ayres-Peres et al. 2011; Baumart et al. 2015; Santos et al. 2015) should promote the isolation and differentiation of *Aegla* populations, increasing the probability of speciation events. Dispersion in South America should thus primarily occur unintentionally, with animals being carried away by water currents during floods (Crivellaro et al. 2017).

Historical factors, such as Pleistocene climatic oscillations and geologic events, also played an important role in shaping the current diversity and distribution of *Aegla* species. For example, Barber et al. (2011) examined two non-sister clades of mostly Chilean *Aegla* species (21 taxa in total) to understand how freshwater assemblages have been formed. They observed that basal divergence in the clades occurred during the Pliocene (between 5.05 and 3.95 Mya) which is consistent with episodes of increased sea levels documented in Chile. However, the majority of cladogenic events date to the Pliocene or Early Pleistocene. Likewise, during Quaternary glaciation events, range restriction of *Aegla* populations near the Coastal Cordillera should have occurred. Range reduction and subsequent postglacial range expansion could have caused population fragmentation and incipient differentiation, explaining the high level of molecular divergence seen in *A. cholchol* (Bracken-Grissom et al. 2011). Climatic oscillations also shaped the current distribution of the highly endemic cave-dwelling species of *Aegla* from southeastern Brazil. In this case, the independent colonization of subterranean habitats by different epigean ancestors was probably followed by vicariance and speciation. Those vicariant events were influenced by Pleistocene climatic fluctuations and drainage changes as well as by the limited potential for dispersion of the species (Fernandes et al. 2013).

In this context, phylogeographic studies helped to understand the routes of dispersion and speciation of aeglids, since they often provide insight into the processes generating biological diversity. For instance, during the glacial advances that occurred in the Pleistocene, the populations of *A. alacalufi* that occurred in Chilean Patagonia were either extirpated or forced into refugia. After the glacial retreats, survivors recolonized these habitats and further expanded their distribution. Due to this expansion, the populations were fragmented and isolated by distance, which probably drove the current population structure (Xu et al. 2009). Likewise, populations of *A. neuquensis* from Argentinean Patagonia were isolated in two river systems due to a fragmentation event that occurred about 137,000 years ago. Although flooding of the region during the Patagonian glaciation would have provided opportunities for gene flow between systems, they were then isolated again following the retreat of glaciers in the region (Barber et al. 2012). The higher genetic diversity, deeper phylogenetic structure, and older divergence times in the Argentinean *A. neuquensis* samples compared to those of the Chilean *A. alacalufi* shows that Pleistocene glacial cycles have impacted these freshwater species differently. The evolution of *A. alacalufi* was greatly influenced by Late Pleistocene glaciations. On the other hand, *A. neuquensis* differentiation was mainly driven by drainage changes established long before the Last Glacial Maximum (Pérez-Losada et al. 2011). While the uplift of the Andes changed forever the South American drainage pattern, generating multiple episodes of vicariance plus drainage system coalescence (Lundberg et al. 1998), Pleistocene glacial cycles remodeled some of these environments, especially on the western slopes of the Andes (Pérez-Losada et al. 2011).

For *Aegla* species found in southern Brazil, speciation and dispersion were probably shaped by other Pleistocene and geological events, with a possible influence of landscape topography on the diversification of populations. For example, alternation between dry and humid phases, with predominance of cold and semi-arid climate during the Late Pleistocene, must have influenced the population dynamics of the species. Specifically, populations of *A. longirostri sensu lato* were possibly isolated during drier periods, which may have reduced or even restricted the gene flow. More recently, when the climate became humid, some of these populations were possibly reconnected, even though they were already reproductively isolated (Crivellaro et al. 2017). Besides that, the presence of deeply divergent groups of *A. longirostri sensu lato* (Meridional Plateau and Central Depression populations) in Rio Grande do Sul state suggests that there were at least two independent events of colonization from the ancestral populations of *Aegla*. That is, one wave of colonization deriving of aeglid representatives coming from the northern continuation of the Meridional Plateau area (clade E) and another originated directly from relatives of Argentina and Uruguay (clade D) (Crivellaro et al. 2017). The origin of *A. platensis*, estimated at 208,000 Ya, was also probably related to events occurring during the Pleistocene. Such events could even have facilitated its dispersion toward southern Brazil (Zimmermann et al. 2018).

Finally, although the area of occurrence of freshwater decapods in southern South America has been mainly attributed to historical factors, such as geoclimatic events, tectonic uplifts, and glaciations, their current distributions could also be strongly influenced by local environmental conditions (Tumini et al. 2016). For instance, latitude, longitude, altitude, aquatic stability, annual temperature range, and conductivity were significant predictors of *Aegla* species occurrence, although interactions among these factors were significant (e.g., the association between altitude and occurrence of aeglids was strong and generally positive, but this relationship depended on latitude and aquatic environmental stability) (Tumini et al. 2016). Thus, the current distribution of *Aegla* species was possibly influenced by modifications in land geography and the formation of continental aquatic environments in southern South America. Thereafter, aeglids could have moved (i.e., a dispersion event) or been restricted (i.e., a vicariant event) to high altitudes; that is, the current high degree of endemism and species differentiation could be a result of the loss of connectivity (Collins et al. 2011), possibly related with limited plasticity and strong adaptation to local habitats (Tumini et al. 2016).

1.4 CRYPTIC SPECIES IN *AEGLA*

High-quality data derived from phylogeographic sampling approaches (i.e., geographically dense and large sample sizes, molecular genetic-based phylogenies) coupled with insights from historical aspects that influenced the current distribution of *Aegla* species (e.g., paleogeography, etc.), have revealed the existence of cryptic species in the group (i.e., species with superficially indistinguishable morphologies that are genetically different from each other) (Moraes et al. 2016; Crivellaro et al. 2017; Zimmermann et al. 2018). Morphological diagnostic characters within *Aegla* are very limited and exhibit low taxonomic variation among related species (Bond-Buckup and Buckup 1994). These subtle phenotypic differences and the problematic distinction between intra- and interspecific variation should generate a discrepancy between the number of species described and the actual number of existing species, with cryptic diversity possibly hidden behind this limited morphological differentiation (Zimmermann et al. 2018). Moreover, as many speciation events in the group are relatively recent, phenotypic evolution may not match the faster genetic divergences of mitochondrial DNA, explaining the cryptic diversity found in the absence of significant differences in diagnostic morphologic characters (Crivellaro et al. 2017). Thus, for a group such as *Aegla* that lacks obvious shared derived morphological characters, systematics and species diagnosis based on molecular data is extremely helpful (Xu et al. 2009).

Growing evidence suggests that several *Aegla* species are assemblages of cryptic species (Bracken-Grissom et al. 2011; Bueno et al. 2016; Santos et al. 2017), and many studies are corroborating these observations. For example, phylogeographic studies carried out with *Aegla* species from Chile and Argentina demonstrated the presence of unrecognized new species that were morphologically similar and geographically close to populations of already known species (Xu et al. 2009; Barber et al. 2012). Analyses of morphological and molecular data revealed that six disjunct populations of *Aegla paulensis* from southern Brazil comprised a species complex, and four new species were described (Moraes et al. 2016). Likewise, surprising results from a phylogeographic study also performed in southern Brazil showed that *A. longirostri* was polyphyletic. From 17 examined populations, species delimitation methods indicated the presence of at least 14 putative cryptic species.

This sum of evidence suggests that cryptic speciation in aeglids is more common than previously thought. Considering this scenario, *Aegla* species with a wider distribution may potentially have a hidden diversity, because candidates for cryptic species complexes are often concealed within broadly distributed taxa. One exception to this pattern seems to be *A. platensis*, which features the widest geographical distribution of all *Aegla* species. Although there is evidence for the presence of at least two cryptic species in this taxon, the vast majority of the populations from Brazil and Argentina (separated by a distance of ~750 km in a straight line) belongs to a single and well-supported monophyletic group. *Aegla platensis* thus defies the phylogeographical pattern observed in most aeglids, which usually have narrow distributions. The wide distribution of *A. platensis* could be related, among other factors, to intrinsic characteristics and/or to historical biogeographical events. However, the mechanisms driving the speciation and spread of *A. platensis* still need to be clarified (Zimmermann et al. 2018).

In summary, anomurans of the genus *Aegla* descend from marine ancestors, which invaded continental waters about 75 Mya and dispersed throughout South America. Since then, the evolution of the group has given rise to approximately 100 species found in most of the main rivers of southern South America. The speciation and dispersion of aeglids were shaped by biological and historical factors, and events that occurred in recent periods, such as the Pleistocene climatic oscillations, played a decisive role.

1.5 PERSPECTIVES

The fascinating history of the family Aeglidae has been partially elucidated, but many issues remain to be addressed. Cryptic species complexes were identified molecularly, raising the number of known species to 100, and probably new species will be soon unveiled as new sampling efforts focus on less studied regions. An update of the family's phylogeny, including the new species and using the same set of genes analyzed by Pérez-Losada et al. (2004), coupled with biogeographic approaches, will allow a better understanding of the diversification of the aeglids in continental waters of southern South America.

The findings about cryptic diversity have a direct and significant effect on the conservation of this group of anomurans. A shocking proportion of almost 70% of the known *Aegla* species are under some level of threat due to their narrow distributions and the rapid degradation of freshwater habitats (Bueno et al. 2016; Santos et al. 2017). Therefore, the discovery of cryptic species assemblages could make them even more critically endangered species with fewer numbers and smaller distributions. In this sense, finding new morphological diagnostic characters to better differentiate among the described species and to allow the description of cryptic species is urgently needed.

Although we do not know the extent of the hidden diversity present in *Aegla*, it is becoming clear that its diversity is probably underestimated and that its species richness must be greater than expected. It is crucial, therefore, to carry out phylogeographic studies and investigate systematic issues for species considered paraphyletic (Pérez-Losada et al. 2004) and for those species that, for example, show intraspecific morphological variation, especially when populations are geographically isolated (Hepp et al. 2012; Fernandes and Bichuette 2013; Giri and Collins 2014; Marchiori et al. 2014, 2015; Trevisan et al. 2016). Since prioritizing habitats for conservation often relies on estimation of species richness and endemism, it is of utmost importance to reveal the cryptic diversity within *Aegla* in order to apply effective conservation measures. We have been focusing our efforts on these issues, hoping to contribute to a better understanding of this unique anomuran family.

Aeglids will face many challenges in the coming years due to anthropogenic interference in freshwater environments, with the worrisome and real possibility that millions of years of evolution and adaptation will be lost in the blink of an eye.

REFERENCES

Ahyong, S. T., and D. O'Meally. 2004. Phylogeny of the Decapoda Reptantia: resolution using three molecular loci and morphology. *The Raffles Bulletin of Zoology* 52(2):673–93.

Anger, K. 2016. Adaptation to life in fresh water by decapod crustaceans: evolutionary challenges in the early life-history stages. In *A Global Overview of the Conservation of Freshwater Decapod Crustaceans*, eds. T. Kawai, and N. Cumberlidge, pp. 127–68. London: Springer.

Ayres-Peres, L., C. Coutinho, J. S. Baumart, A. S. Gonçalves, P. B. Araujo, and S. Santos. 2011. Radio-telemetry techniques in the study of displacement of freshwater anomurans. *Nauplius* 19(1):41–54.

Barber, B. R., P. J. Unmack, M. Pérez-Losada, J. B. Johnson, and K. A. Crandall. 2011. Different processes lead to similar patterns: a test of codivergence and the role of sea level and climate changes in shaping a southern temperate freshwater assemblage. *BMC Evolutionary Biology* 11(1):343.

Barber, B. R., J. Xu, M. Pérez-Losada, C. G. Jara, and K. A. Crandall. 2012. Conflicting evolutionary patterns due to mitochondrial introgression and multilocus phylogeography of the Patagonian freshwater crab *Aegla neuquensis*. *PLoSOne* 7(6):e37105.

Baumart, J. S., M. M. Dalosto, A. S. Gonçalves, A. V. Palaoro, and S. Santos. 2015. How to deal with a bad neighbor? Strategies of sympatric freshwater decapods (Crustacea) for coexistence. *Hydrobiologia* 762(1):29–39.

Berry, E. W. 1922. Outlines of South American geology. *Pan-American geologist*, 38, 187–222, plates 7–15 *apud* Schmitt, W. 1942. The species of *Aegla*, endemic South American fresh-water crustaceans. *Proceedings of the United States National Museum* 91:431–520.

Bloom, D. D., and N. R. Lovejoy. 2011. The biogeography of marine incursions in South America. In *Historical Biogeography of Neotropical Fishes*, eds. J. S. Alberts, and R. E. Reis, pp. 137–44. Berkeley, CA: University of California Press.

Bond-Buckup, G., and L. Buckup. 1994. A família Aeglidae (Crustacea, Decapoda, Anomura). *Arquivos de Zoologia* 32(4):159–346.

Bond-Buckup, G., and L. Buckup. 2000. *Aegla rosanae* Campos Jr., um novo sinônimo de *Aegla* paulensis Schmitt (Crustacea, Aeglidae). *Revista Brasileira de Zoologia* 17(2):385–86.

Bond-Buckup, G., C. G. Jara, M. Pérez-Losada, L. Buckup, and K. A. Crandall. 2008. Global diversity of crabs (Aeglidae: Anomura: Decapoda) in freshwater. *Hydrobiologia* 595(1):267–73.

Bracken-Grissom, H. D., M. E. Cannon, P. Cabezas, et al. 2013. A comprehensive and integrative reconstruction of evolutionary history for Anomura (Crustacea: Decapoda). *BMC Evolutionary Biology* 13:128.

Bracken-Grissom, H. D., T. Enders, C. G. Jara, et al. 2011. Molecular diversity of river versus lake freshwater anomurans in southern Chile (Decapoda: Aeglidae) and morphometric differentiation between species and sexes. In *Phylogeography and Population Genetics in Crustacea*, eds. C. Held, S. Koenemann, and C. D. Schubart, pp. 305–22. Boca Raton, FL: CRC Press.

Bueno, S. L. S., A. L. Camargo, and J. C. B. Moraes. 2017. A new species of stygobiotic Aeglidae from lentic subterranean waters in southeastern Brazil, with an unusual morphological trait: short pleopods in adult males. *Nauplius* 25: e201700021.

Bueno, S. L. S., R. M. Shimizu, and J. C. B. Moraes. 2016. A remarkable anomuran: the taxon *Aegla* Leach, 1820. Taxonomic remarks, distribution, biology, diversity and conservation. In *A Global Overview of the Conservation of Freshwater Decapod Crustaceans*, eds. T. Kawai, and N. Cumberlidge, pp. 23–64. London: Springer.

Bueno, S. L. S., R. M. Shimizu, and S. S. Rocha. 2007. Estimating the population size of *Aegla franca* (Decapoda: Anomura: Aeglidae) by mark-recapture technique from an isolated section of Barro Preto Stream, County of Claraval, state of Minas Gerais, southeastern Brazil. *Journal of Crustacean Biology* 27(4):553–59.

Chablais, J., R. M. Feldmann, and C. E. Schweitzer. 2011. A new Triassic decapod, *Platykotta akaina*, from the Arabian shelf of the northern United Arab Emirates: earliest occurrence of the Anomura. *Paläontologische Zeitschrift* 85:93–102.

Collins, P. A., F. Giri, and V. Williner. 2011. Biogeography of the freshwater decapods in the La Plata basin, South America. *Journal of Crustacean Biology* 31(1):179–91.

Coney, P. J., and C. A. Evenchick. 1994. Consolidation of the American cordilleras. *Journal of South American Earth Sciences* 7(3/4):241–62.

Crivellaro, M. S., B. L. Zimmermann, M. L. Bartholomei-Santos, et al. 2017. Looks can be deceiving: species delimitation reveals hidden diversity in the freshwater crab

Aegla longirostri (Decapoda: Anomura). *Zoological Journal of the Linnean Society* 182(1):24–37.

Dana, J. D. 1852. Crustacea. In *United States Exploring Expedition During the Years 1839, 1840, 1841, 1842, Under the Command of Charles Wilkes, U.S.N.*, pp. 475–78. New Haven, PA: C. Sherman, Atlas Crustacea, 13.

Darriba, D., G. L. Taboada, R. Doallo, and D. Posada. 2012. jModel-Test 2: more models, new heuristics and parallel computing. *Nature Methods* 9:772.

Drummond, A. J., M. A. Suchard, D. Xie, and A. Rambaut. 2012. Bayesian phylogenetics with BEAUti and the BEAST 1.7. *Molecular Biology and Evolution* 29:1969–73.

Edgar, R. C. 2004. MUSCLE: multiple sequence alignment with high accuracy and high throughput. *Nucleic Acids Research* 32:1792–97.

Feldmann, R. M. 1984. *Haumuriaegla glaessneri* n. gen. and sp. (Decapoda, Anomura, Aeglidae) from Haumurian (Late Cretaceous) rocks near Cheviot, New Zealand. New Zealand. *Journal of Geology and Geophysics* 27:379–85.

Feldmann, R. M. 1986. Paleobiogeography of two decapod crustacean taxa in Southern Hemisphere: global conclusions with sparse data. In *Crustaceans Issues 4, Crustacean Biogeography*, eds. R. H. Gore, and K. L. Heck, pp. 5–19. Rotterdam: CRC Press.

Feldmann, R. M., and C. E. Schweitzer. 2006. Paleobiogeography of southern hemisphere decapod Crustacea. *Journal of Paleontology* 80(1):83–103.

Feldmann, R. M., F. J. Vega, S. P. Applegate, and G. A. Bishop. 1998. Early Cretaceous arthropods from the Tlayúa formation at Tepexi de Rodríguez, Puebla, México. *Journal of Paleontology* 72(1):79–90.

Fernandes, C. S., and M. E. Bichuette. 2013. Shape variation of *Aegla schmitti* (Crustacea, Decapoda, Aeglidae) associated to superficial and subterranean stream reaches. *Subterranean Biology* 10:17–24.

Fernandes, C. S., S. L. S. Bueno, and M. E. Bichuette. 2013. Distribution of cave-dwelling *Aegla* spp. (Decapoda: Anomura: Aeglidae) from the Alto Ribeira karstic area in southeastern Brazil based on geomorphological evidence. *Journal of Crustacean Biology* 33(4):567–75.

Gayet, M., T. Sempere, H. Capetta, E. Jaillard, and A. Lévy. 1993. La presence de fossils marins dans le Crétacé terminal des Andes centrales et ses consequences paleogéographiques. *Palaeogeography, Palaeoclimatology, Palaeocology* 102:283–319.

Girard, C. 1855. Description of certain Crustacea, brought home by the U. S. N. Astronomical Expedition. In *The U. S. Naval Astronomical Expedition to the Southern Hemisphere, During the Years 1849, 1851, 1852, Lieutenant James M Gillis, Superintendent*, S. I.: s.n., vol. 2, pp. 154–262. Executive Document number 121, House of Representatives, 33rd Congress, 1st session, serial 729, 15, part 2.

Giri, F., and P. Collins. 2014. Clinal variation in carapace shape in the South American freshwater crab, *Aegla uruguayana* (Anomura: Aeglidae). *Biological Journal of the Linnean Society* 113(4):914–30.

Harries, P. J., E. G. Kauffman, and T. A. Hansen. 1993. Models for biotic survival following mass extinction. In *Biotic Recovery from Mass Extinction Events*, eds. M. B. Hart, pp. 41–60. London: Geological Society, Special Publication, 102.

Hepp, L. U., R. Fornel, R. M. Restello, A. Trevisan, and S. Santos. 2012. Intraspecific morphological variation in a freshwater crustacean *Aegla plana* in southern Brazil: effects of geographical isolation on carapace shape. *Journal of Crustacean Biology* 32(4):511–18.

Jara, C. G., M. Pérez-Losada, and K. A. Crandall. 2018. *Aegla chilota*, new species of anomuran freshwater crab from Chiloé Island, western Patagonia. *Nauplius* 26:e2018029.

Latreille, M. 1818. Crustacés, arachnides et insects. In *Tableau encyclopédique et méthodique des trois règnes de la nature*. Paris: Chez Panckoucke, V. Agasse, 24, plate 308.

Latreille, W. E. 1829. Crustacés, arachnides et partie des insectes. In *Le règne animal distribué d'après son organisation, pour servir de base à l'histoire naturelle des animaux et d'introduction a l'anatomiee comparée*, ed. M. Cuvier, vol. 4, pp. 82–85. Paris: Chez Déterville.

Leach, W. E. 1820. Galatéadées. In *Dictionnaire des sciences naturelles*, eds. F. G. Levrault, pp. 49–56. Strasbourg.

López Greco, L. S., V. Viau, M. Lavolpe, G. Bond-Buckup, and E. M. Rodriguez. 2004. Juvenile hatching and maternal care in *Aegla uruguayana* (Anomura, Aeglidae). *Journal of Crustacean Biology* 24(2):309–13.

Lopretto, E. C. 1978. Las especies de *Aegla* Leach del centro-oeste argentino en base a la morfología comparada del quinto par de pereiópodos (Crustacea, Anomura, Aeglidae). *Neotropica* 24(71):57–68.

Lopretto, E. C. 1979. Estudio comparativo del quinto par de pereiópodos en los representantes del género *Aegla* de la Patagonia Argentina (Crustacea, Anomura). *Neotropica* 25(73):9–22.

Lopretto, E. C. 1980. Análisis de las características del quinto pereiopodo en las especies de *Aegla* del grupo "*platensis*" (Crustacea Anomura Aeglidae). *Physis Sección B* 39(96):37–56.

Lopretto, E. C. 1981. Consideraciones sobre la estructura apendicular vinculada al dimorfismo sexual en los machos de las especies de *Aegla* del noroeste argentino (Crustacea, Anomura, Aeglidae). *Acta Zoologica Lilloana* 36(2):15–35.

Lundberg, J. G., L. G. Marshall, J. Guerrero, B. Horton, M. C. S. L. Malabarba, and F. Wesselingh. 1998. The stage for Neotropical fish diversification: a history of tropical South American rivers. In *Phylogeny and Classification of Neotropical Fishes*, eds. L. R. Malabarba, R. E. Reis, R. P. Vari, Z. M. S. Lucena, and C. A. S. Lucena, pp. 13–48. Porto Alegre: EDIPUCRS.

Marchiori, A. B., M. L. Bartholomei-Santos, and S. Santos. 2014. Intraspecific variation in *Aegla longirostri* (Crustacea: Decapoda: Anomura) revealed by geometric morphometrics: evidence for ongoing speciation? *Biological Journal of the Linnean Society* 112(1):31–39.

Marchiori, A. B., R. Fornel, and S. Santos. 2015. Morphometric variation in allopatric populations of *Aegla platensis* (Crustacea: Decapoda: Anomura): possible evidence for cryptic speciation. *Zoomorphology* 134(1):45–53.

Martin, H. A. 2006. Cenozoic climatic change and the development of arid vegetation in Australia. *Journal of Arid Environments* 66:533–66.

Martin, J. W., and L. G. Abele. 1986. Phylogenetic relationships of the genus *Aegla* (Decapoda: Anomura: Aeglidae), with comments on Anomura phylogeny. *Journal of Crustacean Biology* 6(3):576–616.

Martin, J. W., and L. G. Abele. 1988. External morphology of the genus *Aegla* (Crustacea: Anomura: Aeglidae). *Smithsonian Contributions to Zoology* 453:1–46.

McLaughlin, P. A., R. Lemaitre, and U. Sorhannus. 2007. Hermit crab phylogeny: a reappraisal and its "fall-out". *Journal of Crustacean Biology* 27(1):97–115.

McLaughlin, P. A., and T. Murray. 1990. *Clibanarius fonticola*, new species (Anomura: Paguridea: Diogenidae), from a freshwater pool on Espiritu Santo, Vanuatu. *Journal of Crustacean Biology* 10(4):695.

Moraes, J. C. B., M. Tavares, and S. L. S. Bueno. 2017. Taxonomic review of *Aegla marginata* Bond-Buckup and Buckup, 1994 (Decapoda, Anomura, Aeglidae) with description of a new species. *Zootaxa* 4323(4):519–33.

Moraes, J. C. B., M. Terossi, R. C. Buranelli, M. Tavares, F. L. Mantelatto, and S. L. S. Bueno. 2016. Morphological and molecular data reveal the cryptic diversity among populations of *Aegla paulensis* (Decapoda, Anomura, Aeglidae), with descriptions of four new species and comments on dispersal routes and conservation status. *Zootaxa* 4193(1):1–48.

Morrison, C. L., A. W. Harvey, S. Lavrey, K. Tieu, Y. Huang, and C. W. Cunningham. 2002. Mitochondrial gene rearrangements confirm the parallel evolution of the crab-like form. *Proceedings of the Royal Society of London B* 269:345–50.

Morrone, J. J., and E. C. Lopretto. 1994. Distributional patterns of freshwater Decapoda (Crustacea: Malacostraca) in southern South America: a panbiogeographic approach. *Journal of Biogeography* 21:97–109.

Ortmann, A. E. 1902. The geographical distribution of freshwater decapods and its bearing upon ancient geography. *Proceedings of the American Philosophical Society* 41:267–400.

Oyanedel, A., C. Valdovinos, N. Sandoval, et al. 2011. The southernmost freshwaters anomurans of the world: geographic distribution and new records of Patagonian aeglids (Decapoda: Aeglidae). *Journal of Crustacean Biology* 31(3):396–400.

Páez, F. P., I. C. Marçal, L. Souza-Shibatta, et al. 2018. A new species of *Aegla* Leach, 1820 (Crustacea, Anomura) from the Iguaçu River basin, Brazil. *Zootaxa* 4527(3):335–46.

Pérez-Losada, M., G. Bond-Buckup, C. G. Jara, and K. A. Crandall. 2004. Molecular systematics and biogeography of the southern South American freshwater "crabs" *Aegla* (Decapoda: Anomura: Aeglidae) using multiple heuristic tree search approaches. *Systematic Biology* 53(5):767–80.

Pérez-Losada, M., C. G. Jara, G. Bond-Buckup, and K. A. Crandall. 2002a. Phylogenetic relationships among the species of *Aegla* (Anomura: Aeglidae) freshwater crabs from Chile. *Journal of Crustacean Biology* 22(2):304–13.

Pérez-Losada, M., C. G. Jara, G. Bond-Buckup, M. L. Porter, and K. A. Crandall. 2002b. Phylogenetic position of the freshwater anomuran family Aeglidae. *Journal of Crustacean Biology* 22(3):670–76.

Pérez-Losada, M., J. Xu, C. G. Jara, and K. A. Crandall. 2011. Comparing phylogeographic patterns across the Patagonian Andes in two freshwater crabs of the genus *Aegla* (Decapoda: Aeglidae). In *Phylogeography and Population Genetics in Crustacea*, eds. C. Held, S. Koenemann, and C. D. Schubart, pp. 291–303. Boca Raton, FL: CRC Press.

Porter, M. L., M. Pérez-Losada, and K. A. Crandall. 2005. Model-based multi-locus estimation of decapod phylogeny and divergence times. *Molecular Phylogenetics and Evolution* 37:355–69.

Potter, P. E. 1997. The Mesozoic and Cenozoic paleodrainage of South America: a natural history. *Journal of South American Earth Sciences* 10(5–6):331–44.

Rambaut, A. 2014. Figtree Version 1.4.2. Available at: http://tree.bio.ed.ac.uk/software/figtree/ (accessed April 14, 2018).

Rambaut, A., M. A. Suchard, D. Xie, and A. J. Drummond. 2014. Tracer Version 1.6. Available at: http://beast.bio.ed.ac.uk/Tracer (accessed April 14, 2018)

Rathbun, M. J. 1910. The stalk-eyed crustacea of Peru and the adjacent coast. *Proceedings of the United States National Museum* 38(1766):531–620.

Renne, P. R., A. L. Deino, F. J. Hilgen, et al. 2013. Time scales of critical events around the Cretaceous-Paleogene boundary. *Science* 339:684–87.

Ribeiro, A. C. 2006. Tectonic history and the biogeography from the freshwater fishes of coastal drainages of eastern Brazil: an example of faunal evolution associated with a divergent continental margin. *Neotropical Ichthyology* 4(2):225–46.

Ringuelet, R. A. 1948. Los "cangrejos" Argentinos del género *Aegla* de Cuyo y la Patagonia. *Revista del Museo de La Plata* 5(34):297–347.

Ringuelet, R. A. 1949a. Los anomuros del género *Aegla* del noroeste de la República Argentina. *Revista del Museo de La Plata* 6:1–45.

Ringuelet, R. A. 1949b. Consideraciones sobre las relaciones filogenéticas entre las espécies del género *Aegla* Leach. *Notas del Museo de La Plata* 14:111–18.

Santos, S., G. Bond-Buckup, L. Buckup, et al. 2015. Three new species of Aeglidae (*Aegla*) from Paraná State, Brazil. *Journal of Crustacean Biology* 35(6):839–49.

Santos, S., G. Bond-Buckup, L. Buckup, M. Pérez-Losada, M. Finley, and K. A. Crandall. 2012. Three new species of *Aegla* (Anomura) freshwater crabs from the upper Uruguay River hydrographic basin in Brazil. *Journal of Crustacean Biology* 32(4):529–40.

Santos, S., G. Bond-Buckup, A. S. Gonçalves, M. L. Bartholomei-Santos, L. Buckup, and C. G. Jara. 2017. Diversity and conservation status of *Aegla* spp. (Anomura, Aeglidae): an update. *Nauplius* 25:e2017011.

Santos, S., C. G. Jara, M. L. Bartholomei-Santos, M. Pérez-Losada, and K. A. Crandall. 2013. New species and records of the genus *Aegla* Leach, 1820 (Crustacea, Anomura, Aeglidae) from the west-central region of Rio Grande do Sul, Brazil. *Nauplius* 21(2):211–23.

Schmitt, W. 1942. The species of *Aegla*, endemic South American fresh-water crustaceans. *Proceedings of the United States National Museum* 91:431–520.

Schuldt, M., P. Nuñez, W. Mersing, A. Dell Valle, and M. O. Manceñido. 1988. *Aegla* (Crustacea, Anomura) en el lago Huechulafquen (Neuquen, Argentina) y algunas implicancias filogeneticas para Aeglidae del centro-oeste de Argentina. *Anales de la Sociedad Científica Argentina* 217(50):27–37.

Schweitzer, C. E., and R. M. Feldmann. 2005. Decapod crustaceans, the K/P event, and Paleocene recovery. In *Crustacea and Arthropoda Relationships*, eds. S. Koenemann, and R. Jenner, pp. 17–53. Boca Raton, FL: CRC Press, Crustaceans Issues 16.

Sempere, T., R. F. Butler, D. R. Richards, L. G. Marshall, W. Sharp, and C. C. Swicher III. 1997. Stratigraphy and chronology of Upper-Cretaceous-lower Paleogene strata in Bolivia and northwest Argentina. *The Geological Society of America Bulletin* 109(6):709–27.

Seton, M., R. D. Müller, S. Zahirovic, et al. 2012. Global continental and ocean basins reconstruction since 200 Ma. *Earth-Science Reviews* 113:212–70.

Tan, M. H., H. M. Gan, Y. P. Lee, et al. 2018. Order within the chaos: insights into phylogenetic relationships within the Anomura (Crustacea, Decapoda) from mitochondrial sequences and gene order rearrangements. *Molecular Phylogenetics and Evolution* 127:320–31.

Trevisan, A., M. Z. Marochi, M. Costa, S. Santos, and S. Masunari. 2016. Effects of the evolution of the Serra do Mar mountains on the shape of the geographically isolated populations of *Aegla schmitti* Hobbs III, 1979 (Decapoda: Anomura). *Acta Zoologica* 97(1):34–41.

Tudge, C. C., and D. M. Scheltinga. 2002. Spermatozoal morphology of the freshwater anomurans *Aegla longirostri* Bond-Buckup and Buckup, 1994 (Crustacea: Decapoda: Aeglidae) from South America. *Proceedings of the Biological Society of Washington* 115(1):118–28.

Tumini, G., F. Giri, V. Williner, and P. A. Collins. 2016. The importance of biogeographical history and extant environmental conditions as drivers of freshwater decapod distribution in southern South America. *Freshwater Biology* 61(5):715–28.

Vogt, G. 2013. Abbreviation of larval development and extension of brood care as key features of the evolution of freshwater Decapoda. *Biological Reviews* 88(1):81–116.

Xu, J., M. Pérez-Losada, C. G. Jara, and K. A. Crandall. 2009. Pleistocene glaciation leaves deep signature on the freshwater crab *Aegla alacalufi* in Chilean Patagonia. *Molecular Ecology* 18(5):904–18.

Zimmermann, B. L., M. S. Crivellaro, C. B. Hauschild, et al. 2018. Phylogeography reveals unexpectedly low genetic diversity in a widely distributed species: the case of the freshwater crab *Aegla platensis* (Decapoda: Anomura). *Biological Journal of the Linnean Society* 123:578–92.

CHAPTER 2

Morphology, Taxonomy, and Diversity of Extant Aeglidae

Sandro Santos, Georgina Bond-Buckup, and Carlos G. Jara

CONTENTS

2.1 INTRODUCTION

The aeglids are small anomuran crustaceans (1–6 cm of carapace length in adults), endemic of temperate and subtropical continental waters in southern South America (Bond-Buckup and Buckup 1994), occurring in Argentina, Bolivia, Brazil, Chile, Paraguay, and Uruguay (Martin and Abele 1986), between the latitudes of 20°18'S and 50°34'S (Bueno et al. 2007; Oyanedel et al. 2011). They can be found in a variety of freshwater habitats, such as streams, small creeks, rivers, lakes, lagoons, and caves, in altitudes ranging from sea level to 3,500 meters in northeastern Argentinean cordilleras, as well in depths ranging from very shallow waters in creeks to 320 meters in some Chilean lakes (Jara 1977; Bond-Buckup and Buckup 1994). The uniqueness of the group led Schmitt (1942a) to state that "there are no freshwater Crustacea at all like *Aegla* anywhere else in the world."

The first time an aeglid was acknowledged in the literature, with certainty, was in 1818, when Latreille drew a specimen of a new crustacean, with no description or mention to collection locality, naming it as *Galathea laevis* (*Galathea* is a marine genus). Two years later Leach (1820) recognized it as a new genus, which he called *Aegla*, and proposed a new combination to Latreille's specimen as *Aegla laevis*. It is worth mentioning that the first reference of the occurrence of *Aegla* was made by H. Milne Edwards in 1837, who stated that the genus inhabits the coast of Chile.

The resemblance to galatheids made most naturalists consider the aeglids within this anomuran group. Only in 1852, Dana separated the aeglids in a different sub-tribe ("Aegleidea") from Galatheidea, describing the tribe's features, and in 1855 Girard included the genus *Aegla* within Aeglidae. Nevertheless, the family Aeglidae was included in the superfamily Galatheoidea for a long time. However, the morphological, ecological, and biogeographical unique characteristics of aeglids, as well as molecular evidence, challenged this classification (Martin and Abele 1986, 1988; Pérez-Losada et al. 2002, 2004; Tudge and Scheltinga 2002). McLaughlin et al. (2007) finally elevated the aeglids to the superfamily rank, Aegloidea, based on morphological apomorphies of the group, two of them probably reflecting adaptations to the freshwater environment.

2.2 MORPHOLOGY

The first reference to morphological characters attributed to an "apancora," currently one of the popular names of aeglids in Chile, might go back to Molina (1782). The author mentioned a crustacean, with hairy chelae and a triangular, very long tail, which he named *Cancer apancora*. However, the inaccurate description along with the use of the name "apancora," utilized for several species of crustaceans, led to doubts as to whether the description was indeed of an aeglid (Martin and Abele 1988).

Officially, the first representation of the morphological features of an aeglid was made by Latreille (1818), when he drew a specimen with hairy pereopods and a large lobe on the basis of the movable finger. Leach (1820), when creating the genus *Aegla* referring to Latreille's figure, detailed the morphological characters of the species *Aegla laevis*. A number of early carcinologists reproduced the figure of Latreille, adopting the name proposed by Leach, but with inaccuracies, such as Desmarest (1825), who introduced the misspelling *Aeglea*, which was used for many years. Only in 1910, Rathbun called attention to the correct name given by Leach, *Aegla*.

H. Milne Edwards (1836–1844) published new characterizations of the taxon's morphology, with details of the anterofrontal region, the third maxillipeds and the posterior portion of the abdomen. In 1837, H. Milne Edwards redefined the genus, detailing its morphological aspects. After that, still in the XIX century, three new species were described for the genus. Nicolet (1849) described *Aeglea* (sic) *denticulata* for specimens from the region of Osorno, Chile. Girard (1855) described *Aeglea* (sic) *intermedia* for specimens from the Maypu River, 2000 m above sea level, near Santiago, Chile. Müller (1876) described *Aeglea* (sic) *odebrechtii*, for animals from the Marombas River, near Itajaí in southern Brazil. Dana (1852) presented a detailed study on the morphology of aeglids, including the characterization of gills, and redescribing more precisely the species *Aeglea* (sic) *laevis* (Latreille). Detailed figures of the foregut of *A. laevis* were published by Mocquard (1883). Ortmann (1892) provided a general description of the mouthparts of *A. laevis*. Studying parasites and commensals of aeglids, Mouchet (1931a, b, 1932a, b) described the histology and illustrated parts of the gills.

Schmitt (1942a) described 17 new species, increasing the number of known aeglids from four to 21; he also proposed some characters to be used as diagnostic features, such as form of the cheliped, shape of the orbit, proportion of the carapace and rostrum, development of the anterolateral spines, hepatic lobes, cardiac area, among others. Ringuelet (1948a, b) described two new species using biometric analysis. Snodgrass (1950) made a schematic diagram of the ventral anterior body region, exposing the mandibles of *A. prado* Schmitt (1942). Lopretto (1978a, b, 1979, 1980a, b, 1981) described in detail and compared among patogonian species the morphology of the male fifth pereopod coxa, a structure that the author believed could help in the identification of species.

As from the beginning of the twentieth century, with the description of new species being published, the external morphology of the aeglids has become better characterized and illustrated, allowing more accurate comparisons. Among these publications stand out those from Schmitt (1942a, b), Jara (1977, 1980, 1982, 1986, 1989), Buckup and Rossi (1977), and Bond-Buckup and Buckup (1994). Since then, the morphological characters commonly used in the descriptions are: shape of the rostrum, carapace outline, anterolateral angle of the second abdominal somite, ventral margin of the second pereopod, ornamentation of the fourth thoracic sternite, and several features of the cheliped, such as palmar lobes, basal tooth, and ischium spines. It is known that these characters are extremely variable among individuals; moreover, until 1988 (Martin and Abele), they were seldom illustrated in the studies and often described in an incomplete way or even omitted. Generally speaking, a few descriptions were realized taking into account characters other than those proposed by Schmitt (1942a, b) (Martin and Abele 1988).

2.2.1 Main Morphological Characters Used in the Taxonomy of Aeglids

The most remarkable feature of the genus *Aegla* is its carapace, with several grooves. The distinct cervical groove represents the limit between the anterior and posterior cephalothorax. The anterior central region corresponds to the gastric area (Figure 2.1), delimited by the epigastric prominence at the front and by the cervical groove at its rear end. At both sides of the gastric area, the hepatic areas are found (Figure 2.1). The *linea aeglica dorsalis* is located behind the cervical groove and in postero-lateral direction (Figure 2.2) which, together with the *linea aeglica lateralis*, extended along the carapace border, delimiting the anterior branchial area. On a lateral view of carapace, the *linea aeglica* extends from the antennal sinus up to a point in which it meets the dorsal and lateral *linea aeglica* (Figure 2.2). The *linea aeglica* continues downward as *linea aeglica ventralis* ending at the branchiostegal margin (Figure 2.2). At the rear end of the *linea aeglica*, the epibranchial lobe is found (Figure 2.1); its apex is blunt or acute, carrying a spine, tubercle, or scale.

On the posterior part of carapace two lines are notorious: the dorsal longitudinal and the transverse dorsal (Figure 2.2). The transverse gives rise to the branchial lines

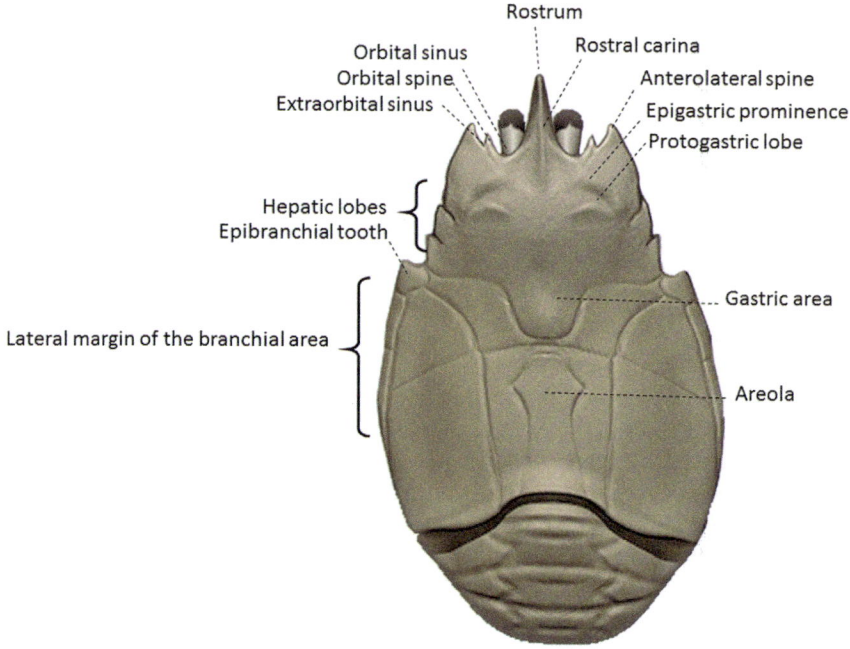

Figure 2.1   A three-dimensional scheme in dorsal view of a crustacean of the genus *Aegla*, with indications of the primary structures and regions of the body.

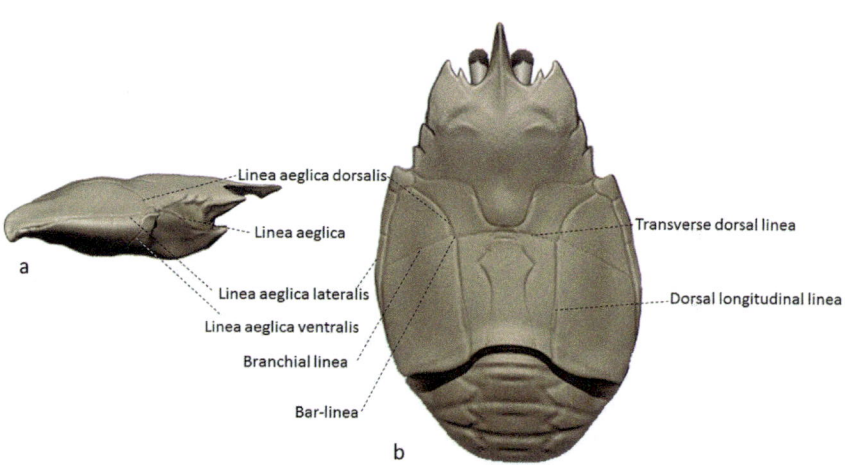

Figure 2.2   A three-dimensional scheme of *Aegla* with the indication of carapace sutures, which are characteristic of the group: (a) lateral view; (b) dorsal view.

extended from one to the opposite carapace margins, delimiting the interior, anterior, and posterior branchial areas. The dorsal longitudinal line extends from the joint between the transverse dorsal linea and the linea aeglica dorsalis to the posterior border of the carapace, parallel to the lateral of the areola.

The carapace presents considerable variation within and among species, just as most characters used in the aeglid taxonomy. However, some features are consistent and unique to the family, such as the carapace, usually dorso-ventrally flattened, giving the animal a flat aspect (Martin and Abele 1988). A cervical groove splits the dorsal surface into two regions: a narrow anterior area and a wider posterior region (Martin and Abele 1988). In the anterior region are some of the main characters used in the taxonomy of the genus *Aegla*, which will be detailed next.

2.2.2 Cephalothoracic Carapace

1. Rostral shape in dorsal view (Figure 2.3): Triangular (broad based gradually narrowing toward the rostral tip)—as in *A. bahamondei*, *A. ringueleti*, and *A. violacea*; linguiform (roughly triangular but lateral borders subparallel along central part of rostrum then converging onto rostral tip); dorsum flattened and carina evanescent as in *A. affinis*, *A. laevis laevis*, and *A. ligulata*; styliform: rostrum narrow, elongate, sharply tipped—as in *A. carinata*, *A. riolimayana*, and *A. uruguayana*.
2. Rostral shape in lateral view (Figure 2.4): From epigastric prominence to rostral tip: straight as in *A. hueicollensis* and *A. parana*; recurved—as in *A. papudo* and *A. spinipalma*; deflected—as in *A. humahuaca* and *A. obstipa*.
3. Rostrum length: Value obtained dividing CL (cephalothorax length, between the rostral tip and the midpoint of the rear border of carapace) by RL (rostral length, between the rostral tip and the midpoint of a line drawn at the height of the deepest point of orbital sinus): lengthy - CL/RL=less than 3.9, as in *A. ludwigi* and *A. sanlorenzo* (representing 20.7% of the known species); medium: CL/RL=4.0 to 4.9, as in *A. franca* and *A. loyolai* (56.1% of the species); short: CL/RL=more

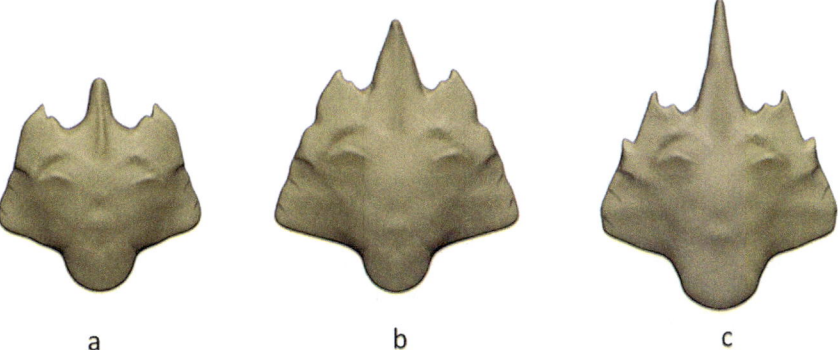

a   b   c

Figure 2.3 Rostral shape in dorsal view: (a) triangular (broad based gradually narrowing toward the rostral tip; e.g., *Aegla violacea*); (b) linguiform (roughly triangular but lateral borders subparallel along central part of rostrum then converging onto rostral tip, e.g., *A. ligulata*); (c) styliform (rostrum narrow, elongate, sharply tipped, e.g., *A. uruguayana*).

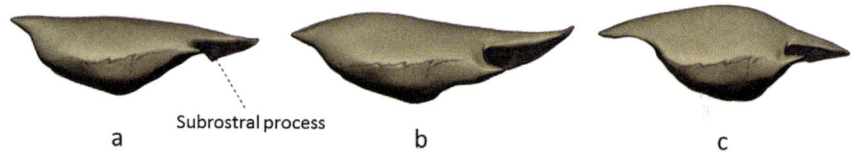

a   Subrostral process   b   c

Figure 2.4   Rostral shape in lateral view in Aeglidae: (a) rectum, as in *Aegla parana*; (b) recurved, as in *A. papudo*; (c) deflected, as in *A. obstipa*. Subrostral process present in *A. paulensis* and absent in *A. spinipalma*.

than 5.0, as in *A. oblata* and *A. strinatii* (23.2% of the species). Although this character is unique, it is worthy to note that the rostrum's dimensions vary significantly among specimens within a single population, mainly when considering the ontogenetic development of the individuals.

4. Scales on the rostral carina: With one row—as in *A. intercalata*; with two rows—as in *A. bahamondei*; with multiple rows—as in *A. carinata* and *A. manuinflata*.

5. Extension of rostral carina: Entire (carina extended all along rostral dorsum)—as in *A. ringueleti*; partial (carina extended scarcely on proximal third)—as in *A. lancinhas*.

6. Subrostral process: The rostrum may have a weak ventral crest that projects basally as a subrostral process. This process may be present (Figure 2.4), as in *A. paulensis* and *A. violacea* or absent, as in *A. camargoi* and *A. spinipalma*.

7. Height of the ventral portion of the rostrum: The extent of ventral portion is compared with the height of rostrum above, being: shallow, if ventral portion rostral is equal or less than the height of rostrum above—as in *A. longirostri* and *A. schmitti*; deep, if the ventral portion rostral is more than the height of rostrum above—as in *A. denticulata denticulata* and *A. serrana*.

8. Orbital sinus: On each side of the rostrum a broad shallow excavation on the margin of the carapace forms the orbital sinus (Figure 2.1), which is flanked ventrolaterally by the orbital spine. This spine may be acute with a cornified tip and nearly equal in length to the anterolateral spine (e.g., *A. platensis*) or it may be reduced and coalesced with the anterolateral spine or absent (e.g., *A. concepcionensis*, *A. papudo*, and *A. serrana*).

9. Extraorbital sinus (Figure 2.1): The extraorbital sinus separating the orbital spine from the anterolateral spine may be of variable depth, or absent as in *A. pomerana*.

10. Anterolateral lobe or spine of the carapace (Figure 2.1): The anterolateral spine is typically acute with a corneous tip and may exceed the length of the eyestalk (e.g., *A. sanlorenzo*), although in most species it extends only as far as the posterior margin of the cornea (e.g., *A. odebrechtii*).

11. Relative width of the front (Figure 2.5): Value obtained dividing PCW (precervical width, as distance between left and right epibranchial margins) and FW (frontal width, as distance between anterolateral angles of carapace); wide: PCW/FW 1.79 to 1.89 (e.g., *A. alacalufi*); narrow: more than 1.90 (e.g., *A. abtao*).

12. Epigastric prominence: At about the level of the first hepatic lobe is a small raised area termed as epigastric prominence (Figure 2.1). This prominence may be highly granulate and obvious (e.g., *A. castro* and *A. platensis*), or it may be inconspicuous or even absent (e.g., *A. parana* and *A. plana*).

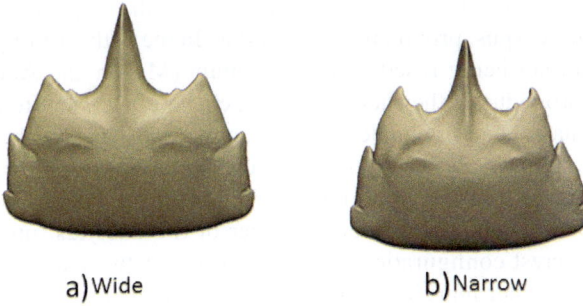

a) Wide   b) Narrow

Figure 2.5 Relative width of the front of Aeglidae: (a) wide: PCW/FW 1.79 to 1.89; (b) narrow: PCW/FW more than 1.90. Value obtained dividing PCW (precervical width, as distance between left and right epibranchial margins) and FW (frontal width, as distance between anterolateral angles of carapace).

13. Protogastric lobes (Figure 2.1): Directly posterior or slightly posterolateral or posteromedial to the epigastric prominences. As in the epigastric prominence, the protogastric lobe may be obvious and granulate (e.g., *A. denticulata*), or indistinguishable from the surrounding carapace (e.g., *A. plana* and *A. rostrata*).
14. Hepatic areas: Both sides of the gastric area are subdivided in three hepatic lobes separated by shallow grooves; their external margins being ornamented with scales or tubercles.
15. Carina along dorsal cephalothorax median line (extended from behind epigastric prominences up to the areola's rear margin): present—as in *A. carinata* and *A. denticulata*; absent—as in *A. jujuyana*, *A. serrana*, and *A. spinipalma*.
16. Epibranchial lobe or tooth: The anterior part of this region is an acute, spine-tipped lobe, usually with a lateral border of smaller spinules. It is separated from the pterygostomial and anterior regions of the carapace by the bifurcated *linea aeglica* (Figure 2.2). Epibranchial tooth ornament: absent (apical portion of lobe blunt and smooth)—as in *A. serrana*; tuberculate (apical portion of lobe with tubercle and scales)—as in *A. itacolomiensis* and *A. violacea*; spiny (apical portion of lobe elongated with spine)—as in *A. denticulata* and *A. jarai*.
17. Lateral margins of posterior branchial areas (Figure 2.1): Not recurved—as in *A. papudo*; recurved—as in *A. parana*.
18. Areola's shape (Figure 2.1): Dimensional relationship between AL (areola length) and AW (areola width): quadratic/subquadratic: AL/AW = less than 1.69, as in *A. manuinflata* and *A. meloi*; subrectangular/rectangular: AL/AW = more than 1.70—as in *A. japi A. pomerana* and *A. rostrata*.

2.2.3 Cephalothoracic Structures

The main cephalothoracic structures used for aeglid taxonomy include the appendages involved in a range of functions such as locomotion, grooming, brooding, feeding, chemoreception, and touching, most of which are still little studied in aeglids (Martin and Abele 1988). The cheliped (first pair of pereopod), as the second, third, and fourth pereopods, is formed by the sequence of coxa, basi-ischium, merus, carpus, propodus, and dactylus. In the fifth pair of pereopods, the basis differs in not being fused with the ischium (Martin and Abele 1988). The dactylus and propodus of the cheliped form a chela (pincers). Often the chelipeds are larger in males than in females, and in many occasions one cheliped is larger than the opposite in males (heterochely, see Chapter 3). Traditionally, the cheliped ornamentation provides important characters for the taxonomy, such as their shape, presence of lobe on the dorsal border of the dactylus, pre-dactylar lobe shape, palmar crest configuration, presence of one or two carpal crests, number of spines on the inner margin of carpus (or dorsal border, see Martin and Abele 1988), type of ornaments on the ventral face of basi-ischium (ventromesial border, see Martin and Abele 1988). Moraes et al. (2016) called attention to the different terms used to describe the taxonomic characters related to the chelipeds of aeglids. According to these authors, Martin and Abele (1988) considered the inward rotation of the first pair of chelipeds, while Schmitt (1942a, b) and Bond-Buckup and Buckup (1994) did not. Martin and Abele (1988) considered the homology of this rotation in relation to other Decapoda, while Schmitt (1942a, b) and Bond-Buckup and Buckup (1994) considered the present condition of the appendices (to check the differences, see Figure 2.6). Pereopods 2 to 4 are similar in shape, not chelate, and locomotor. The sexual pores open on the ventral face of coxae of the female's third pair of pereopods. Sometimes the ornamentation of these appendages provides taxonomical characters, as the presence of setae and scales on the

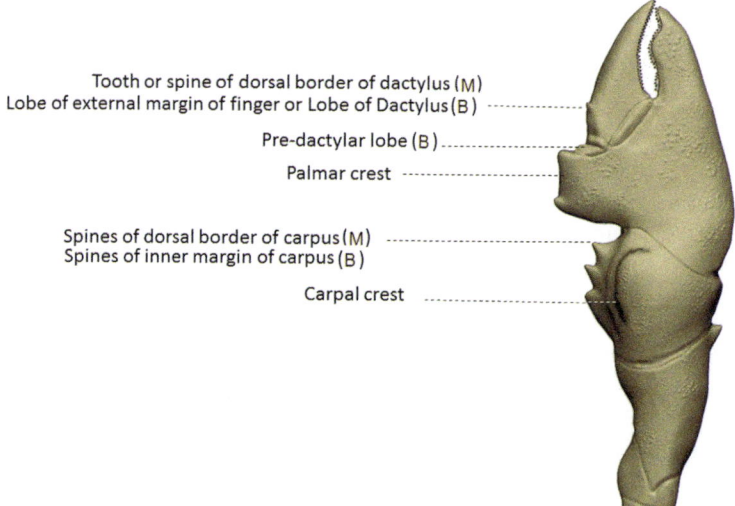

Tooth or spine of dorsal border of dactylus (M)
Lobe of external margin of finger or Lobe of Dactylus (B)
Pre-dactylar lobe (B)
Palmar crest
Spines of dorsal border of carpus (M)
Spines of inner margin of carpus (B)
Carpal crest

Figure 2.6   Right cheliped of an aeglid with indication of the primary structures. (B) Terminology adopted from Bond-Buckup and Buckup (1994); (M) terminology adopted from Martin and Abele (1988).

dorsal margin of propodus and carpus in second, third, and fourth pereopods in *A. meloi* (Santos et al. 2015).

19. Eyestalk: Normal (eyestalks and cornea with normal size, as in most species); reduced (eyestalks and cornea of small size, slender, and with few ommatidia), which occurs in troglobitic species such as *A. cavernicola*, *A. leptochela*, and *A. microphthalma*.
20. Size of adult male chelipeds: Equal (same size)—as in *A. denticulata lacustris* and *A. humahuaca*; unequal (different size)—as in *A. lata* and *A. neuquensis*.
21. Palmar crest: Absent, or rudimentary, as in *A. manuinflata* and *A. uruguayana*, or present, as in most species. When present, the crest may be subrectangular as in *A. longirostri*; subdisciform, as in *A. ludwigi* and *A. schmitti;* or projected in a spine, as in *A. sanlorenzo* (Figure 2.7).
22. Carpal crest or ridge, in dorsal view (one or two ridges): With two ridges, as in *A. denticulata denticulata* and *A. parana*; with one ridge, as in *A. lancinhas* and *A. leachi*.
23. Dorso-distal lobe of palmar crest or pre-dactylar lobe (Figure 2.6): Present, as in *A. affinis* and *A. intercalata*; absent, as in *A. loyolai*.
24. Lobe on proximal end of external margin of finger of chelipeds (Figure 2.6): Present (sometimes rudimentary), as in *A. rostrata* and *A. scamosa*; absent, as in *A. grisella* and *A. serrana*.
25. Dorsal margin of merus of second pereopod: With scales and setae, as in *A. loyolai*; smooth, as in *A. renana*.
26. Ventral margin of merus of second pereopod: With tubercles/scales or spines, as in *A. expansa* and *A. jarai*; smooth, as in *A. georginae* and *A. ludwigi*.
27. Ornaments of fourth thoracic sternum (spine, tubercles, or scale) (Figure 2.8): Absent, as in *A. spinosa*; present, as in *A. carinata* and *A. platensis*.

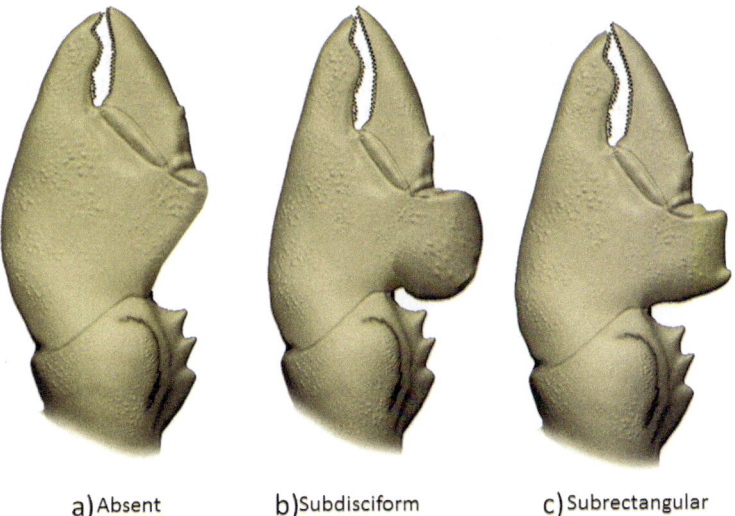

a) Absent   b) Subdisciform   c) Subrectangular

Figure 2.7 Palmar crest types in Aeglidae: (a) absent, or rudimentary, as in *Aegla manuinflata*; (b) subdisciform, as in *A. ludwigi*; (c) subrectangular, as in *A. longirostri*.

2.2.4 Abdomen

The abdomen, formed by six somites, is normally wider and more protuberant in females than in males (see Chapter 3). The first abdominal somite is reduced, and the rear border of the carapace partially covers its tergite; it lacks appendages in both sexes. The second abdominal somite is the largest, and its tergite is ornamented with transverse furrows and protuberances; its laterodorsal pleura extends frontally in an anterolateral angle, its tip being smooth or ornamented with an apical scale or even an acute spine (Figure 2.8), being those features of taxonomic importance. Third to fifth somites are similar in size and do not contribute characters of taxonomic value, but males have vestigial abdominal appendages (except *Aegla charon* Bueno and Moraes, 2017), whereas these appendages are usually well developed from the third to fifth somite of the females. Uropods are the appendages of the sixth abdominal somite, and they stand out of the abdominal ventral cavity where the remaining pleopods (females) are situated; their narrow free margins carry a row of long plumose setae (Figure 2.9). Uropods are paddle-like, and they form, together with the telson plate, a tail fan functioning as a propulsor device mostly employed when crabs escape from menacing situations, swimming retrogradingly.

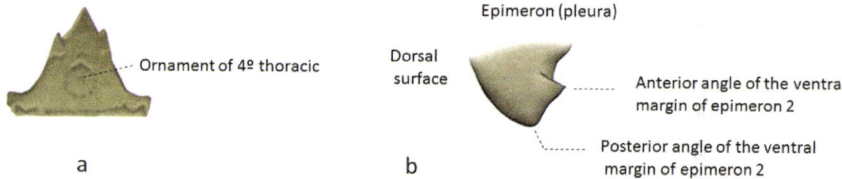

Figure 2.8   (a) Ornament of the fourth thoracic sternite (spine, tubercle, or scale): absent, as in *Aegla spinosa*; present, as in *A. platensis*; (b) dorsum of second abdominal pleura (smooth, as in *A. jujuyana*; carinate, as in *A. rostrata*). Latero-dorsal angle of the second abdominal pleura (blunt, as in *A. intercalata*; acute, with scale/tubercle or spine, as in *A. longirostri*).

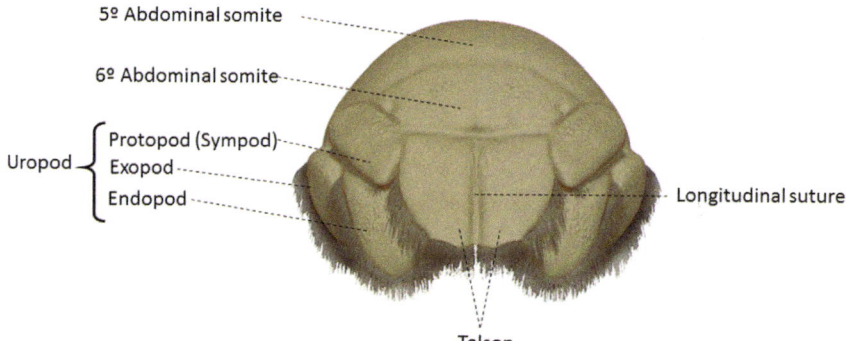

Figure 2.9   Aeglid tail fan, with indication of the telson and uropods.

28. Dorsum of second abdominal pleura (Figure 2.8): Smooth, as in *A. jujuyana* and *A. pomerana*; carinate, as in *A. rostrata* and *A. singularis*.

29. Latero-dorsal angle of second abdominal pleura (Figure 2.8): Blunt, as in *A. intercalata* and *A. oblata*; acute (with scale/tubercle or spine), as in *A. longirostri* and *A. rostrata*.

30. Telson (Figure 2.9): As the last piece of the abdomen, it has a wide shape being subquadrate, oval, or cordiform. In most species the telson plate is formed by two separated plates articulated at the median line (dimerous plate), but in *A. alacalufi*, *A. papudo*, and in some specimens of *A. parva* the plates are fused along the median line, leaving only a furrow where a functional joint exists (monomerous plate) in the remaining species.

2.2.5 Characters More Recently Used and New Proposals

A significant increase in descriptions of new species of the genus *Aegla* has occurred in recent years (Santos et al. 2017). Despite the new records, there are reports in the literature indicating that the cumulative curve of described species of aeglids has not stabilized yet (see Santos et al. 2017). Some aeglids present a restricted distribution, occurring only in a stream or subbasin, while others have a wide distribution. This verification, along with other biological evidence, made some researchers suspect that the group could present some examples of cryptic species (see Chapter 1). Cryptic lineages are difficult to identify morphologically, despite all the evolutionary processes they have gone through during speciation (Bickford et al. 2007). As seen earlier, the diagnostic characters in aeglids, mainly proposed by Schmitt (1942a, b), are limited and probably insufficient to distinguish species encompassing a complex. Facing this scenario, increasing the base of diagnostic characters for the aeglids becomes an utmost necessity.

Moraes et al. (2016), in addition to the shape and size of some body structures traditionally used in the description of aeglid species, proposed other relationships determined by means of algebraic calculations, such as:

31. Shape of rostrum: Calculated as the ratio between RBW (width of the base of the rostrum) and LMR (length of the margin of the rostrum). Shapes recognized are: narrow base triangle (< 1.00) and wide base triangle (≥ 1.00), with the equilateral condition (= 1.00).

32. Shape of areola: Calculated as the ratio of the areolar posterior margin (x) and the areolar anterior demarcation, measured as the distance between the centers of lateral terminal pits (y). If the ratio is ≥ 1.7, then the shape is "trapezoidal." If, however, the ratio is < 1.70, the following algebraic calculation is further required: {h/[(x+y)/2]}, where "h" is the areolar height (Moraes et al. 2016). The following shapes can now be recognized: "rectangular" (> 2.30), "subrectangular" (> 1.60 to 2.30), or "subquadrate" (> 1.00 to 1.60).

33. Shape of cardiac area: Calculated as the ratio between length (as straight line) of transverse dorsal linea (TDL) (a) and length (as straight line) of mesial section of posterior margin at level of intersection with dorsal longitudinal lineae (DLL) (b). Shape is "trapezoidal" when the ratio is ≥ 1.35, or "subrectangular" when the ratio is < 1.35.

34. Shape of the palmar crest of major and minor chelae (outer surface view): It is based on the inflexion (= angle) formed between the posterior margin of the palmar crest and proximal dorsal margin of propodus near the articulation with carpus. Three distinct shapes are recognized: "rudimentary" when inflexion is smooth (angle markedly obtuse), "rectangular" when inflexion tends to orthogonal (right or slightly obtuse angle), and "disciform" when inflexion is convex.
35. Width of the uropods: Based on the ratio between maximum width of the endopod and half the maximum width of telson (taken from anterolateral margin to longitudinal suture). The recognized shapes of uropods are: "wide," when the ratio is $\geq$ 1.00, and "narrow" when the ratio is < 1.00.

Lopretto (1978a, b, 1979, 1980a, b) was one of the first researchers to seek new taxonomic characters for the aeglids, beyond those traditionally used up to then. Based on the taxonomic value of the sexual appendages in many Decapoda, the researcher proposed the utilization of the fifth pair of pereopods as a diagnostic character in aeglids, more specifically, the sexual tube (ventromesial lobe, according to Martin and Abele 1988). Recently, these structures were successfully used to characterize species of the *Aegla paulensis* complex (Moraes et al. 2016). Hereafter, we list some structures that may contribute in the future for the diagnostic of aeglids, mainly for those species composing cryptic complexes. Some of these structures have already been evaluated, as the sexual tube of the fifth pereopod (Moraes et al. 2016), while others still need to be tested:

36. Sexual tube of fifth pereopod (ventromesial lobe: Martin and Abele 1988): A spoon-shaped process that supports the extended vas deferens. Shape of the sexual tube: long and narrow, as in *Aegla platensis* s. lat.; or short and wide, as in *A. leptochela*.
37. Second antenna: Described previously by Martin and Abele (1988) for *A. uruguayana* as much longer than the first antenna and may be twice the length of the body. The peduncle presents five segments, with segments 2 and 3 fused. The basal article (coxa) has a deep ventral groove containing the dorsal border of the pterygostomial region of the carapace, and a large mesial tubercle that bears the aperture of the antennal gland. The mesial border bears many long simple setae and the lateral border has dispersed short setae. The second article is short with a triangular plate extending from its dorsolateral surface; this remnant of an exopod is usually termed a scaphocerite when developed and an antennal scale, squama, or acicle when reduced. The third article is fused with the second, representing the fused basi-ischium. This third segment is longer on the mesial border, which creates an angular articulation; distal segments (four and five) are thereby directed toward the midline so that the antennae appear to originate from under the rostrum. The fourth article or merus is stout and cylindrical; the fifth article or carpus is slightly longer than the fourth and tapers toward the proximal end. The flagellum is long and multi-articulate, each article except the first having a circle of small simple setae on the distal border.
38. Mandibular palp: Described for *A. platensis* by Martin and Abele (1988): two-segmented; the proximal segment bears several simple and plumose setae on the distal half of the dorsal border; the distal segment is flattened and ovate with many simple, pappose, and plumose setae along the entire border; the two segments are approximately equal in length.

39. Third maxilliped: Appendix previously described by Martin and Abele (1988) for
    A. platensis. The dactylus is subcylindrical and armed with simple, stout simple,
    pappose, pore, and serrate setae; most of the serrate setae are located on the distal
    half. The propodus is slightly longer than the dactylus and bears fewer serrate setae;
    a circular field of dense setae on the inner surface contains pappose, sword, serrate,
    and stout serrate setae. The carpus is short and thick with few setae on the outer
    surface but with a circular field of dense pappose, sword, serrate, and stout serrate
    setae on the inner surface.
40. Coxa of female third pereopod: The coxa of the third pereopod in females bears the
    genital aperture on the ventromesial surface.

2.3 TAXONOMY AND DIVERSITY

The family Aeglidae currently presents 87 extant described species and subspecies, all belonging to the genus *Aegla* (Bueno et al. 2017; Santos et al. 2017;
Jara et al. 2018; Páez et al. 2018), but this number will soon exceed 100 since
many new species are to be described (see, for example, Crivellaro et al. 2017, and
Zimmermann et al. 2018). Two monotypic, fossil genera from marine sediments are
also known: *Haumuriaegla glaessneri* and *Protaegla miniscula* (Feldmann 1984;
Feldmann et al. 1998).

If the family Aeglidae is not rich in number of genera, the genus *Aegla* can
be considered very speciose. For the matter of comparison, the known diversity of
crayfish of the genus *Parastacus* is 13 species (Huber et al. 2018); this crayfish is
also endemic to southern South America, occurring in Argentina, Brazil, Chile, and
Uruguay (Ribeiro et al. 2017), and presents a similar distribution to that of aeglids.

The diversity of *Aegla* is astonishing, not only regarding the number of species
but also in the variety of freshwater habitats to which the group has adapted to live.
Most species inhabit lotic environments, but some species also live in lacustrine
habitats, such as *A. rostrata*, which can be found both in rivers and lakes, in shallow and deep waters (Jara 1977). Many species can be found at low to intermediate
altitudes, such as *A. longirostri* (see Zimmermann et al. 2018), and others live at high
altitudes, such as *A. septentrionalis* (see Bond-Buckup 1994).

While most *Aegla* species live in epigean habitats, *A. cavernicola*, *A. charon*,
A. leptochela, and *A. microphtalma* are stygobiotic (obligate cave-dwelling), living
in subterranean waters in southeastern Brazil and presenting some troglomorphic
adaptations to this type of habitat (Türkay 1972; Bond-Buckup and Buckup 1994;
Bueno, Camargo and Moraes 2017). On the other hand, *A. schmitti* and *A. strinatti*
are trogophilic species, meaning that they can be found both inside and outside caves
(Rocha and Bueno 2004, 2011; Fernandes et al. 2013).

The current classification of the extant aeglids, based on Martin and Davis
(2001), and McLaughlin, Lemaitre, and Sorhannus (2007), is as follows:

Order Decapoda Latreille (1802)
Suborder Pleocyemata Burkenroad, 1963
Infraorder Anomura MacLeay (1838)

Superfamily Aegloidea Dana (1852)
Family Aeglidae Dana (1852)
Genus *Aegla* Leach (1820)

Next, we alphabetically list the species described to the present and the diagnosis for each one. Where the diagnosis is not the original, we say who realized it. The geographical distribution area of each species is according to the review in Santos et al. (2017).

2.3.1 *Aegla abtao* Schmitt, 1942

Diagnosis (Bond-Buckup and Buckup 1994): Carapace anterolateral spine not reaching base of cornea; lateral margins of areola subparallel; rostrum triangular, median length, rostral carina slightly sharp in distal third; anterior angle of ventral margin of epimeron 2 with reduced spine; cheliped movable finger with distinguished lobe on outer proximal margin; palmar crest subrectangular, excavated; inner margin of ventral surface of cheliped ischium with tubercles and setae.

Distribution: Chile—Toltén River basin (Colico, Allipen, Toltén, Donguil, Pedregoso rivers, Caburgua, Colico lakes); Valdivia-Cruces River basin (Cruces, Pirén rivers); Valdivia-Calle Calle River basin (Huanehue River); Queule River basin (Queule River); Bueno River basin (Curileufu, Ralitrán, Huilma, Pilmaiquen, Negro, Rahue rivers, Maihue Lake); Maullín River basin (Maullín River, Llanquihue, Todos Los Santos lakes); Chamiza River basin (Chamiza River), Puelo River basin (Puelo River), Chiloé Island (San Juan, Huicha, Butalcura rivers, Huillinco Lake).

2.3.2 *Aegla affinis* Schmitt, 1942

Diagnosis (Bond-Buckup and Buckup 1994): Anterolateral spine of carapace reaching middle of cornea; protogastric lobes absent; transverse dorsal linea sinuous; extra-orbital sinus absent; rostrum very long in adult males, ligulate; rostrum with carina sligthly sharp in distal third; fourth thoracic sternite with several scales; outer proximal margin of movable finger of cheliped without lobe; palmar crest very modest; inner margin of ventral surface of ischium of cheliped only with up to six tubercles; ventral margin of merus of second pereopod with scaliform tubercles.

Distribution: Argentina—Colorado River basin (Colorado, Barrancas, Chico rivers); Desaguadero River basin (Atuel, Tunuyan, Chico, Barrancas, Malargüe rivers, La Matancilla, El Salto, Claro creeks). Chile—Maule River basin (Maule Lake, Maule River).

2.3.3 *Aegla alacalufi* Jara and López, 1981

Diagnosis (Bond-Buckup and Buckup 1994): Anterolateral spine of carapace extending beyond base of cornea; protogastric lobes obsolete; extra-orbital sinus absent; rostrum long in adult males, acuminate, carina absent in distal third; anterior angle of the ventral margin of epimeron 2 unarmed; telson without longitudinal suture; fourth thoracic sternite medially elevated with scale; outer proximal margin of movable finger of cheliped with lobe; cheliped fingers without lobular tooth; palmar crest subrectangular; inner margin of ventral surface of ischium of cheliped only with tubercles and nodes.

Distribution: Chile—Reloncaví River basin (Reloncaví Fiord); Petrohué-Puelo River basin (Cochamó River); Huequi River basin (Huequi River); Chiloé Island (Puchagrán, Chadmo rivers, Huillinco, Tarahuín lakes); Palena River basin (Palena River); Yelcho River basin (El Amarillo River, Yelcho Lake); Madre de Dios Island; Duke of York Island (North, South, Pollux, Thompson rivers).

2.3.4 *Aegla araucaniensis* Jara, 1980

Diagnosis (Bond-Buckup and Buckup 1994): Cephalothorax convex but without longitudinal dorsal carina defined; anterolateral spine of carapace reaching base of cornea; protogastric lobes obsolete; transverse dorsal linea straight; rostrum with medium length in adult males, wide in the base, excavated and moderately ligulate; rostrum with carina slightly sharp in distal third; anterior angle of ventral margin of epimeron 2 with scale; outer proximal margin of movable finger of cheliped with rudimentary lobe; palmar crest subrectangular, slightly excavated, toothed; inner margin of ventral surface of ischium of cheliped only with elevations; ventral margin of merus of second pereipod with scales.

Distribution: Chile—Valdivia River basin (Leufucade, Calle Calle rivers, Challahuin, Chesque, Coihueco, Quilquil, Quilen, Puquiñe, Huillilelfun, Puente Negro, Ñancul creeks, Riñihue lake); Bueno River basin (Curileufu, Puquitre, Coinco, Lumaco, Sagllue, Pichimaule rivers, Cañal creeks); Petrohué River basin (Cayutue River, Todos los Santos Lake); Maullin River basin (Lahuen Ñadi, Machete creeks); Chilóe Island (Caulín, Huicha, Gamboa creeks, Tarahuin Lake).

2.3.5 *Aegla bahamondei* Jara, 1982

Diagnosis (Bond-Buckup and Buckup 1994): Cephalothorax elevated but without longitudinal dorsal carina defined; lateral margins of anterior branchial area of carapace moderately expanded with tubercles; anterolateral spine of the carapace reaching base of cornea; protogastric lobes present; rostrum triangular with medium length, acuminate, carina absent in distal third; anterior angle of ventral margin of epimeron 2 with conical spine; fourth thoracic sternite with conical tubercle; outer proximal margin of movable finger of cheliped with lobe; palmar crest present, toothed, subrectangular; inner margin of ventral surface of ischium of cheliped with a distal tubercle and a small proximal elevation.

Distribution: Chile—Bío Bío River basin (Rafael River); Lebú-Paicavi River basin (Caramávida-Tucapel, Pingueral, Caramávida, Cayucupil, Butamalal rivers).

2.3.6 *Aegla brevipalma* Bond-Buckup and Santos, 2012

Diagnosis: Anterolateral spine of carapace short, not reaching base of cornea; protogastric and epigastric lobes absent; extra-orbital sinus absent, rostrum triangular, deflected, carinate, outer proximal margin of movable finger of cheliped with lobe; palmar crest of cheliped subdisciform; anterior angle of ventral margin of epimeron 2 with modest scales; inner margin of ventral surface of ischium of cheliped with distal scaliform tubercule.

Distribution: Brazil—Uruguay River basin (Matador River).

2.3.7 *Aegla camargoi* Buckup and Rossi, 1977

Diagnosis (Bond-Buckup and Buckup 1994): Anterolateral spine of carapace extending beyond base of cornea; protogastric lobes present; rostrum long in adult males, carinate along entire length; anterior angle of ventral margin of epimeron 2 with spine; outer proximal margin of movable finger of cheliped with lobe ornamented with spine and tubercle; palmar crest of cheliped subdisciform; inner margin of ventral surface of ischium of cheliped with a distal, conical tubercle, a second proximal one, and between them scaliform tubercles; ventral margin of merus of second pereipod with tubercles.

Distribution: Brazil—Uruguay River basin (Apuaê-Inhandava, Pelotas rivers).

2.3.8 *Aegla carinata* Bond-Buckup and Gonçalves, 2014

Diagnosis: Anterolateral spine of carapace extending beyond middle of cornea; protogastric lobes very elevated and with scales; rostrum styliform, carinate along its entire length; cephalothorax with longitudinal dorsal carina on median line, ornamented with scales up to anterior region, meeting the posterior areola; lateral margins of branchial anterior and posterior areas of carapace arched, expanded as a lamina and with tubercles; extra-orbital sinus wide; outer proximal margin of movable finger of cheliped lacking lobe; fingers of cheliped with lobular denticle; palmar crest modest, subrectangular, projected in distal spine; anterior angle of ventral margin of epimeron 2 projected in recurved and robust spine; inner margin of ventral surface of ischium of cheliped with modest distal scaliform tubercle; dorsal margin of carpus of second, third, and fourth pereopods with distal spine followed by scaliform tubercles tipped with tufts of setae.

Distribution: Uruguay—Uruguay River basin (Cuñapiru Creek).

2.3.9 *Aegla castro* Schmitt, 1942

Diagnosis (Bond-Buckup and Buckup 1994): Lateral margins of carapace expanded in branchial region; anterolateral spine of carapace extending beyond base of cornea; protogastric lobes elevated; areola rectangular, long, and narrow; rostrum long in adult males, acuminate, carinate along entire length; anterior angle of ventral margin of epimeron 2 with spine; outer proximal margin of movable finger of cheliped with distinguished lobe; palmar crest of cheliped disciform, excavated; carpal crest of cheliped with tubercles; inner margin of ventral surface of ischium of cheliped with a distal robust spine, a proximal tubercle and up to two scaliform tubercles between them; dorsal margin of merus of second pereipod with tubercles; ventral margin of merus of second pereopod with one or two spines followed by scaliform tubercles.

Distribution: Brazil—Paraná River basin (Upper Paranapanema, Tibagi, Ivai rivers).

2.3.10 *Aegla cavernicola* Türkay, 1972

Diagnosis (Bond-Buckup and Buckup 1994): Anterolateral spine of carapace reaching base of cornea; protogastric lobes obsolete; eyestalks with reduced dimensions; cornea reduced; rostrum triangular, wide in the base, without carina in distal third; outer proximal margin of movable finger of cheliped without lobe; cheliped fingers with lobular tooth; palmar crest of cheliped very modest; inner margin of ventral surface of ischium of cheliped with one distal spine and two scaliform tubercles; pereopods with long segments.

Distribution: Brazil—Southeastern Atlantic system (Ribeira do Iguape River basin, Areias Grot and Areias de Baixo Cave).

2.3.11 *Aegla charon* Bueno and Moraes, 2017

Diagnosis: Pigmented area of cornea slightly reduced; rostrum long with fringe of setae subventrally on mid- and proximal third of lateral margins; tuft of setae behind shaped subrostral process; anterolateral spine curved upward and not reaching basal margin of cornea; cardiac area trapezoidal; areolar area subrectangular; partially developed uniramous, one- or two-segmented pairs of pleopods on pleonal somites 2–5 in adult male specimens; posterolateral margin of telson concave mesially.

Distribution: Brazil—Restricted to the type locality: "Lago Subterrâneo" cave, in the state of São Paulo.

2.3.12 *Aegla chilota* Jara, Pérez-Losada, and Crandall, 2018

Diagnosis: Rostrum neatly triangular, short, scarcely surpassing eyes; orbital sinus wide, shallow, limited by tiny extra-orbital spine and extra-orbital sinus; carpal lobe prominent, triangular, tipped by at least two coalescent acute conical scales; second abdominal epimeron little produced, tipped by acute scale; telson plate roughly pentagonal, with functional median suture; presence of the rectangular palmar crest.

Distribution: Chile—Chiloé Island.

2.3.13 *Aegla cholchol* Jara and Palacios, 1999

Diagnosis: Carapace slightly longer than wide, sparsely setose, moderately expanded at branchial level; rostrum long, subtriangular, acute, styliform at distal half; orbital spine prominent; extraorbital sinus wide, U-shaped; anterolateral angle of first hepatic lobe spiniform; anterior branchial margin subdenticulated, posterior finely serrated, dorsally upturned; anterolateral angle of second abdominal epimeron spiniform; fourth thoracic sternum with blunt medial tubercle, occasionally with one or two scales on frontal border; telson divided; adult males markedly heterochelous, chelae robust; propodus inflated though dorso-ventrally flattened; dactylar lobe blunt; palmar crest subrectangular or arcuate, expanded, dentated; palmar lobe low, blunt, with scales only on frontal edge; medial scale largest; carpal lobe with prominent conical acute spine; mid dorsal carpal ridge tubercles blunt.

Distribution: Chile—Imperial River basin (Pichilumaco, Pichi-Cautín, Traiguén, Colpí, Quillen, Chol Chol, Boroa, Cautin, Quepe rivers); Toltén River basin (Donguil River).

2.3.14 *Aegla concepcionensis* Schmitt, 1942

Diagnosis (Bond-Buckup and Buckup 1994): Anterolateral spine of carapace extending beyond base of cornea; protogastric lobes absent; transverse dorsal linea straight; extra-orbital sinus absent; rostrum long, ligulate, deflected, distally recurved, carina slightly sharp in distal third; anterior angle of ventral margin of epimeron 2 with scaliform tubercle; outer proximal margin of cheliped movable finger with lobe with tubercle; palmar crest subrectangular, excavated; inner margin of ventral surface of ischium of cheliped only with four to six nodes.

Distribution: Chile—Bío Bío River basin (Manantiales, Arabian Stadium creeks); Andalién River basin (unnamed creek at Villa Vergara, Nonguén River, Pineda Lake); Itata River basin (creek in Cerro Cayumanqui).

2.3.15 *Aegla denticulata denticulata* Nicolet, 1849

Diagnosis (Bond-Buckup and Buckup 1994): Cephalothorax with a longitudinal dorsal carina in median line; lateral margins of anterior branchial area of carapace expanded as a lamina, with lobes; anterolateral spine of carapace extending beyond middle of cornea; rostrum long, base wide, carina slightly sharp in distal third; anterior angle of ventral margin of epimeron 2 with robust spine; fourth thoracic sternite medially elevated with tubercles and scales; outer proximal margin of movable finger of cheliped with distinguished lobe; cheliped fingers with lobular tooth; palmar crest rectangular, excavated; inner margin of ventral surface of ischium of cheliped only with modest tubercles; ventral margin of merus of second pereipod with scaliform tubercles.

Distribution: Chile—Bío Bío River basin (Malleco River); Budi River basin (Budi Lake); Toltén River basin (Mahuindanche, Donguil, Huiscapi, Curileufu rivers); Lingue River basin (Lingue River); Valdivia River basin (unnamed creek at Villa Calafquen, Calafquen Lake); Bueno River basin (Bueno, Negro rivers); Maullín River basin (Llanquihue Lake); Chiloé Island (Caulín, Huicha creeks, creeks affluent to Huillinco Lake, Tarahuin Lake).

2.3.16 *Aegla denticulata lacustris* Jara, 1989

Diagnosis (Bond-Buckup and Buckup 1994): Cephalothorax with a dorsal longitudinal carina in median line; lateral margins of anterior branchial area expanded as laminas with spines; anterolateral spine of carapace extending beyond middle of cornea; rostrum long, base wide, carina slightly sharp in distal third; anterior angle of ventral margin of epimeron 2 with robust spine; fourth thoracic sternite medially elevated with tubercle; outer proximal margin of movable finger of cheliped with distinguished lobe; cheliped fingers without lobular tooth; palmar crest rectangular, distinguished; inner margin of ventral surface of ischium of cheliped with one distal spine and up to three tubercles; dorsal magin of merus of second pereopod with anterior spine, followed by tubercles; ventral margin of merus of second pereipod with anterior spine, followed by tubercles.

Distribution: Chile—Bueno River basin (Rupanco Lake).

2.3.17 *Aegla expansa* Jara, 1992

Diagnosis: Rostrum long, broad ligulate, longitudinally excavated both sides of prominent dorsal carina; extraorbital sinus wide U-shaped, separated from orbital sinus by two spines, one above the other; gastric and cardial areas midline markedly convex and protube rant; carapace branchial margins crenulate; propodus of chelae greatly swollen; palmar crest expanded as dorsally concave earlike lamina, its rear end fitting in wide embayment of distodorsal end of carpus; dorsal surface of chelae densely covered by minute subacute scales; venter of fourth thoracic sternum smooth; anterolateral angle of second abdominal epimeron sharply acute.

Distribution: Chile—Bío Bío River basin (Hualqui River).

2.3.18 *Aegla franca* Schmitt, 1942

Diagnosis (Bond-Buckup and Buckup 1994): Anterolateral spine of carapace reaching middle of cornea; protogastric lobes very elevated; areola rectangular, long and narrow; rostrum long, slender, carinate along entire length; anterior angle of ventral margin of epimeron 2 with tubercle; outer proximal margin of movable finger of cheliped with lobe with tubercle; palmar crest subrectangular; inner margin of ventral surface of ischium of cheliped with two spines, one distal and another proximal, and tubercles between them.

Distribution: Brazil—Paraná River basin (Grande River).

2.3.19 *Aegla franciscana* Buckup and Rossi, 1977

Diagnosis (Bond-Buckup and Buckup 1994): Anterolateral spine of carapace extending beyond base of cornea; protogastric lobes absent; rostrum short, triangular, carina slightly sharp in distal third; posterior margin of areola divergent; anterior angle of ventral margin of epimeron 2 with reduced spine; outer proximal margin of movable finger of cheliped with lobe; palmar crest subrectangular; inner margin of ventral surface of ischium of cheliped with one distal spine and up to two tubercles.

Distribution: Brazil—Uruguay River basin (Apuaê-Inhandava, Pelotas rivers); South Atlantic system (Caí, Sinos, Taquari-Tainhas, Tramandaí, Mampituba rivers).

2.3.20 *Aegla georginae* Santos and Jara, 2013

Diagnosis: Anterolateral spine of carapace short, not reaching base of cornea; protogastric lobes present and epigastric lobes very modest; extra-orbital sinus present, U-shaped; rostrum short, carinate; cheliped movable finger without lobe; fingers of both chelipeds with lobular tooth in inner margin; palmar crest of chelipeds subrectangular, both modest and excavated, tending to suboval on the minor cheliped; anterior angle of ventral margin of epimeron 2 with a spine; inner margin of ventral surface of ischium of cheliped with a distal spine, with one to three scales along margin, one proximal tubercle; dorsal margin of dactyle, propod, and carpus of the second, third, and fourth pereopods with scaliform tubercles and setae.

Distribution: Brazil—Uruguay River basin (Perau Creek).

2.3.21 *Aegla grisella* Bond-Buckup and Buckup, 1994

Diagnosis: Cephalothorax very convex; anterolateral spine of carapace reaching middle of cornea; rostrum long, deflected, carinate along entire length; anterior angle of ventral margin of epimeron 2 unarmed; outer proximal margin of movable finger of cheliped without lobe; palmar crest rectangular; inner anterolateral angle of carpus of cheliped obtuse, unarmed; inner margin of ventral surface of ischium of cheliped with two conical spines, one distal and another proximal, with tubercles between them; dorsal and ventral margins of merus of second pereopod only with scales.

Distribution: Brazil—South Atlantic system (Upper Jacuí, Taquari-Tainhas, Ijuí, Passo Fundo rivers).

2.3.22 *Aegla hueicollensis* Jara and Palacios, 1999

Diagnosis: Carapace ovoidal, not expanded at branchial areas. Epibranchial lobe pyramidal, acute, borders scaly. Rostrum short, wide at base, neatly triangular, both sides of low-profiled carina barely troughed, conical at apex, acute. First hepatic lobe blunt. Orbital and extra-orbital sinus present. Branchial margin smooth, slightly nodulated. Second abdominal epimeron acute but not spiniform. Telson divided. fourth thoracic sternum smooth. Male chelae elongated, pyriform, inftated at propodus; palmar crest flabelliform, subrectangular, dentate; dactylar lobe subspiniform; carpal lobe spiniform, apex markedly displaced towards distal end of article, frontal edge with row of tufts of short setae mingled with scales; tubercles of carpal ridge blunt, low, topped by transversal row of scales and setae. Tubercles on dorsal edge of merus of chelipeds blunt, tipped by two to four scales roughly ordered in transversal oblique row. Ventroexternal border of ischium, ventral face of carpus, and ventral borders of chelae with tufts of long setae in row.

Distribution: Chile—Valdivia River basin (Futa, Chaihuín, Colún, Hueicolla rivers).

2.3.23 *Aegla humahuaca* Schmitt, 1942

Diagnosis (Bond-Buckup and Buckup 1994): Anterolateral spine of carapace not reaching base of cornea; protogastric lobes absent; rostrum long, very deflected; rostrum carinate along entire length; outer proximal margin of movable finger of cheliped without lobe; cheliped fingers without lobular tooth; palmar crest of cheliped very modest, rectangular, short; inner margin of ventral surface of ischium of cheliped with one anterior conical spine and up to four tubercles; dorsal margin of merus of second pereopod with scaliform tubercles.

Distribution: Argentina—Paraná River basin (Bermejo, Dulce, Juramento, Grande rivers).

2.3.24 *Aegla inconspicua* Bond-Buckup and Buckup, 1994

Diagnosis: Anterolateral spine of carapace reaching middle of cornea; protogastric lobes absent; rostrum triangular, wide in base, carinate along entire length; subrostral process very developed; anterior angle of ventral margin of epimeron 2 unarmed; outer proximal margin of movable finger of cheliped with lobe; palmar crest of cheliped obsolete; inner margin of ventral surface of ischium of cheliped with one distal spine and three to four smaller elevations.

Distribution: Brazil—South Atlantic system (Guaíba, Caí, Taquari-Antas, Sinos, Gravataí, Tramandaí rivers).

2.3.25 *Aegla inermis* Bond-Buckup and Buckup, 1994

Diagnosis: Anterolateral spine of carapace reaching base of cornea; protogastric lobes present, very elevated; posterior margin of areola divergent; rostrum with medium length in adult males; anterior angle of ventral margin of epimeron 2 unarmed or with a small scale; hands subquadratic; outer proximal margin of movable finger of cheliped without lobe; palm of cheliped not inflated, in median portion, in transversal section, subcylindrical; palmar crest of cheliped absent; ventral margin of merus of second pereopod with scaliform tubercles.

Distribution: Brazil—South Atlantic system (Sinos, Caí, Tramandaí rivers).

2.3.26 *Aegla intercalata* Bond-Buckup and Buckup, 1994

Diagnosis: Anterolateral spine of carapace reaching base of cornea; protogastric lobes moderately distinguished; rostrum very long in adult males, excavated; rostrum carinate along its entire length; anterior angle of ventral margin of epimeron 2 unarmed or with one or more scales; outer proximal margin of movable finger of cheliped without lobe; fingers of cheliped with lobular tooth; palmar crest of cheliped absent or very modest; inner margin of ventral surface of ischium of cheliped with four distinguished tubercles and smaller tubercles between them; ventral margin of merus of second pereopod with scaliform tubercles.

Distribution: Argentina—Mar Chiquita system (Las Lajas River); Dulce River basin (Los Sojas River tributaries); Valle Central River basin (Las Trancas River).

2.3.27 *Aegla intermedia* Girard, 1855 (*Incertae sedis**)

Diagnostics characters listed by Bond-Buckup and Buckup (1994) based on Girard's original description: Rostrum moderate, acute, flattened, and depressed in median portion, with its end slightly recurved. Carapace sides subdenticulated; denticulations more distinguished in stomach region than in branchial region. Carpus with two rows of subconical tubercles on its superior and inner portion. Hand [palmar crest?] with inner process flattened, often denticulated. Lower finger [fixed finger?] has in its margin a series of tubercles arranged in double series in direction to base. Upper finger [movable finger?] has tubercles in margin, the posterior being the largest.

Distribution: Chile—Maipu River basin (Maipu River;* record not confirmed after description).

2.3.28 *Aegla itacolomiensis* Bond-Buckup and Buckup, 1994

Diagnosis: Anterolateral spine of carapace reaching base of cornea; second and third hepatic lobes slightly marked; rostrum long in adult males, triangular, carinate along entire length; anterior angle of ventral margin of epimeron 2 unarmed; outer proximal margin of movable finger of cheliped with lobe; fingers of cheliped with robust lobular tooth; cheliped with palm very inflated; palmar crest of cheliped absent or reduced; inner margin of ventral surface of ischium of cheliped with a distal tubercle and small elevations along the segment.

Distribution: Brazil—South Atlantic system (Gravataí, Sinos rivers).

2.3.29 *Aegla japi* Moraes, Tavares, and Bueno, 2016

Diagnosis: Rostrum triangular, base wide, nearly deflected downward, not reaching distal apex of compound eyes; subrostral process on proximal half, poorly developed, low, lobular, and oriented downward; orbital spines rudimentary; epibranchial area with corneous scales on anterolateral angle only; areola subrectangular; anteromesial region of third thoracic sternite tapered; chelipeds large, palmar crests rectangular with margin entire; uropods narrow; posterolateral margin of telson slightly convex mesially.

Distribution: Brazil—Paraná River basin (Tietê River).

2.3.30 *Aegla jaragua* Moraes, Tavares, and Bueno, 2016

Diagnosis: Rostrum triangular, base wide, straight, extending beyond distal apex of compound eyes. Subrostral process in median position, well developed, high, broad triangular, slightly oriented anteriorly. Orbital spines well developed; epibranchial area with corneous scales on anterolateral angle only; areola subrectangular, anteromesial region of third thoracic sternite truncate; chelipeds large, palmar crests disciform with margin weakly lobulate; uropods narrow; posterolateral margin of telson slightly convex mesially.

Distribution: Brazil—Paraná River basin (Tietê River).

2.3.31 *Aegla jarai* Bond-Buckup and Buckup, 1994

Diagnosis: Anterolateral spine of carapace extending beyond middle of cornea; protogastric lobes discrete; rostrum very long in adult males, styliform, carinate along entire length; anterior angle of ventral margin of epimeron 2 unarmed; outer proximal margin of movable finger of cheliped tipped with tubercle; palmar crest of cheliped disciform, strongly excavated; inner margin of ventral surface of ischium of cheliped with one distal conical spine and up to three tubercles; dorsal margin of merus of second pereopod with spine, followed by tubercles; ventral margin of second pereopod with scaliform tubercles.

Distribution: Brazil—Uruguay River basin (Canoas, Pelotas rivers); South Atlantic system (Itajaí-Açu River).

2.3.32 *Aegla jujuyana* Schmitt, 1942

Diagnosis (Bond-Buckup and Buckup 1994): Anterolateral spine of carapace extending beyond middle of cornea; protogastric lobes obsolete; rostrum very long in adult males, slender, deflected; rostrum carinate along entire length; outer proximal margin of movable finger of cheliped without lobe; fingers of cheliped without lobular tooth; palmar crest of cheliped absent; inner margin of ventral surface of ischium of cheliped with tubercles.

Distribution: Argentina—Paraná River basin (Chico, Grande, Huasamayo rivers, Paco Creek).

2.3.33 *Aegla jundiai* Moraes, Tavares, and Bueno, 2016

Diagnosis: Rostrum triangular, base wide, nearly deflected downward, not reaching distal apex of compound eyes; subrostral process on proximal half, well developed, high, broad triangular, and oriented downward; orbital spines rudimentary; corneous scales at epibranchial area inconspicuous on anterolateral angle only; areola rectangular; third thoracic sternite with anteromesial region truncate; chelipeds small, palmar crests rectangular with margin jagged; uropods narrow; posterolateral margin of telson straight.

Distribution: Brazil—Paraná River basin (Tietê River).

2.3.34 *Aegla laevis laevis* (Latreille, 1818)—accepted as *Aegla laevis* (Latreille, 1818)

Diagnosis (Bond-Buckup and Buckup 1994): Anterolateral spine of carapace extending beyond base of cornea; protogastric lobes obsolete; rostrum long in adult males, ligulate, with carina slightly sharp in distal third; anterior angle of ventral margin of epimeron 2 with small tubercle; outer proximal margin of movable finger of cheliped with small lobe, with scaliform tubercle; palmar crest subrectangular, excavated; inner margin of ventral surface of ischium of cheliped with three to four conical tubercles; ventral margin of merus of second pereopod with scaliform tubercles.

Distribution: Chile—Maipo River basin (Maipo River); Maule River basin (Maule, Putagán rivers).

2.3.35 *Aegla laevis talcahuano* Schmitt, 1942—accepted as *Aegla talcahuano* Schmitt, 1942

Diagnosis (Bond-Buckup and Buckup 1994): Anterolateral spine of carapace reaching base of cornea; rostrum with medium length, ligulate; rostrum with carina absent in distal third; anterior angle of ventral margin of epimeron 2 with spine; outer proximal margin of movable finger of cheliped without lobe; fingers of cheliped with lobular tooth; palmar crest of cheliped subdisciform, distinguished, excavated; inner margin of ventral surface of ischium of cheliped only with two to four tubercles; ventral margin of merus of second pereopod with scales.

Distribution: Chile—Cachapoal River basin (Antivero River); Tinguiririca River basin (Chimbarongo River); Maule River basin (Maule, Lircay, Putagán, Liguay, Perquilauquén rivers).

2.3.36 *Aegla lancinhas* Bond-Buckup and Buckup, 2015

Diagnosis: Anterolateral spine of carapace reaching base of cornea; protogastric lobes weakly marked and epigastric lobes tubercle-shaped, very marked; extraorbital sinus present, rostrum triangular, slightly ligulate, straight, carinate in two proximal thirds, outer proximal margin of movable finger of cheliped without lobe; palmar crest of cheliped modest, subdisciform, slightly concave; anterior angle of ventral margin of epimeron 2 inermis; inner margin of ventral surface of ischium of cheliped with one distal scaliform tubercle and smaller scales.

Distribution: Brazil—Southeastern Atlantic system (Upper Ribeira do Iguape River sub basin, Lancinhas Grot).

2.3.37 *Aegla lata* Bond-Buckup and Buckup, 1994

Diagnosis: Carapace expanded laterally in anterior and posterior branchial areas; anterolateral spine of carapace reaching base of cornea; protogastric lobes present, sometimes obsolete; areola rectangular, long and narrow; rostrum long in adult males, carinate along entire length; anterior angle of ventral margin of epimeron 2 unarmed; outer proximal margin of movable finger of cheliped with modest lobe; palmar crest of cheliped disciform in both chelipeds, slightly excavated; inner margin of ventral surface of ischium of cheliped with four to five subequal tubercles.

Distribution: Brazil—Paraná River basin (Tibagi River, Apertados Creek).

2.3.38 *Aegla leachi* Bond-Buckup and Santos, 2012

Diagnosis: Anterolateral spines of carapace short, reaching base of cornea; protogastric and epigastric lobes absent; extra-orbital sinus absent, rostrum triangular, deflected, carinate, outer proximal margin of movable finger of cheliped with lobe; palmar crest of the major cheliped absent and hand inflated; anterior angle of ventral margin of epimeron 2 unarmed; inner margin of ventral surface of ischium of cheliped without ornamentation.

Distribution: Brazil—Uruguay River basin (Galafre River, Marombas River tributaries, Passo Fundo Creek).

2.3.39 *Aegla leptochela* Bond-Buckup and Buckup, 1994

Diagnosis: Anterolateral spine of carapace not reaching base of cornea; epigastric prominences nodular, ovoidal; protogastric lobes obsolete; eyestalks with reduced dimensions, narrowing distally; cornea reduced; rostrum triangular, with median length, carinate along entire length; anterior angle of ventral margin of epimeron 2 with scale; outer proximal margin of movable finger of cheliped with lobe tipped with a scaliform tubercle; fingers of cheliped without lobular tooth; palmar crest of cheliped absent; inner margin of ventral surface of ischium of cheliped only with tubercles and scaliform tubercles; segments of pereopods very long.

Distribution: Brazil—Southeastern Atlantic system (Ribeira do Iguape River sub basin, Paiva's Grot).

2.3.40 *Aegla leptodactyla* Buckup and Rossi, 1977

Diagnosis (Bond-Buckup and Buckup 1994): Cephalothorax convex with hepatic lobes well delimited; anterolateral spine of carapace extending beyond base of cornea; protogastric lobes very elevated; rostrum with median length, without carina in distal third; outer proximal margin of movable finger of cheliped with lobe with tubercle; hand and fingers thin and long; fingers of cheliped without lobular tooth; palmar crest subrectangular, margin with strong and distinguished spines; inner margin of ventral surface of ischium of cheliped with one distal spine and up to three tubercles; dorsal margin of merus of second pereopod with tubercles; ventral margin of merus of second pereopod with small spines and tubercles.

Distribution: Brazil—Uruguay River basin (Silveira River); South Atlantic system (Taquari-Antas River).

2.3.41 *Aegla ligulata* Bond-Buckup and Buckup, 1994

Diagnosis: Anterolateral spine of carapace extending beyond base of cornea; protogastric lobes absent; rostrum ligulate, low, carina absent in distal third; anterior angle of ventral margin of epimeron 2 unarmed; outer proximal margin of movable finger of cheliped without lobe; palmar crest of cheliped obsolete; inner margin of ventral surface of ischium of cheliped with four to six subequal conical tubercles.

Distribution: Brazil—South Atlantic system (Taquari-Tainhas, Tramandaí rivers).

2.3.42 *Aegla longirostri* Bond-Buckup and Buckup, 1994

Diagnosis: Lateral margins of anterior branchial area slightly expanded with tubercles and scaliform tubercles; anterolateral spine of carapace extending beyond middle of cornea; rostrum very long, styliform, carinate along entire length; anterior angle of ventral margin of epimeron 2 projected by robust spine; outer proximal margin of movable finger of cheliped without lobe; palmar crest rectangular, excavated; inner margin of ventral surface of ischium of cheliped with two robust spines and tubercles between them; ventral margin of merus of second pereopod with anteromedial spine followed by scaliform tubercles.

Distribution: Brazil—Uruguay River basin (Ibicuí River); South Atlantic system (Vacacaí-Mirim, Jacuí, Pardo, Taquari-Tainhas, Antas, Caí, Sinos rivers).

2.3.43 *Aegla loyolai* Bond-Buckup and Santos, 2015

Diagnosis: Anterolateral spine of carapace reaching middle of cornea; protogastric lobes present; extra-orbital sinus present, rostrum elongated, with medium length (4.0 < CL/RL < 4.9), straight, carinate along entire length; outer proximal margin of movable finger of cheliped with pronounced lobe tipped with scaliform tubercle; palmar crest of cheliped subdisciform; anterior angle of ventral margin of epimeron projecting in a spine; inner margin of ventral surface of ischium of cheliped with distal tubercle.

Distribution: Brazil—Paraná River basin (Pajanduvas River).

2.3.44 *Aegla ludwigi* Santos and Jara, 2013

Diagnosis: Anterolateral spine of carapace reaching base of cornea; protogastric lobes absent and epigastric lobes modest; extra-orbital sinus present, U-shaped; rostrum medium, carinate along entire length, recurved distally; movable finger of cheliped without lobe; cheliped fingers with lobular tooth in inner margin; palmar crest of minor cheliped subdisciform and major cheliped subrectangular, both developed and excavated; anterior angle of ventral margin of epimeron 2 unarmed; inner margin of ventral surface of ischium of cheliped with distal spine, one to three scales along margin and proximal tubercle; dorsal margin of dactylus, propodus, and carpus of second, third, and fourth pereopods with scaliform tubercles and setae.

Distribution: Brazil—Uruguay River basin (Cambará Creek, branch of Potiribu River).

2.3.45 *Aegla manni* Jara, 1980

Diagnosis (Bond-Buckup and Buckup 1994): Anterolateral spine of carapace reaching base of cornea; protogastric lobes absent; extra-orbital sinus absent;

rostrum long in adult males, wide, truncated at apex, carinate along entire length; outer proximal margin of movable finger of cheliped with rudimentary lobe; palmar crest subrectangular, excavated; inner margin of ventral surface of ischium of cheliped with a small tubercle and modest elevations.

Distribution: Chile—Valdivia River basin (Futa River, Buenaventura, Joaquines creeks); small coastal basins (Millalafquén, Huiro creeks).

2.3.46 *Aegla manuinflata* Bond-Buckup and Santos, 2009

Diagnosis: Anterolateral spine of carapace reaching base of cornea; protogastric lobes present; extra-orbital sinus U-shaped; rostrum triangular, carinate along entire length, deflected; hand inflated, proximal outer margin of moveable finger of cheliped with lobe; fingers of cheliped without lobular tooth; palmar crest of cheliped absent; anterior angle of ventral margin of epimeron 2 armed; inner margin of ventral surface of ischium of cheliped with one distal tubercle; dorsal margin of carpus of second and third pereopods with short slender setae.

Distribution: Brazil—Uruguay River Basin (Ibicuí-Mirim, Toropi rivers, Taquara, Itaimbé creeks).

2.3.47 *Aegla marginata* Bond-Buckup and Buckup, 1994

Diagnosis: Anterolateral margins of carapace recurved, especially in anterior region; anterolateral spine of carapace reaching base of cornea; protogastric lobes present, elevated; hepatic lobes well marked; rostrum long in adult males, strongly excavated along entire length; anterior angle of ventral margin of epimeron 2 unarmed; hand suboval, palm very inflated; outer proximal margin of movable finger of cheliped without lobe; fingers of cheliped with lobular tooth; palmar crest of cheliped absent; inner margin of ventral surface of ischium of cheliped only with five to six elevations tipped with scales.

Distribution: Brazil—Paraná River basin (Paranapanema, Upper Iguaçu rivers); Southeastern Atlantic system (Ribeira do Iguape River, Paranaguá Bay).

2.3.48 *Aegla meloi* Bond-Buckup and Santos, 2015

Diagnosis: Anterolateral spine of carapace not reaching base of cornea; protogastric lobes flat; extra-orbital sinus reduced. Rostrum styliform, medium length (4.0 < CL/RL < 4.9), carinate. Proximal outer margin of movable finger of cheliped without lobe; palmar crest of cheliped subrectangular; anterior angle of ventral margin of epimeron 2 inermis (unarmed); inner margin of ventral surface of ischium of cheliped with elevation ornamented with scale-shaped distal tubercle.

Distribution: Brazil—Paraná River basin (tributary of Iguaçu River).

2.3.49 *Aegla microphthalma* Bond-Buckup and Buckup, 1994

Diagnosis: Anterolateral spine of carapace acuminate, recurved, reaching the distal third of rostrum; eyestalks with reduced dimensions; cornea absent; rostrum triangular, strongly recurved in distal portion, without carina in distal third; anterior angle of ventral margin of epimeron 2 with tubercle; outer proximal margin of movable finger of cheliped without lobe; fingers of cheliped with lobular tooth; palmar crest of cheliped absent; inner margin of ventral surface of ischium of cheliped with a conical spine and three tubercles.

Distribution: Brazil—Southeastern Atlantic system (Ribeira do Iguape River subbasin, Santana Cave).

2.3.50 *Aegla muelleri* Bond-Buckup and Buckup, 2010

Diagnosis: Anterolateral spine of carapace reaching base of cornea; protogastric and epigastric lobes absent; extra-orbital sinus absent; rostrum triangular, tapered, slightly recurved distally, carinate; proximal outer margin of movable finger of cheliped without lobe; palmar crest of cheliped small, subrectangular; anterior angle of ventral margin of epimeron 2 with small scale; inner margin of ventral surface of ischium of cheliped with distal scaliform tubercle.

Distribution: Brazil—South Atlantic system (Passa Quatro River, Espingarda Creek).

2.3.51 *Aegla neuquensis* Schmitt, 1942

Diagnosis (Bond-Buckup and Buckup 1994): Anterolateral spine of carapace not reaching base of cornea; protogastric lobes absent; transverse dorsal linea straight; rostrum long in adult males, ligulate, without carina in distal third; fourth thoracic sternite medially elevated with scale; outer proximal margin of movable finger of cheliped with lobe; palmar crest subrectangular, margin indented; inner margin of ventral surface of ischium of cheliped with two spines and up to three tubercles.

Distribution: Chile—Simpson River basin (Simpson, Pollux rivers); Argentina—Neuquén River basin (Neuquén River); Negro River basin (Negro, Limay, Aluminé, Collón Curá rivers, Nahuel Huapi, Tromen, Aluminé, Huechulafquen, Lolog lakes); Chubut River basin (Chubut, Mayo, Chico, Tecka, Senguerr rivers, Fontana Lake).

2.3.52 *Aegla oblata* Bond-Buckup and Santos, 2012

Diagnosis: Anterolateral spine not reaching base of cornea; protogastric lobes absent; extra-orbital sinus absent; rostrum subtriangular, short, carinate; outer proximal margin of movable finger of cheliped with lobe; palmar crest of cheliped subdisciform, with lobes; anterior angle of ventral margin of epimeron 2 unarmed; inner margin of ventral surface of ischium of cheliped with elevation ornamented with scaliform distal tubercle.

Distribution: Brazil—Uruguay River basin (Caronas, Lava-Tudo, Périco Redondo rivers, Engenho Velho Creek).

2.3.53 *Aegla obstipa* Bond-Buckup and Buckup, 1994

Diagnosis: Anterolateral spine reaching base of cornea; protogastric lobes absent; transverse dorsal linea straight; rostrum with medium length, deflected, ligulate, without carina in distal third; anterior angle of ventral margin of epimeron 2 unarmed; fourth thoracic sternite flat, subquadratic; outer proximal margin of movable finger of cheliped with modest lobe; palmar crest present, rectangular; inner margin of ventral surface of ischium of cheliped with only four tubercles.

Distribution: Brazil—South Atlantic system (Guaíba Lake, Lower Jacuí, Ratos, Camaquã rivers).

2.3.54 *Aegla occidentalis* Jara, Pérez-Losada, and Crandall, 2003

Diagnosis: Body contour almond-shaped; carapace surface grossly punctated; rostrum triangular to subligulate, short and low profiled, apex with scale surrounded by rosette of short stiff setae and minute acicular scales; orbital spine absent; branchial margins of carapace not expanded, smooth; anterolateral angle of second abdominal epimeron blunt, frequently with small scale hidden among short setae; carpal lobe subpyramidal tipped with row of 1–4 scales in a row mingled with short stiff setae; lobe on proximodorsal end of dactylus of chelae low and blunt; palmar crest subrectangular, slightly expanded and faintly concave, its margin subdenticulate; fourth thoracic sternum flat, unornamented, at most with semicircular swelling ending abruptly at frontal end.

Distribution: Chile—Paicaví River basin (Caramávida, Tucapel rivers, Lanalhue Lake); Lleu Lleu River basin (Lleu Lleu Lake).

2.3.55 *Aegla odebrechtii* Müller, 1876

Diagnosis (Bond-Buckup and Buckup 1994): Anterolateral spine of carapace reaching middle of cornea; protogastric lobes obsolete; rostrum with medium length in adult males, deflected and slightly recurved distally, without carina in distal third; anterior angle of ventral margin of epimeron 2 unarmed; outer proximal margin of movable finger of cheliped with lobe tipped with tubercles; palmar crest of cheliped subdisciform, excavated; inner margin of ventral surface of ischium of cheliped with one distal conical spine, a proximal smaller spine and up to three tubercles between them.

Distribution: Brazil—Uruguay River Basin (Irani, Peixe, Canoas, Pelotas rivers); South Atlantic system (Itajaí-Açu River).

2.3.56 *Aegla okora* Páez and Teixeira, 2018

Diagnosis: Triangular rostrum with narrow base. Subrostral process developed, anteriorly oriented at a 45° with the rostrum. Epigastric prominences and protogastric lobes pronounced, with scales and small setae. Anterolateral spine reaching basal margin of cornea. Branchial region swollen. Areola trapezoidal. Cardiac area trapezoidal. Proximal dorsal margin of movable finger of cheliped without lobe. Palmar crest of major cheliped rectangular. Anterolateral angle of second abdominal epimeron unarmed, with setae. Ventromesial border of ischium of the cheliped ornate with three tubercles, one proximal, one median, and one distal. Uropodal (endopods) wide.

Distribution: Brazil—Paraná River basin (Iguaçu River subbasin, Tapera River) (Páez et al. 2018).

2.3.57 *Aegla papudo* Schmitt, 1942

Diagnosis (Bond-Buckup and Buckup 1994): Carapace very convex, especially in gastric region; anterolateral spine of carapace reaching middle of cornea; protogastric lobes absent; rostrum long, distally ligulate and recurved; rostrum with carina absent in distal third; telson without longitudinal suture; outer proximal margin of movable finger of cheliped with lobe, low, with small tubercle; palmar crest of cheliped very modest; inner margin of ventral surface of ischium of cheliped with three to four subequal tubercles.

Distribution: Chile—Choapa River basin (Choapa, Illapel rivers); Ligua River basin (Ligua River, Papudo Creek); Catapilco River basin (Catapilco River); Aconcagua River basin (Aconcagua River); Marga Marga River basin (Marga Marga River); Maipo River basin (Mapocho River).

2.3.58 *Aegla parana* Schmitt, 1942

Diagnosis (Bond-Buckup and Buckup 1994): Anterolateral spine of carapace long, extending beyond the middle of cornea; rostrum very long, styliform, carinate along entire length; anterior angle of ventral margin of epimeron 2 projected by a spine; outer proximal margin of movable finger of cheliped without lobe; palmar crest discreet, subrectangular; inner margin of ventral surface of ischium of cheliped with two robust spines; dorsal margin of merus of second pereopod with robust spines; ventral margin of merus of second pereopod with spines, followed by tubercles.

Distribution: Brazil—Paraná River basin (Upper and Lower Iguaçu, Timbó, Canoinhas rivers).

2.3.59 *Aegla parva* Bond-Buckup and Buckup, 1994

Diagnosis: Anterolateral spine of carapace reaching middle of cornea; rostrum triangular, with medium length, carinate along entire length; anterior angle of ventral margin of epimeron 2 with spine with reduced dimensions; outer proximal margin of movable finger of cheliped without lobe; palmar crest present, subrectangular; inner margin of ventral surface of ischium of cheliped with two robust spines, one distal, another proximal and tubercles between them; ventral margin of merus of second pereopod with two spines, one distal, another anteromedial, followed by scales.

Distribution: Brazil—Paraná Basin (Upper, Middle and Lower Iguaçu River); South Atlantic system (Itajaí-Açu, Cubatão do Sul, Cedro rivers).

2.3.60 *Aegla paulensis* Schmitt, 1942

Diagnosis (Moraes et al. 2016): Rostrum triangular, base narrow, curved downward, extending beyond distal apex of compound eyes; subrostral process on proximal half, well developed, high, acute triangular, oriented downward; orbital spines well developed; epibranchial area with one corneous scale near anterolateral angle; areola subrectangular; anteromesial region of third thoracic sternite tapered; chelipeds large; palmar crest rectangular with margin lobulated; uropods wide; posterolateral margin of telson slightly concave mesially.

Distribution: Brazil—Southeastern Atlantic system (Cubatão River).

2.3.61 *Aegla perobae* Hebling and Rodrigues, 1977

Diagnosis (Bond-Buckup and Buckup 1994): Anterolateral spine of carapace not reaching base of cornea; areola rectangular, long and narrow; rostrum triangular, short, carinate along entire length; fourth thoracic sternite elevated with tubercle; outer proximal margin of movable finger of cheliped without lobe; palmar crest subrectangular, excavated; inner margin of ventral surface of ischium of cheliped with up to three spines.

Distribution: Brazil—Paraná River basin (Peroba Grot, Tietê-Piracicaba River).

2.3.62 *Aegla pewenchae* Jara, 1994

Diagnosis: Carapace longer than wide; rostrum elongate, styliform; anterolateral angles of carapace acute, slightly divergent; scales on rostral and hepatic borders very small; orbital spine well developed, dorsal to a second smaller one; anterolateral angle of first hepatic lobe spiniform; branchial borders smooth, noticeably arcuate; palmar crest laminar, expanded, its border microdenticulate to dentate; dorsum of propodus of chelae densely covered by minute lens-like scales; anterolateral angle of second abdominal epimeron sharply acute, spiniform.

Distribution: Chile—Rapel River basin (Claro, Cachapoal rivers, Chimbarongo Creek); Mataquito River basin (Mataquito River); Maule River basin (Maule, Lircay, Longaví, Putagán rivers); Itata River basin (Cato, Ñuble, Chillán rivers); Bío Bío River basin (Bío Bío, Laja rivers); Imperial River basin (Traiguén, Quino rivers); Toltén River basin (Donguil River).

2.3.63 *Aegla plana* Buckup and Rossi, 1977

Diagnosis (Bond-Buckup and Buckup 1994): Anterolateral spine of carapace not reaching base of cornea; protogastric lobes absent; rostrum short, carinate along entire length; anterior angle of ventral margin of epimeron 2 unarmed; outer proximal margin of movable finger of cheliped with lobe with scaliform tubercle; fingers of cheliped without lobular tooth; palmar crest subrectangular; inner margin of ventral surface of ischium of cheliped only with three to four elevations with scales; dorsal margin of merus of second pereopod with one spine, followed by scaliform tubercles.

Distribution: Brazil—South Atlantic system (Caí, Taquari-Tainhas rivers).

2.3.64 *Aegla platensis* Schmitt, 1942

Diagnosis (Bond-Buckup and Buckup 1994): Anterolateral spine of carapace reaching middle of cornea; protogastric lobes moderately elevated; rostrum very long in adults, carinate along its entire length; anterior angle of ventral margin of epimeron 2 with spine; fourth thoracic sternite medially elevated with scale; outer proximal margin of movable finger of cheliped with lobe with tubercle; palmar crest rectangular slightly developed; inner margin of ventral surface of ischium of cheliped unarmed.

Distribution: Argentina—La Plata River basin (La Plata River, Garupá Creek); Mar Chiquita system (Dulce River, Tipas, Loro, Singuil rivers); Uruguay River basin (Itacaruaré River, Santa Maria Creek); Martín García Island; Brazil— Uruguay River basin (Parizinho, Passarinhos, Chapecó, Guaraim, Moinho, Ibicuí, Ijuí, Passo Fundo, Piratini, Quarai, Santa Maria, Turvo-Santa Rosa-Santo Cristo, Várzea, Jacutinga rivers); South Atlantic system (Lower Jacuí, Caí, Sinos, Camaquã, Gravataí, Mirim-São Gonçalo, Sinos, Butuí-Icamaquã rivers, Guaíba Lake); Paraguay—Paraguay River basin (Jejuí River); Uruguay—La Plata River Basin (La Plata River, Miguelete Creek); Uruguay River Basin (Uruguay, Negro, Tacuarembó, Quaraí revers); South Atlantic system (Cebollatí River).

2.3.65 *Aegla pomerana* Bond-Buckup and Buckup, 2010

Diagnosis: Anterolateral spine of carapace extending beyond base of cornea; protogastric lobes present; extra-orbital sinus absent; rostrum triangular, slightly ligulate, slightly recurved distally, lacking carina on distal third; outer proximal margin of movable finger of cheliped with lobe; palmar crest of cheliped subrectangular; anterior angle of ventral margin of epimeron 2 unarmed; inner margin of ventral surface of ischium of cheliped with a distal tubercle.

Distribution: Brazil—South Atlantic system (Itajaí-Açu, Itapocu rivers).

2.3.66 *Aegla prado* Schmitt, 1942

Diagnosis (Bond-Buckup and Buckup 1994): Cephalothorax suboval, elevated, but without dorsal longitudinal carina; lateral margins of anterior branchial area of carapace expanded with scaliform tubercles; epigastric prominences obsolete to absent; protogastric lobes elevated; hepatic lobes well marked; rostrum long in adult males, styliform, carinate along entire length; anterior angle of ventral margin of epimeron 2 projected by a spine; fourth thoracic sternite medially elevated with a spine; outer proximal margin of movable finger of cheliped with lobe tipped by tubercle; palmar crest subrectangular; inner margin of ventral surface of ischium of cheliped with two spines, one distal, another proximal, and tubercles between them.

Distribution: Brazil—South Atlantic system (Mirim Lagoon, Jaguarão River).

2.3.67 *Aegla quilombola* Moraes, Tavares, and Bueno, 2017

Diagnosis: Rostrum wide at base; extra-orbital sinus shallow, anterolateral spines straight; epigastric prominences pronounced; protogastric lobes pronounced with a set of corneous scales; anterior margin of first hepatic lobe and axis of rostrum oriented with respect to each other less than 90°; hepatic lobes poorly defined; transverse dorsal linea rather straight; areola subrectangular; cardiac area trapezoidal; anterolateral angle and lateral margin of epibranchial area with small corneous scales; anteromesial region of third thoracic sternite tapered; major cheliped propodus with palmar crest rudimentary, low, serrated, outer surface not excavated; anterolateral angle of second abdominal epimeron with corneous scales; ventral angles of third and fourth abdominal epimera unarmed; sexual tube short, wide; uropods narrow-shaped; telson with anterolateral and posterolateral margins well differentiated, posterolateral margin straight.

Distribution: Brazil—Southeastern Atlantic system (Ribeira de Iguape Basin: several localities in the Intervales State Park and one locality in the Alto Ribeira Touristic State Park—PETAR).

2.3.68 *Aegla riolimayana* Schmitt, 1942

Diagnosis (Bond-Buckup and Buckup 1994): Anterolateral spine of carapace not reaching base of cornea; posterior margins of areola convergent; rostrum with medium length in adult males, styliform, carinate along its entire length; anterior angle of ventral margin of epimeron 2 with spine with reduced dimensions; outer proximal margin of movable finger of cheliped with rudimentary lobe; palmar crest present, subrectangular, excavated, with tubercles; inner margin of ventral surface of ischium of cheliped only with two tubercles; dorsal magin of merus of second pereopod unarmed with setae.

Distribution: Argentina—Negro River basin (Limay, Aluminé, Chimehuin rivers, Moquehue, Huechlafquen, Aluminé lakes, Jones Creek); Valdivia River basin (Lácar Lake). Chile—Valdivia River basin (Cruces, Leufucade, Antilhue, Reyehueico, San Pedro-Calle Calle rivers); Bueno River basin (Lake Puyehue); Chilóe Island (Huicha, Butalcura rivers, Huillinco Lake).

2.3.69 *Aegla renana* Bond-Buckup and Santos, 2010

Diagnosis: Anterolateral spine of carapace short, not exceeding base of cornea; protogastric lobes absent; extra-orbital sinus absent; rostrum ligulate, slightly recurved distally, lacking carina on distal third; outer proximal margin of movable finger of cheliped with lobe absent but with scales; palmar crest of cheliped subrectangular, with scales on its margin; anterior angle of ventral margin of epimeron 2 unarmed; inner margin of ventral surface of ischium of cheliped with four tubercles. Dorsal margin of dactylus, propodus, and carpus of second, third, and fourth pereopods with rows of setae, scales.

Distribution: Brazil—South Atlantic system (Caí River).

2.3.70 *Aegla ringueleti* Bond-Buckup and Buckup, 1994

Diagnosis: Anterolateral spine of carapace reaching base of cornea; protogastric lobes absent; rostrum triangular, carinate along its entire length; anterior angle of ventral margin of epimeron 2 with small tubercle; fourth thoracic sternite elevated with tubercle; outer proximal margin of movable finger of cheliped without lobe; fingers of cheliped without lobular tooth; palmar crest of cheliped very modest, narrow; inner margin of ventral surface of ischium of cheliped with up to three tubercles; dorsal margin of merus of second pereopod unarmed with long setae; ventral margin of merus of second pereopod with scaliform tubercles; inner margin of coxa of cheliped with one conical spine.

Distribution: Argentina—Paraná River basin (Salado River).

2.3.71 *Aegla rosanae* Campos Jr., 1998

Diagnosis (Moraes et al. 2016): Rostrum triangular, base narrow, curved upward distally, extending beyond distal apex of compound eyes; subrostral process on proximal half, well developed, low, broad, triangular, and tip oriented anteriorly; orbital spines well developed; epibranchial area with corneous scales on anterolateral angle and on lateral margin; areola rectangular; anteromesial region of third thoracic sternite abrupt; chelipeds moderately large, palmar crests rectangular with margin lobulate; uropods narrow; posterolateral margin of telson straight mesially.

Distribution: Brazil—Southeastern Atlantic system (Paraíba do Sul River).

2.3.72 *Aegla rossiana* Bond-Buckup and Buckup, 1994

Diagnosis: Anterolateral spine of carapace reaching base of cornea; rostrum long in adult males, carinate along its entire length; anterior angle of ventral margin of epimeron 2 with modest spine; outer proximal margin of movable finger of cheliped with rudimentary lobe; palmar crest of cheliped absent or very modest; inner margin of ventral surface of ischium of cheliped only with one distal tubercle.

Distribution: Brazil—South Atlantic system (Araranguá, Tramandaí rivers).

2.3.73 *Aegla rostrata* Jara, 1977

Diagnosis (Bond-Buckup and Buckup 1994): Lateral margins of anterior branchial area of carapace moderately expanded with robust spines; anterolateral spine of carapace reaching base of cornea; protogastric lobes distinguished; rostrum very long, styliform, carinate along its entire length; anterior angle of ventral margin of epimeron 2 projected by a spine; outer proximal margin of movable finger of cheliped with lobe with tubercle; palmar crest subrectangular, excavated; inner margin of ventral surface of ischium of cheliped with one distal conical spine, low and small proximal elevation; dorsal margin of merus of second pereopod with scaliform tubercles.

Distribution: Chile—Toltén River basin (Caburga, Colico, Villarica lakes); Valdivia River basin (Huanehue, San Pedro, Cau Cau rivers, Calafquén, Neltume, Panguipulli, Riñihue lakes).

2.3.74 *Aegla saltensis* Bond-Buckup and Jara, 2010

Diagnosis: Anterolateral spine of carapace extending beyond base of cornea; protogastric lobes absent; extra-orbital sinus narrow, V-shaped; rostrum triangular, carinate along its entire length, strongly deflected; proximal outer margin of moveable finger of cheliped without lobe; fingers of cheliped without lobular tooth; palmar crest of cheliped absent; anterior angle of ventral margin of epimeron 2 unarmed; inner margin of ventral surface of ischium of cheliped with one distal spine and a small proximal tubercle; dorsal margin of carpus of second and third pereopods with one distal spine followed by scale-shaped tubercles.

Distribution: Argentina—Paraná River basin (Pasaje, Juramento rivers).

2.3.75 *Aegla sanlorenzo* Schmitt, 1942

Diagnosis (Bond-Buckup and Buckup 1994): Anterolateral spine of carapace reaching distal portion of cornea; protiogastric lobes elevated; rostrum long, acuminate, deflected; rostrum carinate along its entire length; anterior angle of ventral margin of epimeron 2 with acute spine; outer proximal margin of movable finger of cheliped without lobe; fingers of cheliped without lobular tooth; palmar crest of cheliped absent; inner margin of ventral surface of ischium of cheliped only with two robust spines; ventral margin of merus of second pereopod with distal spines followed by tubercles.

Distribution: Argentina—Paraná River basin (San Lorenzo River, Los Berros Creek).

2.3.76 *Aegla scamosa* Ringuelet, 1948

Diagnosis (Bond-Buckup and Buckup 1994): Anterolateral spine of carapace reaching and sometimes extending beyond base of cornea; protogastric lobes absent; transverse dorsal linea sinuous; rostrum triangular, long, carinate with scales slightly sharp in distal third; anterior angle of ventral margin of epimeron 2 with two or more scales; fourth thoracic sternite slightly elevated medially with scale; outer proximal margin of movable finger of cheliped without lobe; palmar crest present, subrectangular, excavated; inner margin of ventral surface of ischium of cheliped only with five tubercles; dorsal and ventral margins of second pereopod with scaliform tubercles and scales.

Distribution: Argentina—Colorado River basin (Mendoza, San Juan rivers, Uspallata, Villa, El Infiernillo, Água Negra creeks).

2.3.77 *Aegla schmitti* Hobbs III, 1979

Diagnosis (Bond-Buckup and Buckup 1994): Anterolateral spine of carapace extending beyond middle of cornea; protogastric lobes present, elevated; rostrum long in adult males, carinate along its entire length; anterior angle of ventral margin of epimeron 2 with spine; outer proximal margin of movable finger of cheliped with lobe tipped by tubercles and scales; palmar crest of minor cheliped strongly disciform, excavated; inner anterolateral angle of carpus of cheliped obtuse, flattened, with scaliform tubercle; inner margin of ventral surface of ischium of cheliped with two spines, one distal, another proximal, and up to two tubercles between them; dorsal margin of merus of second pereopod with scaliform tubercles.

Distribution: Brazil—Paraná River basin (Paranapanema, Upper Iguaçu, Paraná, Tibagi, Ivai, Timbó rivers); Southeastern Atlantic system (Ribeira do Iguape River).

2.3.78 *Aegla septentrionalis* Bond-Buckup and Buckup, 1994

Diagnosis: Anterolateral spine of carapace reaching middle of cornea; protogastric lobes absent; extra-orbital sinus absent; rostrum long in adult males, ligulate, with carina slightly sharp in distal third; anterior angle of ventral margin of epimeron 2 unarmed; fourth thoracic sternite without ornamentation, flat; outer proximal margin of movable finger of cheliped without lobe; fingers of cheliped with robust lobular tooth; palmar crest of cheliped absent; inner margin of ventral surface of ischium of cheliped only with three to four nodes.

Distribution: Bolivia—Paraná River basin (Salo, Sella rivers); Argentina— Paraná River basin (Bermejo, Pilcomayo, Arenales rivers).

2.3.79 *Aegla serrana* Buckup and Rossi, 1977

Diagnosis (Bond-Buckup and Buckup 1994): Cephalothorax very convex; anterolateral spine of carapace reaching base of cornea; protogastric lobes absent; extra-orbital sinus narrow, sometimes very reduced; rostrum long in adult males, carinate along its entire length; subrostral process very developed; anterior angle of ventral margin of epimeron 2 unarmed; outer proximal margin of movable finger of cheliped without lobe; palmar crest of cheliped subdisciform, slightly excavated; inner margin of ventral surface of ischium of cheliped with one anterior conical tubercle, a second posterior tubercle and up to two tubercles or nodes between them.

Distribution: Brazil—Uruguay River basin (Peixe, Santa Rita rivers); South Atlantic system (Caí, Taquari-Tainhas, Sinos, Tramandaí rivers).

2.3.80 *Aegla singularis* Ringuelet, 1948

Diagnosis (Bond-Buckup and Buckup 1994): Lateral margins of anterior branchial area of carapace with scaliform tubercles; anterolateral spine of carapace reaching middle of cornea; epigastric prominences absent; protogastric lobes elevated; rostrum with medium length, styliform, carinate along its entire length; anterior angle of ventral margin of epimeron 2 projected by a spine; fourth thoracic sternite elevated with scale; outer proximal margin of movable finger of cheliped with lobe tipped by a robust spine; anterolateral angle of carpus of cheliped acute and projected by a spine turned to the palmar carpal articulation; palmar crest rectangular with lobes and spines; inner margin of ventral surface of ischium of cheliped with one distal spine and up to two medial tubercles; dorsal margin of merus of second pereopod with tubercles and scaliform tubercles.

Distribution: Argentina—Uruguay River basin (Pindapoy Creek); Brazil— Uruguay River basin (Apuaê-Inhandava, Turvo, Ijui, Passo Fundo, Várzea, Jacutinga, Peixe rivers).

2.3.81 *Aegla spectabilis* Jara, 1986

Diagnosis (Bond-Buckup and Buckup 1994): Cephalothorax elevated but without dorsal longitudinal carina defined; lateral margins of anterior branchial area of carapace expanded with robust spines; anterolateral spine of carapace reaching middle of cornea; rostrum very long, styliform, carinate along its entire length; anterior angle of ventral margin of epimeron 2 projected by a robust spine; outer proximal margin of movable finger of cheliped with small tubercle; fingers of cheliped without lobular tooth; palmar crest projected only by a unique spine; inner margin of ventral surface of ischium of cheliped only with small elevations; dorsal margin of merus of second pereopod only with one robust spine; ventral margin of merus of second pereopod with one robust distal spine followed by scales.

Distribution: Chile—Imperial River basin (Chol Chol, Perquenco, Quepe, Cautín rivers); Toltén River basin (Donguil River).

2.3.82 *Aegla spinipalma* Bond-Buckup and Buckup, 1994

Diagnosis: Anterolateral spine of carapace extending beyong middle of cornea; rostrum with medium length, acuminate, distally recurved, without carina in distal portion; anterior angle of ventral margin of epimeron 2 with robust spine; outer proximal margin of movable finger of cheliped without lobe; palmar crest subrectangular, excavated and with dorsal spine; inner margin of ventral surface of ischium of cheliped with two robust subequal spines, one distal, another proximal; dorsal margin of merus of second pereopod with one small distal spine, followed by scales; ventral margin of merus of second pereopod with a distal tubercle and scales.

Distribution: Brazil—South Atlantic system (Jacuí, Taquari-Tainhas, Sinos, Upper and Lower Jacuí rivers); Uruguay River basin (Ijuí River).

2.3.83 *Aegla spinosa* Bond-Buckup and Buckup, 1994

Diagnosis: Lateral margins of anterior branchial area of carapace with tubercles and scales; anterolateral spine of carapace reaching middle of cornea; protogastric lobes obsolete; rostrum with medium length, styliform, carinate along its entire length; anterior angle of ventral margin of epimeron 2 with robust spine; outer proximal margin of movable finger of cheliped with lobe tipped by tubercles and scales; palmar crest of cheliped subdisciform, not excavated; inner margin of ventral surface of ischium of cheliped with one robust distal spine, a second smaller, proximal spine, and up to smaller spines between them; dorsal and ventral margins of merus of second pereopod with robust spines.

Distribution: Brazil—Uruguay River basin (Apuaê-Inhandava, Jacutinga, Canoas, Pelotas rivers).

2.3.84 *Aegla strinatii* Türkay, 1972

Diagnosis (Bond-Buckup and Buckup 1994): Carapace depressed; anterolateral spine of carapace not reaching base of cornea; protogastric lobes present, sometimes obsolete; transverse dorsal linea very sinuous; rostrum short, carinate along its entire length; anterior angle of ventral margin of epimeron 2 unarmed; outer proximal margin of movable finger of cheliped without lobe; palmar crest of minor cheliped more disciform; inner margin of ventral surface of ischium of cheliped only with two tubercles, one distal and one proximal, and scaliform tubercles between them.

Distribution: Brazil—Southeastern Atlantic system (Ribeira do Iguape River, Diabo Cave, Tapagem Grot).

2.3.85 *Aegla uruguayana* Schmitt, 1942

Diagnosis (Bond-Buckup and Buckup 1994): Anterolateral spine of carapace extending beyond middle of cornea; rostrum long, tending to styliform, carinate along its entire length; anterior angle of ventral margin of epimeron 2 projected by a spine; fourth thoracic sternite elevated with a spine; outer proximal margin of movable finger of cheliped without lobe; fingers of cheliped with lobular tooth; palmar crest of cheliped absent; inner margin of ventral surface of ischium of cheliped with one robust spine and up to three elevations along the segment; dorsal margin of merus of second pereopod with one distal spine and scaliform tubercles; ventral margin of merus of second pereopod with scales.

Distribution: Argentina—Paraná River basin (Areco, Dulce, Tercero, Paraná, Baradero rivers); Uruguay River basin (Yeruá, Molino creeks); Martín García Island; La Plata River basin (Cepeda, Manantiales creeks); Brazil—Uruguay River basin (Ibicuí, Quaraí, Santa Maria rivers); South Atlantic system (Guaíba, Lower Jacuí, Negro, Mirim, São Gonçalo, Icamaquã rivers); Uruguay—Uruguay River basin (Uruguay River); La Plata River basin (Rosario River, Las Vacas, San Carlos creeks); South Atlantic system (Cebollatí River).

2.3.86 *Aegla vanini* Moraes, Tavares, and Bueno, 2016

Diagnosis: Rostrum triangular, base wide, nearly deflected downward, extending beyond distal apex of compound eyes; subrostral process on proximal half, well developed, high, broad triangular, and oriented downward; orbital spines absent; corneous scales at epibranchial area present on anterolateral angle only; areola subrectangular; anteromesial region of third thoracic sternite truncate; chelipeds moderately large, palmar crests rectangular with margin lobulate; uropods narrow; posterolateral margin of telson slightly convex mesially.

Distribution: Brazil—Southeastern Atlantic system (Claro River).

2.3.87 *Aegla violacea* Bond-Buckup and Buckup, 1994

Diagnosis: Anterolateral spine of carapace exending beyond base of cornea; protogastric lobes elevated; transverse dorsal linea straight; rostrum triangular, moderately elevated, carinate; anterior angle of ventral margin of epimeron 2 with spine with reduced dimensions; outer proximal margin of movable finger of cheliped with lobe; fingers of cheliped with robust lobular tooth; palmar crest of cheliped absent; inner margin of ventral surface of ischium of cheliped only with one distal spine and one proximal tubercle; ventral margin of merus of second pereopod only with scales.

Distribution: Brazil—South Atlantic system (Lower Jacuí, Lake Guaíba tributaries).

2.4 PERSPECTIVES

Most of the current 87 species and subspecies (about 70%) described for the genus *Aegla* are under some level of threat (Santos et al. 2017). Although *Aegla* is a speciose genus, its general morphology is conservative. The limited number of morphological features with systematic value has difficulted both the construction of morphology-based phylogenies for the group (Bond-Buckup and Buckup 1994) and the identification of cryptic species, which do not present differences in their diagnostic characters. *Aegla longirostri* and *A. platensis* are two examples of cryptic diversity recently revealed by phylogeographic approaches (Crivellaro et al. 2017; Zimmermann et al. 2018). It might be speculated that isolated populations of the same taxon in some cases are actually cryptic species, which need to be described in the near future. Molecular methods have helped to unveil part of the cryptic diversity in aeglids, and integrative approaches are needed to understand the taxonomy of the group. It is of utmost importance to describe the cryptic species so that they can be included in biodiversity assessments and conservation programs. This way, before describing these new species, it is necessary to find new diagnostic characters that are able to distinguish among species. We are now concentrating our efforts on identifying additional morphological features of systematic importance for aeglids.

REFERENCES

Bickford, D., D. J. Lohman, N. S. Sodhi, et al. 2007. Cryptic species as a window on diversity and conservation. *Trends in Ecology and Evolution* 22(3):148–55.

Bond-Buckup, G., and L. Buckup. 1994. A Família Aeglidae (Crustacea, Decapoda, Anomura). *Arquivos de Zoologia* 32:1–346.

Buckup, L., and A. Rossi. 1977. O gênero *Aegla* no Rio Grande do Sul, Brasil (Crustacea, Decapoda, Anomura, Aeglidae). *Revista Brasileira de Biologia* 37(4):879–92.

Bueno, S. L. S., A. L. Camargo, and J. C. B. Moraes. 2017. A new species of stygobiotic Aeglidae from lentic subterranean waters in southeastern Brazil, with an unusual morphological trait: short pleopods in adult males. *Nauplius* 25:e201700021.

Bueno, S. L. S., R. M. Shimizu, and S. S. Rocha. 2007. Estimating the population size of *Aegla franca* (Decapoda: Anomura: Aeglidae) by mark-recapture technique from an isolated section of Barro Preto Stream, County of Claraval, state of Minas Gerais, southeastern Brazil. *Journal of Crustacean Biology* 27(4):553–59.

Crivellaro M. S., B. L. Zimmermann, M. L. Bartholomei-Santos, et al. 2017. Looks can be deceiving: species delimitation reveals hidden diversity in the freshwater crab *Aegla longirostri* (Decapoda: Anomura). *Zoological Journal of the Linnean Society* 182:24–37.

Dana, J. D. 1852. Crustacea. In *United States Exploring Expedition During the Years 1839, 1840, 1841, 1842, Under the Command of Charles Wilkes, U.S.N.*, pp. 475–78. New Haven, PA: C. Sherman. Atlas Crustacea, 13.

Desmarest, A. G. 1825. *Considérations générales sur la Classe des Crustáces et description des espècesde cesanimaux qui vivent dans la mer, sur les côtes, ou dans les eaux douces de la France.* Paris: F.G. Levrault.

Feldmann, R. M. 1984. *Haumuriaegla glaessneri* n. gen. and sp. (Decapoda, Anomura, Aeglidae) from Haumurian (Late Cretaceous) rocks near Cheviot, New Zealand. New Zealand. *Journal of Geology and Geophysics* 27:379–85.

Feldmann, R. M., F. J. Vega, S. P. Applegate, and G. A. Bishop. 1998. Early Cretaceous arthropods from the Tlayúa formation at Tepexi de Rodríguez, Puebla, México. *Journal of Paleontology* 72(1):79–90.

Fernandes, C. S., S. L. S. Bueno, and M. E. Bichuette. 2013. Distribution of cave-dwelling *Aegla* spp. (Decapoda: Anomura: Aeglidae) from the Alto Ribeira karstic area in southeastern Brazil, based on geomorphological evidence. *Journal of Crustacean Biology* 33:567–75.

Girard, C. 1855. Description of certain Crustacea, brought home by the U. S. N. Astronomical Expedition. In *The U. S. Naval Astronomical Expedition to the Southern Hemisphere, During the Years 1849, 1851, 1852, Lieutenant James M Gillis, Superintendent*, S. I.: s.n., vol. 2, pp. 154–262. Executive Document number 121, House of Representatives, 33rd Congress, 1st session, serial 729, 15, part 2.

Huber, A. F., F. B. Ribeiro, and P. B. Araujo. 2018. New endemic species of freshwater crayfish *Parastacus* Huxley, 1879 (Crustacea: Decapoda: Parastacidae) from the Atlantic forest in southern Brazil. *Nauplius* 26:e2018015.

Jara, C. G. 1977. *Aegla rostrata* n. sp., (Decapoda, Aeglidae), nuevo crustáceo dulceacuicola del sur de Chile. *Studies on Neotropical Fauna and Environment* 12:165–75.

Jara, C. G. 1980. Dos nuevas especies de *Aegla* Leach (Crustacea, Decapoda, Anomura) del sistema hidrografico del Rio Valdivia. *Anales del Museo de Historia Natural de Valparaiso* 13:255–66.

Jara, C. G. 1982. *Aegla bahamondei*, new species (Crustacea: Decapoda: Anomura) from the Coastal Mountain Range of Nahuelbuta, Chile. *Journal of Crustacean Biology* 2(2):232–38.

Jara, C. G. 1986. *Aegla spectabilis*, a new species of freshwater crab from the eastern slope of the Nahuelbuta Coastal Cordillera, Chile. *Proceedings of the Biological Society of Washington* 99(1):34–41.

Jara, C. G. 1989. *Aegla denticulata lacustris*, new subspecies, from Lake Rupanco, Chile (Crustacea, Decapoda, Anomura, Aeglidae). *Proceedings of the Biological Society of Washington* 102(2):385–93.

Jara, C. G., M. Pérez-Losada, and K. A. Crandall. 2018. *Aegla chilota*, new species of anomuran freshwater crab from Chiloé Island, western Patagonia. *Nauplius*, 26:e2018029.

Latreille, M. 1818. Crustacés, arachnides et insectes. In *Tableau encyclopédique et méthodique des trois règnes de la nature*, p. 24. Paris: Chez Mme. Veuve Agasse, plate 308.

Leach, W. E. 1820. Galatéadées. In *Dictionnaire des sciences naturelles*, ed. F. G. Levrault pp. 49–56. Strasbourg.

Lopretto, E. C. 1978a. Estructura exosqueletaria y miologica del quinto par de pereiópodos del macho de la familia Aeglidae (Crustacea, Anomura). *Limnobios* 1(8):284–98.

Lopretto, E. C. 1978b. Las especies de *Aegla* Leach del centro-oeste Argentino en base a la morfologia comparada, del quinto par de pereiópodos (Crustacea, Anomura, Aeglidae). *Neotropica* 24(71):57–68.

Lopretto, E. C. 1979. Estudio comparativo del quinto par de pereiópodos en los representantes del genero *Aegla* de la Patagonia Argentina (Crustacea Anomura). *Neotropica* 25(73):9–22.

Lopretto, E. C. 1980a. Analisis de las caracteristicas del quinto pereiopodo en las especies de *Aegla* del grupo "platensis" (Crustacea, Anomura, Aeglidae). *Physis* 39(96): 37–56.

Lopretto, E. C. 1980b. Clave para la determinación de las especies del genero *Aegla* de la República Argentina en base al estudio comparativo del quinto par de pereiópodos masculinos (Crustacea, Anomura, Aeglidae). *Limnobios* 1(10):431–36.

Lopretto, E. C. 1981. Consideraciones sobre la estructura apendicular vinculada al dimorfismo sexual en los machos de las especies de *Aegla* del noroeste Argentino (Crustacea, Anomura, Aeglidae). *Acta Zoologica Lilloana* 36(2):15–35.

Martin, J. W., and L. G. Abele. 1986. Phylogenetic relationships of the genus *Aegla* (Decapoda, Anomura, Aeglidae), with comments on anomuran phylogeny. *Journal of Crustacean Biology* 6(3):576–616.

Martin, J. W., and L. G. Abele. 1988. External morphology of the genus *Aegla* (Crustacea: Anomura: Aeglidae). *Smithsonian Contributions to Zoology* 453:1–46.

Martin, J. W., and G. E. Davis. 2001. An updated classification of the recent Crustacea. *Natural History Museum of Los Angeles County, Science Series* 39:1–124.

McLaughlin, P. A., R. Lemaitre, and U. Sorhannus. 2007. Hermit crab phylogeny: a reappraisal and its "fall-out". *Journal of Crustacean Biology* 27(1):97–115.

Milne Edwards, H. 1836–1844. Les crustacés. In *Le règne animal, distribué d'après son organisation, pour servir de base a l'histoire naturelle des animaux, et d'introduction à l'anatomie comparée.* ed. G. Cuvier, 3rd edn, vol. 8, 278 pages. Paris, atlas, 86 plates.

Milne Edwards, H. 1837. *Histoire naturelle des crustacés, comprenant L' Anatomic, la physiologie et la classification de ces animaux*, vol. 2, pp. 258–60. Paris: Librarie Encyclopédique de Roret.

Mocquard, M. F. 1883. Recherches anatomiques sur l'estomac des Crustacés Podophthalmaires. *Annales des Sciences Naturelles* (Paris; Zoologie et Paleontologie), series 6, 16(1): 311 pages, plates 1–11.

Molina, J. I. 1782. *The Geographical, Natural and Civil History of Chili*. New York: AMS Press (1973: Reprint from the 1808 edition, based on the original version).

Moraes, J. C. B., M. B. R. C. Terossi, M. Tavares, F. L. Mantelatto, and S. L. S. Bueno. 2016. Morphological and molecular data reveal the cryptic diversity among populations of

Aegla paulensis (Decapoda, Anomura, Aeglidae), with descriptions of four new species and comments on dispersal routes and conservation status. *Zootaxa* 4193:1–48.

Mouchet, S. 1931a. Sur la branchie d'*Aeglea laevis* (Latreille) et son parasite *Lagenophrys aeglea* nov. sp. *Comptes Rendus Hebdomadaires des Séances et Mémoires de la Société de Biologie et de Ses Filiales* (Montevideo) 109:148–50.

Mouchet, S. 1931b. Note préliminaire sur l'étude de la branchie de *Aeglea laevis* (Latreille). *Archivos Société de Biologie de Montevideo* 3(2):188–200.

Mouchet, S. 1932a. Phénomènes pathologiques dans la branchie d'*Aeglea laevis* (Latreille). *Comptes Rendus Hebdomadaires des Seances et Mimoires de la Société de la Biologie et de Ses Filiales et Associées* (Montevideo) 110:861–62.

Mouchet, S. 1932b. Notes sur la biologie du galathéide *Aeglea laevis* (Latr.). *Bulletin de la Société zoologique de France* 57(59):316–40.

Müller, F. 1876. *Aeglea odebrechtii* n. sp. *Jenaische Zeitschrift für Naturwissenschaft* 10(3):13–24, plate 1, new series.

Nicolet, H. 1849. Crustaceos. In *Historia Fisica y* Politica de Chile: Zoologia, ed. C. Gay, pp. 198–201. Paris: Maulde y Renou.

Ortmann, A. E. 1892. Die Dekapoden-Krebse des Strassburger Museums, Theil IV: Die Abtheilungen Galatheidea und Paguridea. *Zoologische Jahrbücher Abtheilung für Systematik, Geographie und Biologie der Thiere* 6:241–26, plates 11 and 12.

Oyanedel, A., C. Valdovinos, N. Sandoval, et al. 2011. The southernmost freshwaters anomurans of the world: geographic distribution and new records of Patagonian aeglids (Decapoda: Aeglidae). *Journal of Crustacean Biology* 31(3):396–400.

Páez, F. P., I. C. Marçal, L. Souza-Shibatta, et al. 2018. A new species of *Aegla* Leach, 1820 (Crustacea, Anomura) from the Iguaçu River basin, Brazil. *Zootaxa* 4527(3):335–46.

Pérez-Losada, M., G. Bond-Buckup, C. G. Jara, and K. A. Crandall. 2004. Molecular systematics and biogeography of the southern South American freshwater "crabs" *Aegla* (Decapoda: Anomura: Aeglidae) using multiple heuristic tree search approaches. *Systematic Biology* 53(5):767–80.

Pérez-Losada, M., C. G. Jara, G. Bond-Buckup, M. L. Porter, and K. A. Crandall. 2002. Phylogenetic position of the freshwater anomuran family Aeglidae. *Journal of Crustacean Biology* 22(3):670–76.

Rathbun, M. J. 1910. The stalk-eyed crustacea of Peru and the adjacent coast. *Proceedings of the United States National Musuem* 38(1766):531–620.

Ribeiro, F. B., A. F. Huber, C. D. Schubart, and P. B. Araujo. 2017. A new species of *Parastacus* Huxley, 1879 (Crustacea, Decapoda, Parastacidae) from a swamp forest in southern Brazil. *Nauplius* 25:e2017008.

Ringuelet, R. A. 1948a. Una nueva *Aegla* del nordeste Argentino (Decapoda, Anomura). *Notas del Museo de La Plata* 13(Zoologia 111):203–08.

Ringuelet, R. A. 1948b. Los "cangrejos" Argentinos del genero *Aegla* de Cuyo y La Patagonia. *Revista del Museo de La Plata* 5(Zoologia 34):297–347.

Rocha, S. S., and S. L. S. Bueno. 2004. Crustáceos decápodes de água doce com ocorrência no Vale do Ribeira de Iguape e rios costeiros adjacentes, São Paulo, Brasil. *Revista Brasileira de Zoologia* 21:1001–10.

Rocha, S. S., and S. L. S. Bueno. 2011. Extension of the known distribution of *Aegla strinatii* Türkay, 1972 and a checklist of decapod crustaceans (Aeglidae, Palaemonidae and Trichodactylidae) from the Jacupiranga State Park, south of São Paulo State, Brazil. *Nauplius* 19:163–67.

Santos, S., G. Bond-Buckup, L. Buckup, et al. 2015. Three new species of Aeglidae (*Aegla*) from Paraná State, Brazil. *Journal of Crustacean Biology* 35(6):839–49.

Santos S., G. Bond-Buckup, A. S. Gonçalves, M. L. Bartholomei-Santos, L. Buckup, and C. G. Jara. 2017. Diversity and conservation status of *Aegla* spp. (Anomura, Aeglidae): an update. *Nauplius* 25:e2017011.

Schmitt, W. L. 1942a. The species of *Aegla*, endemic South American freshwater crustaceans. *Proceedings of the United States National Museum* 91(3132):431–524.

Schmitt, W. L. 1942b. Two new species of *Aeglea* from Chile. *Revista Chilena de Historia Natural* 44(1940):25–31.

Snodgrass, R. E. 1950. Comparative studies on the jaws of mandibulate arthropods. *Smithsonian Miscellaneous Collections* 116(1): 1–85.

Tudge, C. C., and D. M. Scheltinga. 2002. Spermatozoal morphology of the freshwater anomurans *Aegla longirostri* Bond-Buckup and Buckup, 1994 (Crustacea: Decapoda: Aeglidae) from South America. *Proceedings of the Biological Society of Washington* 115(1):118–28.

Türkay, M. 1972. Neue Höhlendekapoden aus Brasilien (Crustacea). *Revue Suisse de Zoologie* 79(15):415–18.

Zimmermann B.L, M. S. Crivellaro, C. B. Hauschild, et al. 2018. Phylogeography reveals unexpectedly low genetic diversity in a widely distributed species: the case of the freshwater crab *Aegla platensis* (Decapoda: Anomura). *Biological Journal of the Linnean Society* 123:578–92.

CHAPTER 3

Population Structure and Morphological Maturity

Setuko Masunari

CONTENTS

3.1 INTRODUCTION

Studies on aeglids date back to the early 1800s, when Latreille described the first species, *Aegla laevis*, which was still classified under the name of *Galathea laevis* (Schmitt 1942) at the time. Three species were discovered in that century, but most of the more than 80 currently known species (Santos et al. 2017) have been described from the middle of the last century onwards, especially in the work of Schmitt (1942), Bond-Buckup and Buckup (1994), and Santos et al. (2012, 2013, 2015).

Despite the long-established systematic studies, population studies of aeglids began only in the 1960s, with the pioneering work by Bahamonde and López (1961) on *A. laevis* from el Monte, Chile, in which they described various biological aspects such as sexuality, reproduction, growth, migration, and population composition. The subsequent studies on aeglid populations focused on species from the state of São Paulo, Brazil: López (1965) described a population of *A. paulensis* occurring in the biological reserve at the Boraceia Biological Station in the municipality of

Salesópolis, and Rodrigues and Hebling (1978) analyzed a population of *A. perobae* from São Pedro. However, most of the existing studies on aeglid populations have been conducted since 2000 [see review (in Bueno et al. 2016a)] and have generated a large volume of descriptive data, especially for those species occurring in southern Brazil, the Paraguay-Paraná River basin and the Uruguay River basin.

This chapter provides information on the habitat, distribution, density, structure, and morphological maturity of epigean populations of aeglids. The only currently available record on the density of cave-dwelling aeglids (Maia et al. 2013) will be referred to in the section on population density estimates (Section 3.2).

3.2 HABITAT, DISTRIBUTION, AND DENSITY

Most epigean and cave-dwelling aeglids inhabit lotic environments composed not only of rapids but also of backwaters (Bond-Buckup 2003); they can also live in lentic ecosystems (Bond-Buckup and Buckup 1994).

Temperature is considered one of the most important variables in these freshwater environments. The temperature of stream water is driven by the interaction of natural environmental processes (e.g., air temperature, solar radiation, and conduction from soil) and anthropogenic disturbances of the natural thermal regime, such as deforestation and hydroelectric development. Water temperature regulates a wide range of biological processes in a river system, such as the spawning periods, growth rates, and mortality rates of the aquatic inhabitants. As a typical ectothermic organism, aeglids can thrive only when water temperatures are within their preferred thermal range [see review (in Harvey et al. 2011)].

Another important variable in lotic environments is the concentration of dissolved oxygen (DO) present in the water. This oxygen is produced during daylight hours through photosynthesis and is consumed 24 hours a day through respiration and the decomposition of organic matter by bacteria and fungi. As a result, the DO concentration in a river will be higher during the day and lower during the night, with the lowest levels occurring just before sunrise. The DO concentration is also influenced by the altitude and water temperature; oxygen is more soluble in colder than in warmer water. Under natural conditions, these two variables are actually always interrelated: high temperatures with low oxygen content and low temperatures with high oxygen content (Aldstadt III et al. 2010). The maximum solubility of oxygen in water at 1 atm pressure (standard air pressure at sea level) ranges from approximately 6.41 mg.L^{-1} at 40°C to 14.62 mg.L^{-1} at 0°C; that is, ice-cold water can hold more than twice as much DO as warm water (Cole 1983).

Aeglid populations are distributed in subtropical and temperate climates (southern South America) where aquatic habitats have relatively low temperatures. Although most authors describe the waters inhabited by aeglids as well oxygenated, few studies have accurately measured DO. Rodrigues and Hebling (1978) were the first authors to measure DO during their study of the biology of *Aegla perobae* from the Peroba stream (São Pedro, São Paulo, Brazil); they reported a concentration of 10.8 mg.L^{-1} in waters with a mean temperature of 16.5°C (ranging from 11 to 22°C), the highest DO value recorded in streams inhabited by aeglids and very close to the maximum oxygen solubility at that temperature (Cole 1983). The remaining aeglid populations whose habitat temperature and DO are known live in waters with temperatures ranging from 10.8 to 22.8°C and DO ranging from 5.0 to 9.7 mg.L^{-1} (Figure 3.1). The Quebra-Perna River (Ponta Grossa, Paraná, Brazil), where *A. castro* lives, showed the narrowest DO range (8.16 to 8.96 mg.L^{-1}) (Swiech-Ayoub and Masunari 2001a), while the Taquara stream (São Pedro do Sul, Rio Grande do Sul, Brazil), where *A. manuinflata* lives, showed the most variable DO values (5.30 to 9.70 mg.L^{-1}) (Trevisan 2008). Considering that the habitats of aeglid populations are nearly always above sea level (see Table 3.1), these data confirm that aeglids live in well-oxygenated waters. The lower values from the Taquara stream are certainly related to the time of the day at which the DO measurement was taken, perhaps at sunrise, when the lowest DO is typically observed (Aldstadt III et al. 2010). The ranges of water temperature and dissolved oxygen content in the streams inhabited by *Aegla* species are graphically represented in Figure 3.1. For comparative purposes, a population of *Macrobrachium potiuna*, a freshwater palaemonid shrimp species from southern Brazil (Blumenau, Santa Catarina), was included in the graph. This shrimp lives in slightly warmer waters (15 to 25°C) and can tolerate lower DO (4.0 to 9.6 mg.L^{-1}) than aeglids (Boos and Althoff 2002).

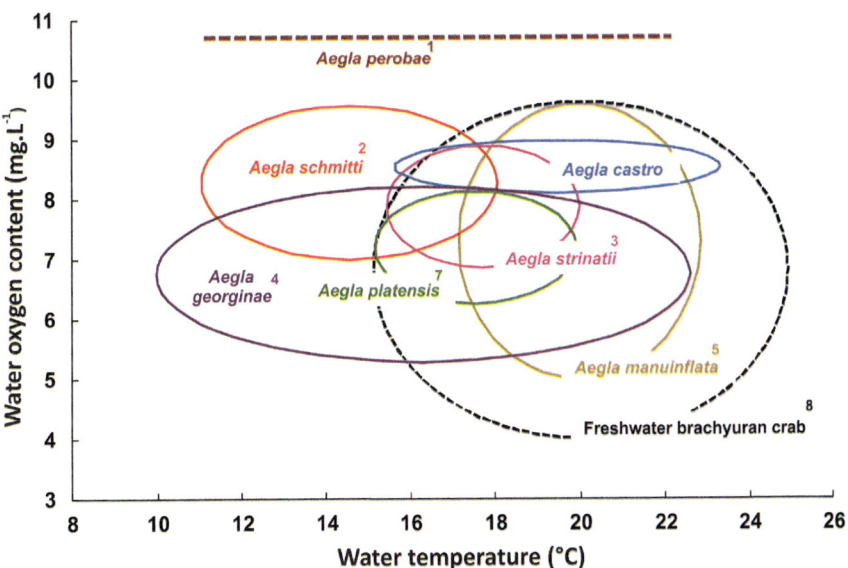

Figure 3.1   Range of water temperature and dissolved oxygen content in the streams inhabited by some aeglid species. For comparative purposes, data of these two abiotic variables from habitat of southern Brazilian freshwater shrimp were included. References: [1]Rodrigues and Hebling (1978), [2]Teodósio and Masunari (2009), [3]Rocha (2007), [4]Copatti et al. (2016a), [5]Trevisan (2008), [6]Swiech-Ayoub and Masunari (2001a), [7]Dalosto et al. (2014), [8]Giri and Collins (2014), [42]Boos and Althoff (2002).

Table 3.1 Geographic Position, Local Average Air Temperature, Longevity, and Locality of Aeglids

	Latitude	Altitude (m)	Temperature (°C)	Longevity (months) Male	Longevity (months) Female	Locality (Brazil)	References
Aegla georginae	29°29'	290*	19.5**	19.0	24.0	Jaguari, RS	Copatti et al. (2016b)
Aegla longirostri	29°24'	400*	18.2**	21.0	24.0	Itaára, RS	Silva-Castiglioni et al. (2006)
Aegla jarai	27°01'	330*	19.1**	24.0	24.0	Serra Itajaí National Park	Boos et al. (2006)
Aegla itacolomiensis	29°33'	18*	19.7**	26.0	30.0	Igrejinha, RS	Silva-Gonçalves et al. (2009)
Aegla marginata	24°23'	575	18.5**	29.0	30.0	Intervales State Park	Silva et al. (2016)
Aegla strinatii	24°38'	515*	23.0	33.5	34.0	Jacupiranga State Park	Rocha (2007)
Aegla jaragua	23°27'	790	18.5**	33.9	40.2	São Paulo, SP	Cohen et al. (2011)
Aegla schmitti	24°31'	351*	18.9	34.0	38.4	Intervales State Park	Chiqueto-Machado et al. (2016)
Aegla manuinflata	29°36'	176*	19.3**	36.0	30.0	São Pedro do Sul, RS	Trevisan and Santos (2011)
Aegla perobae	22°31'	673	20.6**	36.0	36.0	São Pedro, SP	Bueno et al. (2014)
Aegla laevis	33°40'	389*	16.1**	37.0	37.0	El Monte (Chile)	Bahamonde and López (1961)

Source of data: Google Earth Pro (2018).
**Source of data:* Climate-data.org (2018).

Dalosto and Santos (2011) evaluated the relationship between the oxygen consumption and daily activity of the aeglid *A. longirostri* under limited and constant O_2 conditions. This species has high and oxygen-independent metabolism but does not have a true oxygen-regulator pattern. This indicates a high metabolic rate compared to those of other crustaceans, which may improve its success in environments where oxygen saturation levels are high. In addition, it is probably incapable of maintaining this metabolic rate for prolonged periods when oxygen availability is restricted. These data strongly suggest that the need for well-oxygenated water is an intrinsic characteristic of the species. If this is a pattern of all aeglids, their distribution is likely to be related to cold water because of its ability to carry large amounts of dissolved oxygen.

The flow velocity of the streams inhabited by aeglids ranges from 0.2 to 1.2 m.s^{-1} (heavy rain excluded) (Swiech-Ayoub and Masunari 2001a; Bueno and Shimizu 2008; Dalosto et al. 2014; Giri and Collins 2014; Copatti et al. 2016a), indicating constant water movement, which facilitates the dissolution of oxygen. Other known water parameters of the environments inhabited by aeglids include pH ranging from 4.8 to 7.7 and varying levels of ammonia, sulfates, and phosphates (Swiech-Ayoub and Masunari 2001a; Teodósio and Masunari 2009; Dalosto et al. 2014; Giri and Collins 2014; Copatti et al. 2016a).

The substrates of the habitats occupied by aeglids have been described by several authors as composed of rocks of various sizes (Figure 3.2), interspersed with foliage and sand in varied proportions. Although aeglids have strong swimming abilities, they are primarily benthic animals (Ayres-Peres et al. 2011a). Therefore, such substrates benefit these animals by providing shelter from strong currents and predators (fish and birds).

Figure 3.2 Natural substrate composed of pebbles and boulders in a creek inhabited by *Aegla schmitti* at Mananciais da Serra, Paraná State, Brazil.

According to Tokeshi and Arakaki (2012), the more abundant the physical elements in an aquatic habitat, the more abundant the organisms living in that system. The results of a study on *A. manuinflata* from Arroio Passo Taquara (São Pedro do Sul, Rio Grande do Sul, Brazil) suggest a positive correlation between abundance and pebble density (Trevisan 2008). However, the diameter of the pebbles, which ranged from 32 to 256 mm, did not influence abundance. Further studies are required for an improved understanding of the influence of habitat architecture on the abundance of all aeglids.

Artificial substrates such as man-made concrete structures can fully replace the natural habitats of aeglids, as demonstrated by a study by Teodósio and Masunari (2009) with *A. schmitti* at Mananciais da Serra (Piraquara, Paraná, Brazil). At this site, aged submerged concrete walls, which once accommodated water filters for the Curitiba city water supply, offered an unusual biotope: their cracks and holes amplified the capacity of the substrate in providing shelter and enabling aeglids to occur in higher abundance as compared to natural substrates. In addition to the enhanced carrying capacity of aged concrete walls, they provided protection against strong water currents and against natural predators such as amphibians and birds. This relationship has already been described by Downes et al. (1998, 2000) who experimentally demonstrated that habitat structure could influence the invertebrate population abundance in stony upland streams. Habitat structural features such as large crevices, roughness, and presence of macroalgae promoted an increase in the density of individuals.

The presence of aeglids in artificial substrates was also observed by López (1965), who described a predominance of large adult individuals of *A. paulensis* in a cement tank connected to natural streams. These adult aeglids migrated up to 300 m between the stream and the cement tank, overcoming barriers up to 1.5 m high, as shown by tracking tagged individuals.

In addition to the structural elements in the streams, varied water velocity additionally contributes to increasing the habitat complexity. Water flow in a stream is primarily related to the stream's gradient, but it is also controlled by the geometry of the stream channel (Wohl 2010). Water flow velocity is decreased by friction along the streambed and the edges; therefore, it is slowest at the bottom and edges and fastest near the surface and in the middle of the stream (Wohl 2010). The pebbles present in the bottom of the aeglids' habitat weaken the water flux and make multiple, varied shelter spaces available for them (Downes et al. 2000). Aquatic plants, which are typically more frequent along the edges of the stream than in the middle, also provide shelter. For example, smaller individuals of both sexes of *A. castro* were more frequent than larger individuals in narrower stretches of the creek, where the substrate was composed by pebbles and boulders and where aquatic vegetation was present along its margins. These smaller aeglids seek out substrates that have more complex architecture and offer appropriate spacing for their shelter. These biotopes were approximately 225–300 m distant from the widest areas and sandy substrate, where larger individuals of both sexes predominated, especially males (Swiech-Ayoub and Masunari 2001a).

Juveniles in the early stages of development were either absent entirely or present in low numbers in samples of most aeglid populations that were studied (see Section 3.3.3). This fact strongly indicates that these very small juveniles live in different habitats than the other demographic categories; perhaps they seek out even more complex habitats far from adults in order to evade cannibalism. These observations constitute important data for the future recovery of aeglid habitats. Future studies are certain to reveal further important information regarding the space occupation of aeglids during their different life cycle phases.

Unfortunately, healthy habitats are being altered at levels that seriously threaten the populations of aeglids, a subject that has been highlighted by several authors [e.g., (Bueno et al. 2016a, b)], but no efficient strategy is known to avoid local or total population extinction. Bueno et al. (2014) observed how vulnerable aeglid populations are, even to naturally occurring events. The authors recorded a drastic decline in the population density of *A. perobae* after an extraordinarily heavy rainfall, from 9.05 ind.m^2 to numbers so low that a reliable estimate of the density could not be obtained. The beginning of population recovery was verified only after seven months. In this context, studies that return live aeglids to the environment after obtaining morphometric and biological data (see Silva-Gonçalves et al. 2009; Dalosto et al. 2014; Chiqueto-Machado et al. 2016; Silva et al. 2016) are highly appreciated.

Density has been estimated for only five species of aeglids [*A. franca, A. jaragua* (formerly *A. paulensis*), *A. manuinflata, A. perobae*, and *A. platensis*] using the mark-recapture method of Schumacher-Eschmeyer or Petersen (Bueno et al. 2007; Cohen et al. 2013; Bueno et al. 2014; Dalosto et al. 2014; Trevisan and Santos 2014). The mean densities estimated for these species vary from 1.52 ind.m^{-2} for a population of *A. manuinflata* from the Taquara stream (Trevisan and Santos 2014) to 11.5 ind.m^{-2} for a population of *A. platensis* occurring in the Lajeado Bonito stream (Dalosto et al. 2014). The densities of aeglids living in caves and related habitats, estimated with the mark-recapture and triple-catch methods, are considered apart from this range because they are very low, with a mean of 0.33 ind.m^{-2} (Maia et al. 2013).

Density is a population dynamic parameter, and its variation depends on both extrinsic (climate, food, nutrients, predators, pathogens, and shelter) and intrinsic factors (behavioral, physiological, and genetic changes associated with social interactions within the population) (Krebs 2014). Although experiments and observations on the factors influencing the density of aeglids have not been performed, it can be inferred that the low densities recorded for cave dwellers are caused mainly by food scarcity and limited space (Barr 1967).

The population density of animals is inversely related to their body mass in several animal groups, including aquatic invertebrates (Krebs 2014). This relationship was also demonstrated for arthropods by Williamson and Lawton (1991) in a study of fractal geometry in ecological habitats. This negative relationship was observed for the five aeglid species for which density was estimated, with males having a maximum carapace length (CL) range of 21.62–31.75 mm (Figure 3.3). Although this figure uses body length instead of body mass, the two parameters are correlated. The population of *A. platensis* from the Mineiro stream (Bueno and Bond-Buckup 2000) was excluded from this analysis because the authors did not indicate whether the carapace measurement included the rostrum.

Silva et al. (1997), in a study of population density and energy use as a function of body mass of mammals and birds, pointed out the relationship between the

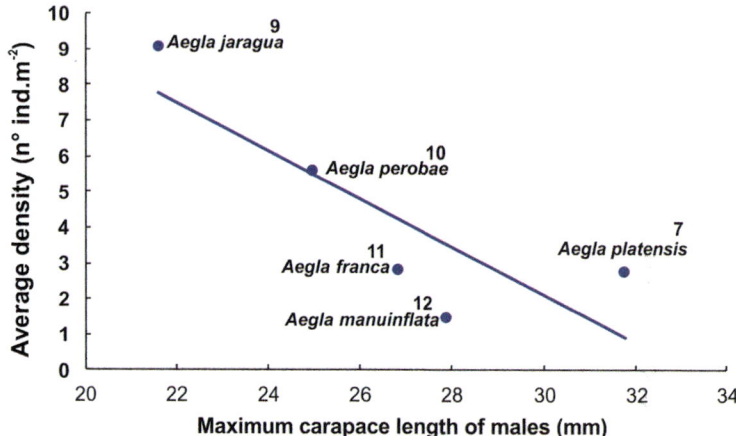

Figure 3.3   Relationship between the maximum carapace length of males of some aeglid species and density of the respective populations. References: [7]see caption of Figure 3.1, [9]Cohen et al. (2013), [10]Bueno et al. (2014), [11]Bueno et al. (2007), [12]Trevisan and Santos (2014).

morphology, physiology, and behavior of individual organisms and the ecological performance of these endothermic vertebrates. However, the cause of the negative relationship between population density and body size (or mass) among invertebrates, including aeglids, remains unknown.

3.3 POPULATION STRUCTURE

This section presents the sexual dimorphism of aeglid populations in terms of size, sex ratio, growth, longevity, and size composition.

Currently, more than 20 populations of aeglid species have had their size structure described. The measurement adopted as representative of body size in all cases is the carapace length (CL). Some authors (Rodrigues and Hebling 1978; Swiech-Ayoub and Masunari 2001a; Trevisan and Santos 2012 and others) defined CL as the distance between the tip of the rostrum and the posterior margin of the carapace, while others (Boos et al. 2006; Silva-Gonçalves et al. 2009; Silva et al. 2016 and others) excluded the rostrum from this procedure, most probably due to the fragility of the rostrum. On the other hand, Cohen et al. (2011) made both measures available. Due to this discrepancy in the measurement procedure, all CL values measured without the rostrum were converted to CL including the rostrum in this chapter, based on measurements taken from the figures of the species descriptions.

3.3.1 Sexual Dimorphism and Sex Ratio

Sexual dimorphism in aeglids is usually recognized by certain secondary sexual characteristics: the presence of pleopods and a genital opening in the third pair of pereopods in females and longer carapace and chelipeds and a larger palmar crest in males (Bahamonde and López 1961; Martin and Abele 1988). In addition, at least in *A. schmitti*, abdominal somites 2, 3, and 4 are longer in females than in males (Bueno et al. 2016a). Several authors have also detected sexual dimorphism in carapace shape with the aid of geometric morphometry techniques [review (in Diawol et al. 2015)]. Females have a wider posterior carapace than males, a characteristic that the authors consider a morphological adaptation to accommodate the egg mass during the incubation period, similar to what is observed in brachyuran crabs.

Sexual dimorphism in carapace length is widely seen among aeglids [review (in Bueno et al. 2016a)], with males having longer length than females. This pattern is evident in Figure 3.4, where we compared the maximum CL value found for males and females of each population.

The population of *A. marginata* in the Barrinha River (Tunas do Paraná, Paraná, Brazil) showed the lowest maximum CL values, with males and females measuring 20.4 mm and 17.2 mm, respectively (Adam et al. 2018). On the other hand, the population of *A. parana* living in the Negro River, Upper Iguaçu Basin, Brazil, attained the largest dimensions, with the male and female measuring 50.2 mm and 38.7 mm, respectively (A. Schafaschek unpublished data).

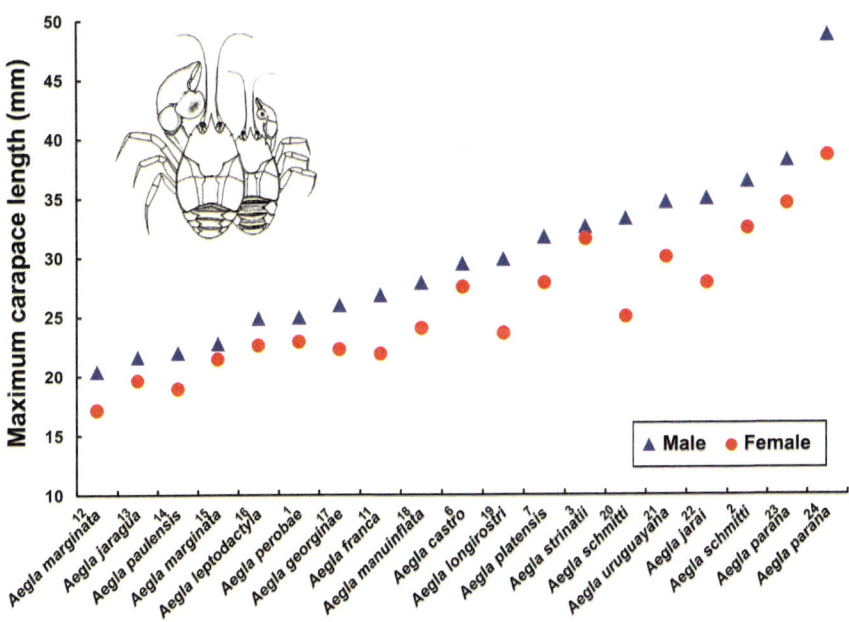

Figure 3.4 Sexual size dimorphism in aeglid species discerned through maximum carapace length of males and females. References: [1–12]see caption of Figures 3.1 and 3.3, [13]Adam and Masunari (2018), [14]Cohen et al. (2011), [15]López (1965), [16]Silva et al. (2016), [17]Noro and Buckup (2003), [18]Copatti et al. (2016b), [19]Silva-Castiglioni et al. (2006), [20]Chiqueto-Machado et al. (2016), [21]Viau et al. (2006), [22]Boos et al. (2006), [23]Grabowski et al. (2013), [24]Schafaschek (unpublished data).

Among the 21 species for which maximum CL values for males and females are available, only in *A. platensis* from the Mineiro creek (Taquara, Rio Grande do Sul, Brazil) (Bueno and Bond-Buckup 2000) and *A. franciscana* from the Rolante stream (São Francisco de Paula, Rio Grande do Sul, Brazil) (Gonçalves et al. 2006) are males smaller than females (these populations were not included in Figure 3.4). The sampled areas for these populations, stretches of 18 m and 10 m, respectively, may have influenced this reverse result: these areas were probably not large enough to sample all size classes. In the other studies, the sampled stream stretches ranged from 30 m to 1000 m. The finding of larger males than females in a population of the same species—*A. platensis*—from the Lajeado Bonito River (Frederico Westphalen, Rio Grande do Sul, Brazil), recorded by Oliveira and Santos (2011) and Dalosto et al. (2014), supports this assumption.

It is known that *A. manuinflata* males can migrate up to 175 m in eight days, as monitored by radiotelemetry (Ayres-Peres et al. 2011b). Moreover, selective migrations of aeglids were recorded by López (1965) in *A. paulensis*, where larger adults of both sexes can migrate up to 300 m and overcome barriers up to 1.5 m high to reach concrete tanks. Certainly, the increased mobility of larger aeglids is related to reproduction.

The trend of male aeglids attaining larger maximum CL than females might be under the major evolutionary forces of sexual selection in favor of larger male size improving intrasexual combat success, increasing reproductive output [see review (in Liao et al. 2013)]. On the other hand, females tend to allocate more energy for egg production than to somatic growth and thus attain smaller body sizes than males (Hartnoll 1978, 1982, 1985).

The estimated sex ratio of aeglid populations is closely related to the sampling method. Aeglids have been captured manually with the aid of hand nets or traps baited with beef or chicken liver. Most populations that were sampled with hand nets showed sex ratios that did not differ significantly from the expected 1:1, or, when the difference was significant, the proportions of males and females were close, indicating that the parents' expenditures on male and female offspring were similar (Fisher 1930). On the other hand, populations sampled with traps always showed a marked predominance of males, resulting in sex ratios significantly different from the expected 1:1 (Figure 3.5). A timely study by Trevisan and Santos (2014) tested the influence of the two collection methods on the same *A. manuinflata* population living in the Taquara stream, southern Brazil. In the samples obtained with a hand net, the sex ratio was very close to 1:1 (238 males and 235 females), whereas in those collected with a trap, the ratio was 1:0.19 (334 males and 64 females). Dalosto et al. (2014) and Copatti et al. (2016b) also confirmed this influence of the sampling technique.

According to Dalosto et al. (2014) and Copatti et al. (2016b), sex ratio varies with the sampling method due to the differential size and behavior of males and females. The aggressiveness of males is well documented by Ayres-Peres et al. (2011a) in laboratory studies of *A. longirostri*. As males are larger than females in

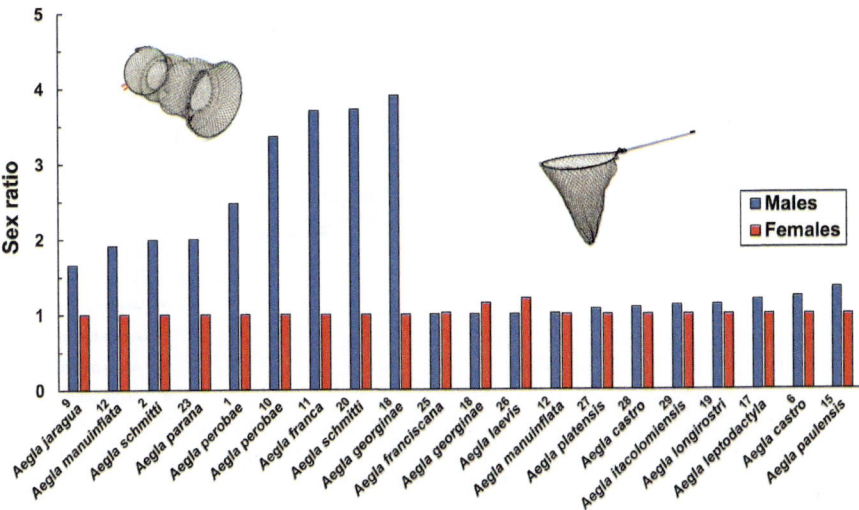

Figure 3.5 Sex ratio of aeglid populations sampled with traps (first nine populations at left) and with hand nets (remaining eleven populations at right). References: [1-24]see caption of Figures 3.1, 3.2, 3.3, and 3.4, [25]Gonçalves et al. (2006), [26]Bahamonde and López (1961), [27]Bueno and Bond-Buckup (2000), [28]Fransozo et al. (2003), [29]Silva-Gonçalves et al. (2009).

most aeglid populations (see Section 3.4), females and juveniles apparently remain outside the traps to avoid agonistic encounters with males. This assumption was confirmed by filming a group of *A. schmitti* occurring in a submerged concrete wall with cracks and holes at Mananciais da Serra, Piraquara, southern Brazil (Teodósio and Masunari 2009), with a camera coupled to a baited trap set up near one of these cracks. Large males readily entered this trap; however, smaller individuals (smaller females or juveniles) failed to do so, despite apparent attraction to the bait (S. Masunari unpublished data). Therefore, hand nets are the recommended sampling method to estimate the sex ratio in aeglid populations. However, additional collections using traps are essential for determining the maximum sizes of males and females within a given population or species.

In the four populations for which the relationship between sex ratio and size class has been described, the proportion of males increased from intermediate to larger classes, although ratios were variable in the smaller classes (Figure 3.6). In addition, males consistently predominated in the larger size classes in the 21 species for which size structure were analyzed. These results suggest that aeglids fit the anomalous pattern in accordance to the classification proposed by Wenner (1972), which is characterized by a significant deviation from a 1:1 ratio among the older or larger individuals.

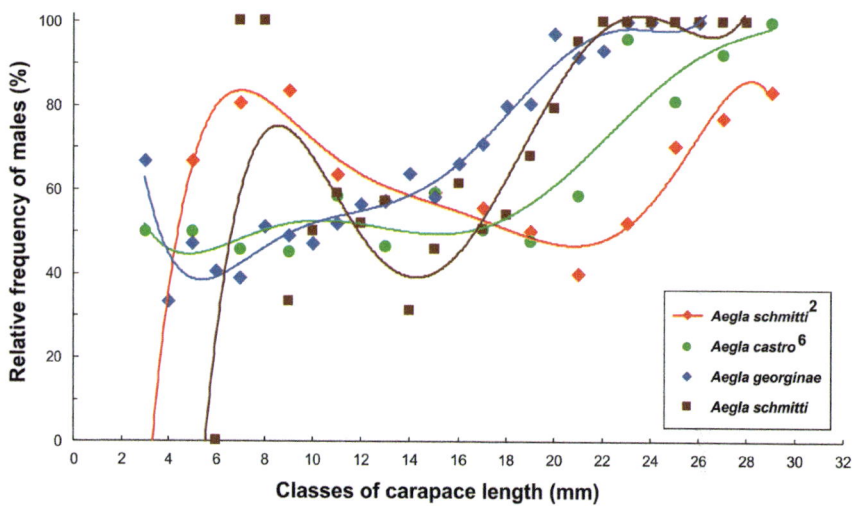

Figure 3.6 Probability curves based on the distribution of the sex ratio within various size classes for four aeglid species. References: [2-6]see caption of Figure 3.1, [20]see caption of Figure 3.4.

3.3.2 Growth and Longevity

Aeglids exhibit the energy budget pattern of ectothermic animals, in which no energy is allocated for production of heat and the distribution of total energy intake varies during their ontogeny (Werner et al. 2018). Prior to puberty, energy is allocated exclusively to maintenance and growth, and after this stage, energy is also allocated to reproduction [see revision (in Werner et al. 2018)]. The aeglid body continues to grow even after the onset of sexual maturity (see also Section 3.4), another common pattern among ectothermic animals (Werner et al. 2018). In this context, the most suitable growth model for the aeglids is the von Bertalanffy function, which predicts infinite theoretical growth (Kozlowski 1992; Kozlowski and Teriokin 1999). This growth model establishes a relationship between body size (carapace length) and age and therefore allows longevity to be estimated from large individuals of the sampled population (von Bertalanffy 1938). Growth patterns and all longevity estimates discussed in this item were based on this model.

Of the four basic techniques described by Hartnoll (2001) for determining the growth, age, and longevity of crustaceans, the one based on size-frequency distribution analysis has been adopted in studies of aeglid populations, e.g., the population of *A. jaragua* (formerly *A. paulensis*) from São Paulo, southeastern Brazil (Cohen et al. 2011). It depends on the identification of modes in the distribution that correspond to year classes or to recruitment cohorts. A single sample may be analyzed to show a series of year classes, or serial samples may be collected to follow the progression of modes with time. When the modes are clear, the analysis and interpretation are straightforward. However, this is rarely the case, and efforts have been made to develop means to analyze less well-defined distributions (Hartnoll 2001).

Somatic growth has been described and longevity estimated for 11 out of more than 80 *Aegla* species (Santos et al. 2017): *Aegla platensis* (see Bueno et al. 2000), *A. leptodactyla* (see Noro and Buckup 2003), *A. jarai* (see Boos et al. 2006), *A. longirostri* (see Silva-Castiglioni et al. 2006), *A. strinatii* (see Rocha 2007), *A. itacolomiensis* (see Silva-Gonçalves et al. 2009), *A. jaragua* (see Cohen et al. 2011), *A. manuinflata* (see Trevisan and Santos 2014), *A. schmitti* (see Chiquetto-Machado et al. 2016), *A. georginae* (see Copatti et al. 2016a), and *A. marginata* (see Silva et al. 2016). Growth rates are diverse among aeglid species and between the sexes, resulting in evident discrepancies in the growth curves for each species and each sex. This indicates that procedure standardization is needed in order to minimize these differences between estimates of von Bertalanffy equation parameters. However, males reach a higher maximum CL (see Section 3.3.1) and grow at a faster rate than females. Additionally, males reach sexual maturity with a larger CL than females (see Section 3.4) due to their faster growth and shorter lifespan than females, except that *A. manuinflata* shows the inverse pattern, and *A. laevis*, *A. perobae*, and *A. jarai* show equal longevity in both sexes (Table 3.1).

Longevity within these aeglids ranged from 19.0 to 36.0 months among males and from 24.0 to 40.2 months among females (Table 3.1). In *A. laevis* (see Bahamonde and López 1961), longevity was assessed by the identification of cohorts without discriminating the sexes. Therefore, the longevity of aeglids (1.58–3.35 years) is within the range of most freshwater decapod species (approximately 63.3%), which live from one to five years (Vogt 2012).

According to Vogt (2014), the longevity of decapod crustaceans depends on the taxonomic affiliation, geographic latitude, altitude, habitat, and water depth, although for freshwater decapods, the latter factor (water depth) is less important. Freshwater decapods that live at high latitudes and high altitudes tend to live longer (Vogt 2014). The altitude distribution of aeglids whose longevity is currently known varied from 18 to 790 m in latitudes from 22°31′ S to 33°40′ S (see references in Table 3.1). Although *A. jaragua* is the species with the longest lifespan (with males living 2.8 years and females living 3.3 years) and recorded at the highest altitude (790 m) (Cohen et al. 2011), this relationship is not clear when the entire set of species are considered (Table 3.1).

The variation in longevity is not related to latitude either: species recorded at close latitudes showed very different longevities (Table 3.1), e.g., *A. longirostri* and *A. manuinflata* in localities that were less than one degree of latitude apart (29°24′19″ S and 29°36′01″ S, respectively) showed very distinct longevities (mean of 1.9 and 2.8 years, respectively). This lack of relationship has already been verified by Silva et al. (2016). It is probable that the range of latitudinal distributions of these aeglids (just over 11°) is not sufficient to make this relation evident.

Considering that local air temperature is the primary determinant of the temperature of the aquatic systems of a region (Webb et al. 2003) and that it directly reflects the geographical position (altitude and latitude) and local climatic peculiarities, a graph of the relationship between the local average annual air temperature and the longevity of aeglids, as estimated by the von Bertalanffy model, is shown (Figure 3.7). Species that occur in habitats with similar local average annual air temperatures, such as *A. marginata* and *A. jaragua* (18.5°C), can have very different

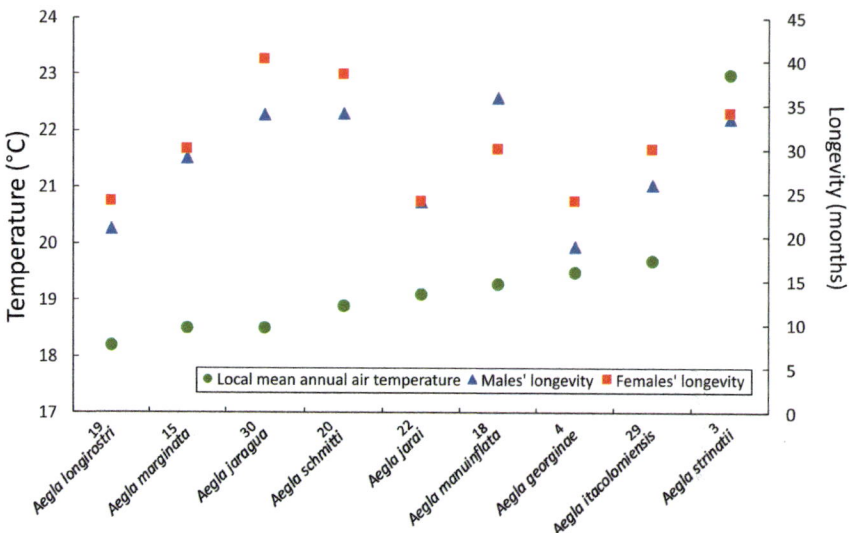

Figure 3.7 Relationship between longevity of nine aeglid species and average annual air temperature of the respective localities of occurrence. References: [3–4]see caption of Figure 3.1, [13–22]see caption of Figure 3.4, [29]see caption of Figure 3.5, [30]Trevisan and Santos (2011).

lifespans (2.5 years and 3.1 years, respectively) (Cohen et al. 2011; Silva et al. 2016) (Figure 3.4).

The discussion in this section strongly suggests that factors related to the microclimate and other local environmental variables influence the longevity of aeglids more than the geographic localization of populations. Additionally, the longevity of aeglids shows no relationship with the maximum CL attained by the respective species, since those with different maximum CLs can have the same longevity [see partial review (in Trevisan and Santos 2014)].

3.3.3 Size Composition

The overall frequency distributions in size classes (CL = carapace length) of aeglid populations show a unimodal distribution in *A. franca* and *A. jaragua* (Bueno et al. 2007 and Cohen et al. 2011, respectively) and a bimodal distribution in *A. castro* and *A. georginae* (Fransozo et al. 2003 and Copatti et al. 2016b, respectively). However, most species have multimodal distributions due to the multiple cohorts that periodically enter the population; these cohorts usually constitute the main source of individual size variation within populations (Fransozo et al. 2003; Bueno et al. 2007; Cohen et al. 2011; Copatti et al. 2016).

These cohorts represent various ages and therefore various individual sizes. The number of cohorts can be equal for both sexes, as in *A. schmitti* from Alto da

Ribeira State Park (São Paulo State, Brazil) (Chiqueto-Machado et al. 2016), which was estimated as four cohorts per year. On the other hand, different cohort numbers between the sexes are recorded for *A. jaragua* from Jaraguá State Park (four cohorts in males and five in females) (Cohen et al. 2011), for *A. perobae* from the São Pedro River (three and four, respectively) (Bueno et al. 2014), and for *A. marginata* from Intervales State Park (eight and six, respectively) (Silva et al. 2016).

Sexual dimorphism regarding body size (Section 3.3.1) is the second most important factor that causes size diversity among aeglids. Males reach a larger CL size than females in 19 out of 21 species whose size composition data are available.

Samples of aeglid populations rarely include hatchlings that measure approximately 1.66 to 2.27 mm CL (smallest population group in Figure 3.8). Most of the dimension data available for these individuals have been measured on specimens hatched in the laboratory (Francisco et al. 2007; Teodósio and Masunari 2007; Moraes and Bueno 2013, 2015; Silva et al. 2017). Only *A. longirostri* samples included very small juveniles measuring less than 2.00 mm CL (rostrum included) (Colpo et al. 2005; Silva-Castiglioni et al. 2006), which may be juveniles in the early stages of development (no descriptions or meristic data are available for this species).

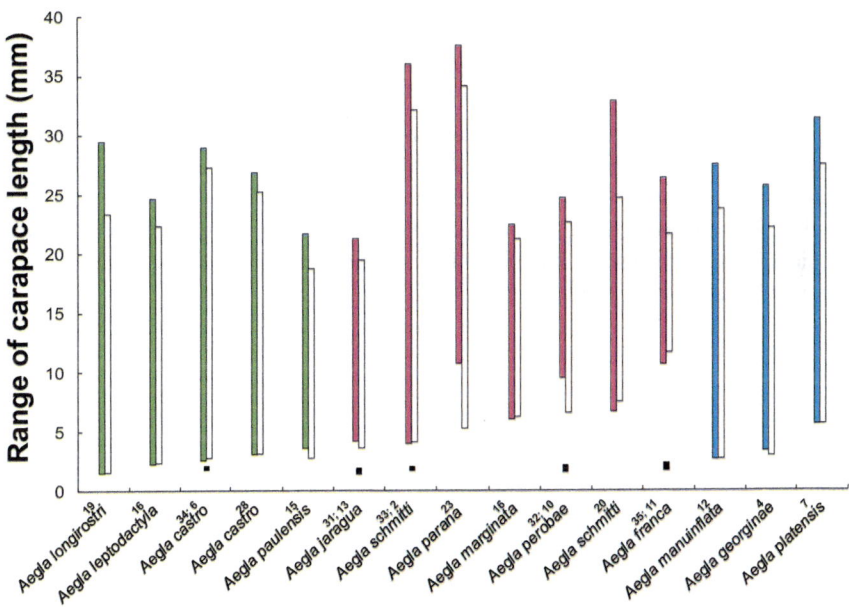

Figure 3.8 Range of carapace length of aeglid populations. Except for the newly hatched juveniles that were reared in the laboratory (black bars), the individuals were collected in the nature with hand nets (green bars), with traps (purple bars), or with both devices (blue bars). Filled bars: males, empty bars: females. References: [2–7]see caption of Figure 3.1, [10–12]see caption of Figure 3.3, [13–23]see caption of Figure 3.4, [28]see caption of Figure 3.5, [31]Moraes and Bueno (2013), [32]Moraes and Bueno (2015), [33]Teodósio and Masunari (2007), [34]Silva et al. (2017), [35]Francisco et al. (2007).

The exclusion of newly hatched juveniles from population samples is a consequence of the protective behavior of females in which offspring are kept underneath the abdomen, as observed by Rodrigues and Hebling (1978) in the laboratory. Additionally, Bueno et al. (2014) recorded hatchlings of A. *perobae* that were still attached to the surface of the female's body that was collected in the nature, but these juveniles were excluded from the trap sampling.

These data indicate that the methodologies usually employed in the study of aeglid populations are adequate only to sample individuals larger than 2.00 mm CL. The low frequency in the smaller size classes in all size composition studies (see references in the captions of Figures 3.4, 3.5, and 3.8) confirms this assumption. However, it is possible that very small juveniles do not live in the same biotope as the other larger individuals due to low structural complexity (see Section 3.2) and to avoid cannibalism.

A careful sampling in the headwater of a tributary of the Negro River (Mafra, Santa Catarina, Brazil) resulted in a sample of individuals composed predominantly of juvenile A. *parana* in the early stages of development, measuring from 4.48 to 9.13 mm CL (A. Schafaschek unpublished data). This biotope was approximately 2.0 m wide, and its substrate was predominantly composed of coarse sand with some pebbles (A. Schafaschek unpublished data). Future research should clarify whether this juvenile segregation was motivated by the availability of suitable spaces between sand grains (Tokeshi and Arakaki 2012) or whether it was due to the evasion of cannibalism and predation.

Once again, it is noted that the sampling device influenced the size composition of the collected individuals (Figure 3.8). Very small juveniles were better represented in those populations and species that were sampled with a hand net (the green block of five species), while in those sampled with a trap (the purple block of seven species), they were absent. The blue block, composed of three species (both device types), showed varied results.

3.4 MORPHOLOGICAL MATURITY

The attainment of sexual maturity in decapods involves three types of maturation: 1) external morphology, 2) internal physiology, and 3) functional aspects (Waiho et al. 2017). However, because of various governing factors, such as genetic selection and abiotic variables of the environment, the three types of maturation do not always synchronize with one another [see review (in Waiho et al. 2017)]. This asynchrony has been recorded in populations of A. *castro* (see Takano et al. 2016), A. *franca* (see Bueno and Shimizu 2008, 2009), and A. *uruguayana* (see Viau et al. 2006).

Morphological sexual maturity (MSM) is typically recognized by the sudden change in secondary sexual characters at the pubertal molt in most decapods (Hartnoll 1974, 1978). In the prepubertal phase, body structures gradually grow at each molt in both sexes. After the pubertal molt, males and females undergo distinct changes in the proportional dimensions of body parts, especially those related to reproduction, in relation to carapace size. However, the onset of MSM is not always easy to recognize in decapod crustaceans due to the possible occurrence of more than one pubertal molt (Hartnoll 2001).

Males acquire heterochelous chelipeds that are larger and more contrasting than those of females and contribute to success in intrasexual combat (Hartnoll 1974, 1978). On the other hand, females have a visibly wider abdomen than males, a pattern interpreted as an adaptation for accommodating the largest possible number of eggs adhered to the well-developed pleopods (Hartnoll 1974, 1978). For these reasons, most studies on the onset of MSM in decapods are based on the relative growth of the chelipeds (length or height of propodus) in males and on the abdomen width in females, which show an abrupt change in shape and size after the pubertal molt [see review (in Hartnoll 1985)].

As in decapods, the onset of MSM in aeglid males is characterized by a positive allometric growth of the chelipeds after the pubertal molt, i.e., they start to grow at a faster rate than females' chelipeds. Based on this allometric growth of the cheliped dimensions, the size at onset of MSM has been estimated for males of nine populations belonging to eight species: *A. castro* (Swiech-Ayoub and Masunari 2001b; Takano et al. 2016), *A. franca* (see Bueno and Shimizu 2009), *A. longirostri* (see Colpo et al. 2005), *A. georginae* (see Copatti et al. 2016b), *A. manuinflata* (see Trevisan and Santos 2012), *A. marginata* (see Adam and Masunari 2018), *A. platensis* (see Oliveira and Santos 2011), and *A. strinatii* (see Rocha 2007). On the other hand, aeglids do not show sexual dimorphism in the posterior portion of the abdomen (fifth and sixth abdominal somites plus telson) as observed in brachyuran decapods, and the females do not show any sudden morphological change, even with sexual maturation. These facts make it difficult to estimate the size of females at the onset of MSM based on the relative growth of this dimension (posterior portion of the abdomen). Bahamonde and López (1961) and other authors have estimated this dimension based on the size of the smallest ovigerous female present in the population.

More recently, Viau et al. (2006), Trevisan and Santos (2012), Copatti et al. (2016a), Adam et al. (2018), and Marçal et al. (2018) were successful in estimating the size of females at the onset of MSM using the relative growth of the width of the second abdominal somite as measured dorsally. In total, 16 species/populations have had the size at onset of MSM estimated, based on the smallest ovigerous female (seven species), the relative growth of the second abdominal segment (six species), or gonad inspection (three species).

Aeglid males and females of the same species attain MSM at different CLs, with males reaching this condition at a longer CL than females (Figure 3.9). Furthermore, the values of the onset of MSM show a direct relationship with the maximum CL reached by the particular species/population, both in males and in females (Figure 3.9). The equations for this relation are $y = 0.7871x - 6.8443$ ($r^2 = 0.8493$) for males and $y = 0.6249x - 1.813$ ($r^2 = 0.8177$) for females, where $y = $ CL at onset of MSM and $x = $ maximum CL of the species.

This relationship was also recorded by Waiho et al. (2017) for 133 populations of 55 species of brachyuran crabs. Although the maximum CL of the species can undergo intraspecific variations according to the oscillation of abiotic variables affecting their populations (Krebs 2014), the relationship between size at the onset

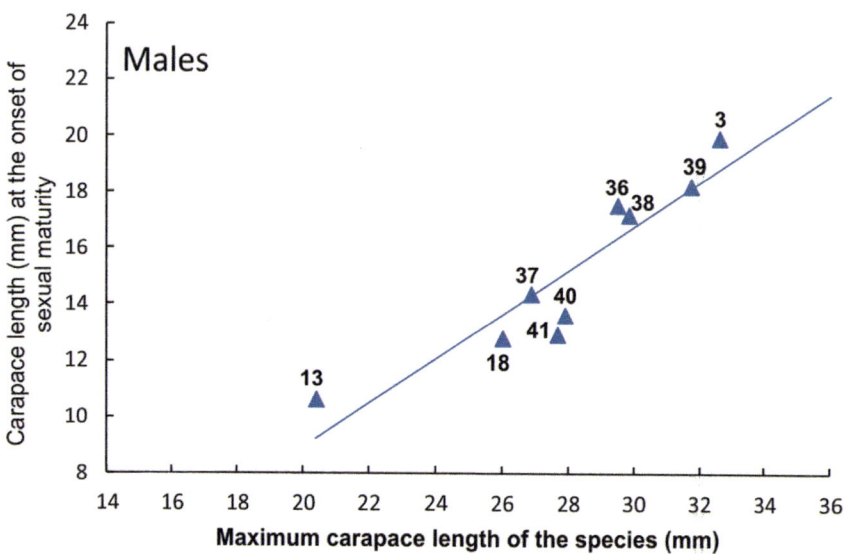

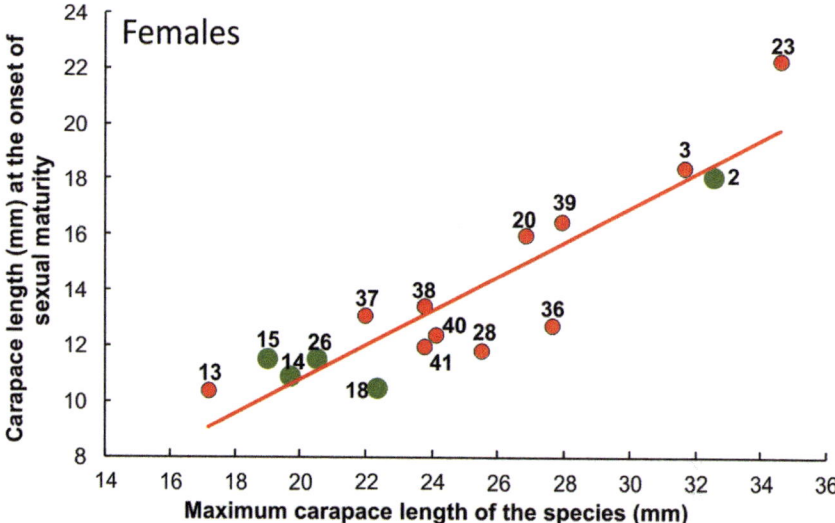

Figure 3.9 Relationship between the carapace length at the onset of sexual maturity and the maximum carapace length of the species for males and females. Red circles: estimate from relative growth of abdomen or gonad inspection, green circles: smallest ovigerous female in the population. References: [2–3]see caption of Figure 3.1, [13–23]see caption of Figure 3.4, [26 and 28]see caption of Figure 3.5, [36]Swiech-Ayoub and Masunari (2001b), [37]Bueno and Shimizu (2009), [38]Colpo et al. (2005), [39]Oliveira and Santos (2011), [40]Trevisan and Santos (2012), [41]Takano et al. (2016).

of MSM and the maximum CL reached by the population seems to be an intrinsic factor of the population and species.

ACKNOWLEDGMENTS

Thanks are due to Dr. Donato Seiji Abe from the Associação Instituto Internacional de Ecologia e Gerenciamento Ambiental, São Carlos, SP, for providing software for the calculation of the dissolved oxygen concentration in the water from the saturation percentage data, and to Dr. Roberto Munehisa Shimizu from São Paulo University, São Paulo, SP, for critical reading. American Journal Experts revised the English of an early version of the manuscript.

REFERENCES

Adam, C. L., M. Z. Marochi, and S. Masunari. 2018. Ontogenetic shape changes and sexual dimorphism in *Aegla marginata* Bond-Buckup and Buckup, 1994. *Anais da Academia Brasileira de Ciências* 90(2):1521–32.

Aldstadt III, J. H., H. A. Bootsma, and J. L. Ammerman. 2010. Chemical properties of water. In *River Ecosystem Ecology*, ed. G. E. Likens, pp. 8–15. Cambridge: Academic Press.

Ayres-Peres, L., P. B. Araújo, and S. Santos. 2011b. Description of the agonistic behavior of *Aegla longirostri* (Decapoda: Aeglidae). *Journal of Crustacean Biology* 31:379–88.

Ayres-Peres, L., C. Coutinho, J. S. Baumart, A. S. Gonçalves, P. B. Araújo, and S. Santos. 2011a. Radio-telemetry techniques in the study of displacement of freshwater anomurans. *Nauplius* 19:41–54.

Bahamonde, N., and M.T. López. 1961. Estudios biológicos en la población de *Aegla laevis laevis* (Latreille) de el Monte (Crustacea, Decapoda, Anomura). *Investigaciones Zoológicas Chilenas* 7:19–58.

Barr, T. C. 1967. Observations on the ecology of caves. *American Naturalist* 101:475–92.

Bond-Buckup, G. 2003. Família Aeglidae. In *Manual de identificação dos Crustacea Decapoda de água doce do Brasil*, ed. G. A. S. Melo, pp. 21–116. São Paulo: Loyola.

Bond-Buckup, G., and L. Buckup. 1994. A família Aeglidae (Crustacea, Decapoda, Anomura). *Arquivos de Zoologia* 32:159–347.

Boos, H., Jr., and S. L. Althoff. 2002. Biologia reprodutiva de *Macrobrachium potiuna* (Müller, 1880) (Crustacea, Decapoda, Palaemonidae) no Parque Natural Municipal São Francisco de Assis, Blumenau, SC. *Revista Estudos de Biologia – Champagnat* 24(48):45–50.

Boos, H., Jr., D. Silva-Castiglioni, K. Schacht, L. Buckup, and G. Bond-Buckup. 2006. Crescimento de *Aegla jarai* Bond-Buckup and Buckup (Crustacea, Anomura, Aeglidae). *Revista Brasileira de Zoologia* 23:490–96.

Bueno, A. A. P., and G. Bond-Buckup. 2000. Dinâmica populacional de *Aegla platensis* Schmitt (Crustacea, Decapoda, Aeglidae). *Revista Brasileira de Zoologia* 17:43–49.

Bueno, A. A. P., G. Bond-Buckup, and L. Buckup. 2000. Crescimento de *Aegla platensis* Schmitt em ambiente natural (Crustacea, Decapoda, Aeglidae). *Revista Brasileira de Zoologia* 17:51–60.

Bueno, S. L. S., S. Santos, S. S. Rocha, K. M. Gomes, E. C. Mossolin, and F. L. Mantelatto. 2016b. Avaliação dos Eglídeos (Decapoda: Aeglidae). In *Livro vermelho dos crustáceos do Brasil: Avaliação 2010–2014*, org. M. A. A. Pinheiro and H. Boos, pp. 35–63. Porto Alegre: Sociedade Brasileira de Carcinologia. Available at: https://www.researchgate. net/publication/309864568_Avaliacao_dos_Eglideos_Decapoda_Aeglidae

Bueno, S. L. S., and R. M. Shimizu. 2008. Reproductive biology and functional maturity in females of *Aegla franca* (Decapoda: Anomura: Aeglidae). *Journal of Crustacean Biology* 28:652–62.

Bueno, S. L. S., and R. M. Shimizu. 2009. Allometric growth, sexual maturity, and adult male chelae dimorphism in *Aegla franca* (Decapoda: Anomura: Aeglidae). *Journal of Crustacean Biology* 29:317–28.

Bueno, S. L. S., R. M. Shimizu and J. C. B. Moraes. 2016a. A remarkable anomuran: the taxon *Aegla* Leach, 1820. Taxonomic remarks, distribution, biology, diversity and conservation. In *A Global Overview of the Conservation of Freshwater Decapod Crustaceans*, eds. T. Kawai and N. Cumberlidge, pp. 23–64. London: Springer.

Bueno, S. L. S., R. M. Shimizu, and S. S. Rocha. 2007. Estimating the population size of *Aegla franca* (Decapoda: Anomura: Aeglidae) by mark-recapture technique from an isolated section of Barro Preto stream, county of Claraval, state of Minas Gerais, southeastern Brazil. *Journal of Crustacean Biology* 27:553–59.

Bueno, S. L. S., B. F. Takano, F. P. A. Cohen et al. 2014. Fluctuations in the population size of the highly endemic *Aegla perobae* (Decapoda: Anomura: Aeglidae) caused by a disturbance event. *Journal of Crustacean Biology* 34:165–73.

Chiquetto-Machado, P. I., L. C. M. Vieira, R. M. Shimizu, and S. L. S. Bueno. 2016. Life cycle of the freshwater anomuran *Aegla schmitti* Hobbs, 1978 (Decapoda: Anomura: Aeglidae) from southeastern Brazil. *Journal of Crustacean Biology* 36(1):39–45.

Climate-model by climate-data.org and Location data by openstreetmap.org. 2018. Available at: https://pt.climate-data.org/america-do-sul/brasil-114 (accessed January to April, 2018).

Cohen, F. P. A., B. F. Takano, R. M. Shimizu, and S. L. S. Bueno. 2011. Life cycle and population structure of *Aegla paulensis* (Decapoda: Anomura: Aeglidae). *Journal of Crustacean Biology* 31:389–95.

Cohen, F. P. A., B. F. Takano, R. M. Shimizu, and S. L. S. Bueno. 2013. Population size of *Aegla paulensis* (Decapoda: Anomura: Aeglidae). *Latin American Journal of Aquatic Research* 41:746–52.

Cole, G.A. 1983. *Textbook of limnology*. Saint Louis, MO: Mosby.

Colpo, K. D., L. O. Ribeiro, and S. Santos. 2005. Population biology of the freshwater anomuran *Aegla longirostri* (Aeglidae) from south Brazilian streams. *Journal of Crustacean Biology* 25:495–99.

Copatti, C. E., R. P. Legramanti, A. Trevisan, and S. Santos. 2016a. Growth, sexual maturity and sexual dimorphism of *Aegla georginae* Santos and Jara, 2013 (Decapoda: Anomura: Aeglidae) in a tributary of the Ibicuí River in southern Brazil. *Zoologia* 33(3): e20160010.

Copatti, C. E., R. P. Legramanti, A. Trevisan, and S. Santos. 2016b. Method of capture and population structure of *Aegla georginae* Santos and Jara, 2013 (Decapoda: Anomura: Aeglidae) in a tributary of the Ibicuí River in southern Brazil. *Brazilian Journal of Biology* 7(4):1035–42.

Dalosto, M. M., A. V. Palaoro, D. Oliveira, E. Samuelsson, and S. Santos. 2014. Population biology of *Aegla platensis* (Decapoda: Anomura: Aeglidae) in a tributary of the Uruguay River, state of Rio Grande do Sul, Brazil. *Zoologia* 31(3):215–22.

Dalosto, M. M., and S. Santos. 2011. Differences in oxygen consumption and diel activity as adaptations related to microhabitat in Neotropical freshwater decapods (Crustacea). *Comparative Biochemistry and Physiology, Part A* 160:461–66.

Diawol, V. P., F. Giri, and P. A. Collins. 2015. Shape and size variations of *Aegla uruguayana* (Anomura, Aeglidae) under laboratory conditions: a geometric morphometric approach to the growth. *Iheringia, série Zoologia* 105(1):76–83.

Downes, B. J., P. S. Lake, E. S. G. Schreiber, and A. Glaister. 1998. Habitat structure and regulation of local species diversity in a stony, upland stream. *Ecological Monographs* 68(2):237–57.

Downes, B. J., P. S. Lake, E. S. G. Schreiber, and A. Glaister. 2000. Habitat structure, resources and effects of surface roughness and macroalgae on stream invertebrates. *Oecologia* 123:569–81.

Fisher, R. A. 1930. *The Genetical Theory of Natural Selection.* Oxford: Oxford University Press.

Francisco, D. A., S. L. S. Bueno, and T. C. Kitahara. 2007. Description of the first juvenile of *Aegla franca* Schmitt, 1942 (Crustacea, Decapoda, Aeglidae). *Zootaxa* 1509:17–30.

Fransozo, A., R. C. Costa, A. L. D. Reigada, and J. M. Nakagaki. 2003. Population structure of *Aegla castro* Schmitt, 1942 (Crustacea: Anomura: Aeglidae) from Itatinga (SP), Brazil. *Acta Limnologica Brasiliensia* 15:13–20.

Giri, F., and P. Collins. 2014. Clinal variation in carapace shape in the South American freshwater crab, *Aegla uruguayana* (Anomura: Aeglidae). *Biological Journal of the Linnean Society* 113:914–30.

Gonçalves, R. S., D. S. Castiglioni, and G. Bond-Buckup. 2006. Ecologia populacional de *Aegla franciscana* (Crustacea, Decapoda, Anomura) em São Francisco de Paula, RS, Brasil. *Iheringia, série Zoologia* 96:109–14.

Google Earth Pro. V7.3.2.5495. (October 31, 2018). Various locations and various coordinates. Google LLC 2018. Open GL. Microsoft Windows. NVIDIA Corporation. Available at: https://www.google.com.br/intl/pt-BR/earth/(accessed January to April, 2018).

Grabowski, R. C., S. Santos, and A. L. Castilho. 2013. Reproductive ecology and size of sexual maturity in the anomuran crab *Aegla parana* (Decapoda: Aeglidae). *Journal of Crustacean Biology* 33:1–7.

Hartnoll, R. G. 1974. Variation in growth pattern between some secondary sexual characters in crabs (Decapoda Brachyura). *Crustaceana* 27(2):131–36.

Hartnoll, R. G. 1978. The determination of relative growth in Crustacea. *Crustaceana* 34(3):281–93.

Hartnoll, R. G. 1982. Growth. In *The Biology of Crustacea: Embryology, Morphology and Genetics*, ed. L. G. Abele, vol. 2, pp. 111–96. New York: Academic Press.

Hartnoll, R. G. 1985. Growth, sexual maturity and reproductive output. In *Factors in Adult Growth*, ed. A.A. Wenner, pp. 101–28. Boston: A.A. Balkema.

Hartnoll R. G. 2001. Growth in Crustacea: twenty years on. *Hydrobiologia* 449(1):111–22.

Harvey, R., L. Lye, A. Khan, and R. Paterson. 2011. The Influence of air temperature on water temperature and the concentration of dissolved oxygen in Newfoundland rivers. *Canadian Water Resources Journal* 36 (2):171–92.

Kozlowski, J. 1992. Optimal allocation of resources to growth and reproduction: implications for age and size at maturity. *Trends on Ecology and Evolution* 7(1):15–19.

Kozlowski, J., and A.T. Teriokin. 1999. Allocation of energy between growth and reproduction: the Pontryagin Maximum Principle solution for the case of age- and season-dependent mortality. *Evolutionary Ecology Research* 1:423–41.

Krebs, C. J. 2014. *Ecology: The Experimental Analysis of Distribution and Abundance.* Harlow: Pearson.

Liao, W. B., Y. Zeng, and J. D. Yang. 2013. Sexual size dimorphism in anurans: roles of mating system and habitat types. *Frontiers in Zoology* 10:65.

López, M. T. 1965. Estudios biológicos en *Aegla odebrechtti paulensis*, Schmitt. *Boletim da Faculdade de Filosofia, Ciências e Letras da Universidade de São Paulo, Série Zoologia* 25:301–14.

Maia, K. P., S. L. S. Bueno, and E. Trajano. 2013. Ecologia populacional e conservação de eglídeos (Crustacea: Decapoda: Aeglidae) em cavernas da área cárstica do Alto Ribeira, em São Paulo. *Revista da Biologia* 10:40–45.

Marçal, I. C., L. M. Ioshimura, J. J. S. Rosa, and G. M. Teixeira. 2018. Population structure and sexual maturity of *Aegla castro* (Decapoda, Anomura), an endemic freshwater crab from Brazil. *Invertebrate Reproduction and Development* 62(1):35–42.

Martin, J. W., and L. G. Abele. 1988. External morphology of the genus *Aegla* (Crustacea: Anomura: Aeglidae). *Smithsonian Contribution to Zoology* 453:1–46.

Moraes, J. C. B., and S. L. S. Bueno. 2013. Description of the newly-hatched juvenile of *Aegla paulensis* (Decapoda, Anomura, Aeglidae). *Zootaxa* 3635(5):501–19.

Moraes, J. C. B., and S. L. S. Bueno. 2015. Description of the newly-hatched juvenile of *Aegla perobae* (Decapoda, Anomura, Aeglidae). *Zootaxa* 3973(3):491–510.

Noro, C. K., and L. Buckup. 2003. O crescimento de *Aegla leptodactyla* Buckup & Rossi (Crustacea, Anomura, Aeglidae). *Revista Brasileira de Zoologia* 20:191–98.

Oliveira, D., and S. Santos. 2011. Maturidade sexual morfológica de *Aegla platensis* (Crustacea, Decapoda, Anomura) no Lajeado Bonito, norte do estado do Rio Grande do Sul, Brasil. *Iheringia, série Zoologia* 101(1–2):127–30.

Rocha, S. S. 2007. Biologia reprodutiva, estrutura e dinâmica populacional e avaliação do grau de risco de extinção de *Aegla strinatii* Türkay, 1972 (Crustacea, Decapoda, Aeglidae). Ph.D. diss., São Paulo University.

Rodrigues, W., and N. J. Hebling. 1978. Estudos biológicos em *Aegla perobae* Hebling & Rodrigues, 1877 (Decapoda, Anomura). *Revista Brasileira de Biologia* 38:383–90.

Santos, S., G. Bond-Buckup, L. Buckup et al. 2015. Three new species of Aeglidae (Aegla Leach, 1820) from Paraná State, Brazil. *Journal of Crustacean Biology* 35:839–49.

Santos, S., G. Bond-Buckup, L. Buckup, M. Pérez-Losada, M. Finley, and K.A. Crandall. 2012. Three new species of *Aegla* (Anomura) freshwater crabs from Upper Uruguay River hydrographic basin in Brazil. *Journal of Crustacean Biology* 32:529–40.

Santos, S., G. Bond-Buckup, A. S. Gonçalves, M. L. Bartholomei-Santos, L. Buckup, and C. G. Jara. 2017. Diversity and conservation status of *Aegla* spp. (Anomura, Aeglidae): an update. *Nauplius* 25:e2017011.

Santos, S., C. G. Jara, M. L. Bartholomei-Santos, M. Pérez-Losada, and K. A. Crandall. 2013. New species and records of the genus *Aegla* Leach, 1820 (Crustacea, Anomura, Aeglidae) from the West-Central region of Rio Grande do Sul, Brazil. *Nauplius* 21(2):211–23.

Schmitt, W. L. 1942. The species of *Aegla*, endemic South American fresh-water crustaceans. *Proceedings of the United States National Museum* 91:431–520.

Silva, A. R., M. R. Wolf, and A. L. Castilho. 2016. Reproduction, growth and longevity of the endemic South American crab *Aegla marginata* (Decapoda: Anomura: Aeglidae). *Invertebrate Reproduction & Development* 60(1):59–72.

Silva, L. S. A., C. M. Guerrero-Ocampo, M. L. Negreiros-Fransozo, and G. M. Teixeira. 2017. Description of the newly-hatched juvenile of *Aegla castro* Schmitt, 1942 (Crustacea, Anomura, Aeglidae). *Zootaxa* 4237(1):167–80.

Silva, M., J. H. Brown, and J. A. Downing. 1997. Differences in population density and energy use between birds and mammals: a macroecological perspective. *Journal of Animal Ecology* 66:327–40.

Silva-Castiglioni, D., D. F. Barcelos, and S. Santos. 2006. Crescimento de *Aegla longirostri* Bond-Buckup & Buckup (Crustacea, Anomura, Aeglidae). *Revista Brasileira de Zoologia* 23:408–13.

Silva-Gonçalves, R., G. Bond-Buckup, and L. Buckup. 2009. Crescimento de *Aegla itacolomiensis* (Crustacea, Decapoda) em um arroio da Mata Atlântica no sul do Brasil. *Iheringia, Série Zoologia* 99:397–402.

Swiech-Ayoub, B. P., and S. Masunari. 2001a. Flutuações temporal e espacial de abundância e composição de tamanho de *Aegla castro* Schmitt (Crustacea, Anomura, Aeglidae) no Buraco do Padre, Ponta Grossa, Paraná, Brasil. *Revista Brasileira de Zoologia* 18:1003–17.

Swiech-Ayoub, B. P., and S. Masunari. 2001b. Biologia reprodutiva de *Aegla castro* Schmitt (Crustacea, Anomura, Aeglidae) no Buraco do Padre, Ponta Grossa, Paraná, Brasil. *Revista Brasileira de Zoologia* 18:1019–30.

Takano, B. F., F. P. A. Cohen, A. Fransozo, R. M. Shimizu, and S. L. S. Bueno. 2016. Allometric growth, sexual maturity and reproductive cycle of *Aegla castro* (Decapoda: Anomura: Aeglidae) from Itatinga, state of São Paulo, southeastern Brazil. *Nauplius* 24 e2016010.

Teodósio, E. A. F. M. O., and S. Masunari. 2007. Description of first two juvenile stages of *Aegla schmitti* Hobbs III, 1979 (Anomura: Aeglidae). *Nauplius* 15(2):73–80.

Teodósio, E. A. F. M. O., and S. Masunari. 2009. Estrutura populacional de *Aegla schmitti* (Crustacea: Anomura: Aeglidae) nos reservatórios dos Mananciais da Serra, Piraquara, Paraná, Brasil. *Revista Brasileira de Zoologia* 26:19–24.

Tokeshi, M., and S. Arakaki. 2012. Habitat complexity in aquatic systems: fractals and beyond. *Hydrobiologia* 685:27–47.

Trevisan, A. 2008. Biologia populacional de *Aegla* sp. n. (Crustacea, Decapoda, Aeglidae) no Arroio Passo Taquara, São Pedro do Sul, RS. M.Sc. diss., Federal University of Santa Maria.

Trevisan, A., and S. Santos. 2011. Crescimento de *Aegla manuinflata* (Decapoda, Anomura, Aeglidae) em ambiente natural. *Iheringia, Série Zoologia* 101:336–42.

Trevisan, A., and S. Santos. 2012. Morphological sexual maturity, sexual dimorphism and heterochely in *Aegla manuinflata* (Anomura). *Journal of Crustacean Biology* 32 (4):519–27.

Trevisan, A., and S. Santos. 2014. Population dynamics of *Aegla manuinflata* Bond-Buckup and Santos 2009 (Decapoda: Aeglidae), a threatened species. *Acta Limnologica Brasiliensia* 26(2):154–62.

Viau, V. E., L. S. L. Greco, G. Bond-Buckup, and E. M. Rodríguez. 2006. Size at the onset of sexual maturity in the anomuran crab, *Aegla uruguayana* (Aeglidae). *Acta Zoologica* (Stockholm) 87:253–64.

Vogt, G. 2012. Ageing and longevity in the Decapoda (Crustacea): a review. *Zoologischer Anzeiger* 251(1):1–25.

Vogt, G. 2014. Life span, early life stage protection, mortality, and senescence in freshwater Decapoda. In *Advances in Freshwater Decapod Systematics and Biology*, eds. D.C.J. Yeo, S. Klaus and N. Cumberlidge, pp. 17–52. Leiden: Koninklijke Brill NV, Crustaceana Monographs 19.

Von Bertalanffy, L. 1938. A quantitative theory of organic growth. *Human Biology* 10:181–213.

Waiho, K., H. Fazhan, J. C. Baylon et al. 2017. On types of sexual maturity in brachyurans, with special reference to size at the onset of sex maturity. *Journal of Shellfish Research* 36(3):807–39.

Webb, B. W., P. D. Clack, and D. E. Walling. 2003. Water-air temperature relationships in a Devon river system and the role of flow. *Hydrological Processes* 17:3069–84.

Wenner, A. M. 1972. Sex ratio as a function of size in marine Crustacea. *The American Naturalist* 106(949):321–50.

Werner, J., N. Sfakianakis, A. D. Rendall, and E. M. Griebeler. 2018. Energy intake functions and energy budgets of ectotherms and endotherms derived from their ontogenetic growth in body mass and timing of sexual maturation. *Journal of Theoretical Biology* 444:83–92.

Williamson, M. H., and J. H. Lawton. 1991. Fractal geometry of ecological habitats. In *Habitat Structure: The Physical Arrangement of Objects in Space*, eds. S. S. Bell, E. D. McCoy and H. R. Mushinsky, pp. 69–86. London: Chapman and Hall.

Wohl, E. 2010. Hydrology: streams. In *River Ecosystem Ecology*, ed. G.E. Likens, pp. 23–31. Cambridge: Academic Press.

CHAPTER 4

Trophic Ecology

Pablo Collins

CONTENTS

4.1 HISTORICAL RELEVANCE OF TROPHIC RELATIONSHIPS

The trophic ecology of species is a dynamic process that needs to be observed from the species history and its joint evolution to shifts in the environments and its organisms through mutual interactions. From their ancestors, all species bring a series of tools that serve to capture and process food (e.g., appendages and specialized organs, behaviors, and physiological and biochemical aspects) (Castro et al. 2015), allowing organisms to develop in their occupied habitat. Over time, species have adjusted these tools to the environmental conditions or have modified them accordingly (Greenaway 2003). This adaptation is observed in relation to nutritional requirements, cycles (e.g., molting, reproduction), and the evolution of the resources in the environment together with abiotic and biotic factor cycles (Figure 4.1).

The extant Aeglidae is the unique family of Anomura that lives in hypoosmotic environments of continental aquatic systems, while the remaining anomuran

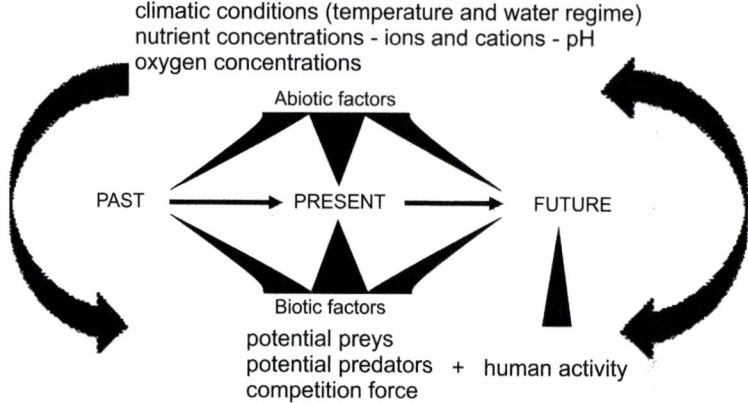

Figure 4.1   Factors that affect the trophic ecology of the family Aeglidae in the evolutionary context. The human activity is a new factor that affects the nutrient concentrations, pH, ions-cations, and more globally, the climatic conditions together to prey available and the predator risks.

families reside in marine systems (see Chapter 1). Fossil species of aeglids in Mexico (*Protaegla minuscula*, Feldmann et al. 1998, approximately 110 million years ago—Mya) and New Zealand [*Haumuriaegla glaessneri* Feldman 1984, approximately 74 Mya (Maastrichtian age)] from marine sediments suggest a marine origin for the group (Feldmann 1984; Feldmann et al. 1998). A colonization history was proposed for the ancestor of the extant aeglids, reaching South America from the Pacific Ocean via Chilean shores, approximately 75 Mya (Pérez-Losada et al. 2004). Since then, the trophic relationships of aeglids have changed towards the current conditions, according to the characteristics of each biogeographic region where the species live today.

In that time, when the ancestors of aeglids might have come from the Australasia region (Feldmann 1984), important tectonic events occurred in the Southern Hemisphere. The Antarctic and Australian continents moved towards the east, and oceanic currents had a new arrangement, a drift change (Kennet 1977; Dalziel 1991). The temperature was more elevated, and the sea level increased until terrestrial South American was converted into a plain inundated by the sea. Furthermore, the marine water was diluted by rain, provoking modifications in the salinity (Poulsen et al. 2003). During some periods, only approximately 20% of the land was above the sea level (Figure 4.2). In these seas, the oceanic plankton and benthos evolved until the manifestation of the current characteristics, in which diatoms, dinoflagellates, foraminifera, calcareous nannoplankton, and ammonites represented the base for the rest of the food web (Kelly 2002; Gibbs et al. 2006). The brachiopods diminished in importance, and the Asteroidea and Hexacorallia increased, while the Bryozoa experienced a considerable expansion together with the heterodont bivalve molluscs (rudists) (Lehmann and Hillner 1983). All the groups developed complex adaptations to avoid becoming prey (e.g., burial behavior, greater shell thickness, and forms of

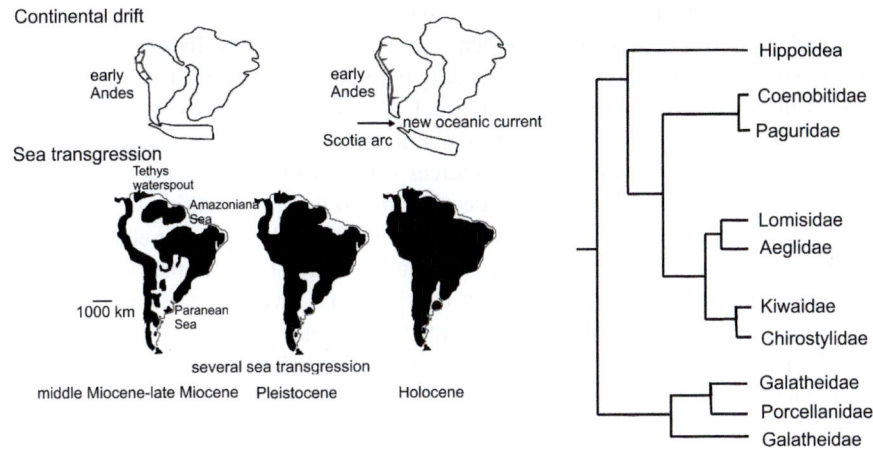

Figure 4.2 Some main factors that influenced the current distribution of the family Aeglidae in the South American continent, and a possible tree of relationships among the families of Anomura groups. [Modified from Schnabel et al. (2011).]

protection) (Lehmann and Hillner 1983). Vegetation with flowers (e.g., angiosperms) had an explosive diversification, signifying the emergence of a new resource in the trophic web (Francis et al. 2006).

In this context, it is necessary to note the occurrence of events that provoked the massive species extinctions during the Cretaceous and Paleogene (T/K extinctions, 65 Mya) (Schweitzer and Feldmann 2005). In this turbulent setting, aeglids were adjusting their trophic expression towards current interactions with the new environment and resources (see Chapter 1).

One way to understand the current trophic ecology of a species or a group of related species is to recognize how sister groups or species from close lineages that had a similar origin were derived and to recognize how the force of the environment molded the new interactions.

The phylogenetic relationships of the Aegloidea with other anomuran taxa is still an ongoing discussion (see Chapter 1), depending on the type of data considered (i.e., traditional morphology, sperm morphology, and molecular data). The interpretation of these data have indicated close phylogenetic affinities with Paguroidea Latreille 1802, Hippoidea Latreille 1825, and Lomisoidea Bouvier 1895 (Martin and Abele 1988; Pérez-Losada et al. 2002; Tudge and Scheltinga 2002; McLaughlin et al. 2007; Schnabel and Ahyong 2010; Tsang et al. 2011).

Closely related groups of the aeglids (e.g., Galatheoidea) can feed from resources in suspension and elements in deposits, are algal grazers, scavengers, predators, and occasional cannibals (Lovrich and Thiel 2011). Within the food spectrum, some species consume macroalgae or phytodetritus, including epibiotic bacteria as an important resource (Lovrich and Thiel 2011; Zwirglmaier et al. 2015). In the stomachs of multiple species from Galatheoidea (e.g., *Cervimunida* spp., *Galathea* spp., *Munida* spp., *Munidopsis* spp., *Porcellana* spp.) and Kiwaoidea (e.g., *Kiwa* spp.), diatoms, small macroalgae, unicellular metazoans, bacteria, sediment, and remains of organic particulate were encountered (Lovrich and Thiel 2011; Zwirglmaier et al. 2015). These species display an omnivore feeding type with strong tendencies toward herbivory in larval, juvenile, and adult stages. Furthermore, the closely related families of Galatheoidea and Kiwaoidea that inhabit the ocean can be considered filter-feeders, scrapers, and collectors, using resources with low mobility such as polychaetes, molluscs, and epibionts, which provide protein and lipids in addition to micronutrients required for development (Lovrich and Thiel 2011; Zwirglmaier et al. 2015).

Current conditions can be assumed to be similar to those in the past era when Aeglidae entered the South American continent, i.e., when the environment was stable and in the absence of important climatic or geological events that could modify the landscape and macro factors (Potter 1997). Under these conditions, and as a characteristic that forces a feeding type, the new environment harbored a high diversity of species with slow movements and that live on the bottom. Thus, aeglids had adjusted and adapted to the continental environment, according to the unstable characteristics of the new habit (i.e., freshwater environments) and that differed from their original environment in the basic conditions (e.g., water presence, oxygen, and ion and cation concentrations), using new food resources to obtain energy and essential elements to live.

The environments occupied by aeglid species differed from those where their recent marine ancestors lived, mainly in relation to the stability indicated by the water presence that defined continental aquatic cycles [e.g., temperature, hydrosedimentology, heliophany (the duration in hours of sunlight over a given place on the earth's surface, between dawn and dusk)], oxygen variations, and ion as well as cation concentration differences. All these fluctuations should be sustained energetically by each animal, with a nutrient matrix capable of developing and permitting the presence of crustacean populations (Collins et al. 2007). In contrast, from the initial colonization of the continental habitats, these environments experienced various changes, including the rise of the Andes together with the directional change of the existing basins caused by the movement of tectonic plates (Ribeiro 2006). The separation of Africa from the Americas and, subsequently, the Antarctic continent from Australasia, provoked variations of the direction of marine currents, diverse marine ingressions, periods of extreme cold, and ice ages (del Río and Martínez Chiappara 1998), resulting in habitats that were not occupied by the species or that presented harsh environmental conditions, which were difficult to survive (Figure 4.2). The presence of dry and humid periods, in addition to the moments of intense cold mainly affecting the aquatic environments in southern South America, should have formed relictual sites, maintained at higher temperatures caused by the actions of active volcanoes or because they were ice-free (Oyanedel Pérez 2015) where the trophic web was already secured.

All the aforementioned history of the marine ancestors of the aeglids and, more recently, of current populations in continental aquatic systems provoked changes in the type of environment with different trophic frames of its specific components and in energy needs that served the individual and population requirements. A combination of the modifications in the basins and the environmental macro factors (e.g., temperature, salinity, pH, and ion concentrations) as well as changes in the trophic web contributed to adaptations that allowed the permanence and present species diversity of Aeglidae.

The food items consumed must provide the necessary nutrients for the survival and development of the species and populations. Additionally, their amino acids, fatty acids, and carbohydrates must assure the incorporation of the elements needed to live and occupy a specific habitat (Oliveira et al. 2003, 2011). Moreover, the species must have the availability of enzymes and substrates for the biochemical reactions, as well as the absorption and digestion capacity for the reserve mobilization as developed in other groups of decapods (Rosas et al. 2006; Oliveira et al. 2007; Buckup et al. 2008; Dutra et al. 2008; Musin et al. 2017, 2018; Sacristán et al. 2017).

4.2 TROPHIC SPECTRA AND RELATED VARIATIONS

4.2.1 Stomach Structure

The stomach of aeglids was described in *A. camargoi* (see Castro 2003), *A. leptodactyla* (see Castro 2003), *A. platensis* (see Castro and Bond-Buckup 2003), and *A. uruguayana* (see Williner 2010) indicating two parts, the cardiac and pyloric chambers, with functions of trituration and filtration (Figure 4.3). Each part has specialized structures that permit the efficient processing of food and use a great diversity of trophic items (Dall and Mortiary 1983; Felgenhauer and Abele 1989). The gastric mill is the most conspicuous structure in the stomach with a series of ossicles, silk, teeth, and thorns that permit food movement (Icely and Nott 1992). The hardness and prominence of the teeth and thorns of the zygocardiac and urocardiac ossicles provide evidence of the intensity of the stomach's mechanical action. Even more, the macrophagy is indicated by the sharpness of the teeth, thorns of the ossicles, and by the number of crests that present the cardiopyloric valve. However, the aeglids stomach has silks along the ridges of the cardiopyloric valve that indicate

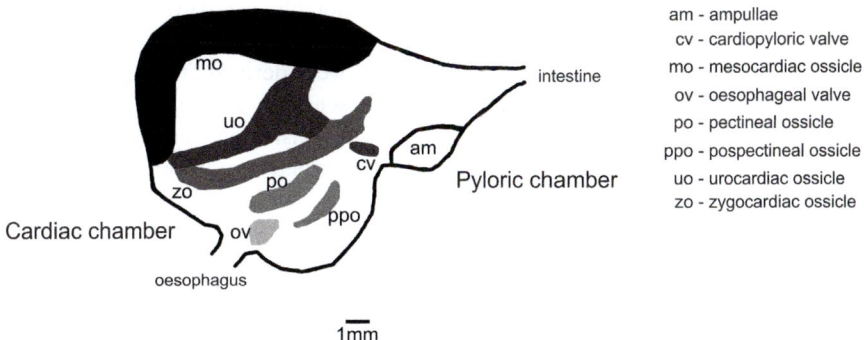

am - ampullae
cv - cardiopyloric valve
mo - mesocardiac ossicle
ov - oesophageal valve
po - pectineal ossicle
ppo - pospectineal ossicle
uo - urocardiac ossicle
zo - zygocardiac ossicle

1mm

Figure 4.3 Stomach scheme of aeglids, indicating the main structures of the gastric mill and others of cardiac and pyloric chambers. [Modified from Williner (2010).]

the ability of filtration and processing of micro-organisms (unicellular algae and other planktonic animals) (Williner 2010). The presence of silk and teeth in the cardiopyloric valve suggests multiple functions for retention and movement of particles.

In the cardiac chamber, the ingested particles are stored until they pass through the gastric mill, where they are crushed and triturated (Castro and Bond-Buckup 2003; Williner 2010). When the size of the particles decreases, they are passed to the pyloric chamber. During this process, aeglids begin an extracellular digestion performed by the exocytosed enzymes from the hepatopancreas cells to the lumen therein and from there to the pyloric chamber of the stomach (Al-Mohanna and Nott 1987; Vogt 1994). This process takes approximately 5 hours in aeglids (e.g., *A. longirostri*) but depends on the food type in relation to the amount of soft and hard parts (Sokolowicz et al. 2007).

The size of the stomach combined with the presence of internal structures that imprint a certain pace on the feeding process, in addition to other factors, is one of the elements which may act as structuring causes. In turn, the cycle presented by F, R, and B cells in the hepatopancreas completes the complex interaction of elements (Sacristán et al. 2013).

4.2.2 Feeding Ecology Introduction

Once aeglids started the colonization of the freshwater environment, adaptive radiation could have been facilitated by the presence of a considerable number and wide range of ecological niches (Simpson 1953; Jackson et al. 2016). The groups that successfully invaded the limnic habitat may have been pre-adapted morphologically and physiologically, as was observed and discussed for freshwater crabs of the families Trichodactylidae and Pseudothelphusidae (Bott 1969). The diversification of aeglids was assisted by the heterogeneity of the habitat and numerous ecological niches and inhabited microhabitats, resulting from frequently changing topographical and hydrological characteristics of their environments (i.e., lakes, streams, swamps, and caves; or aquatic environments in tropical, temperate and cold regions; or aquatic environments with permanent or temporary water presence; or aquatic environments in mountains, plains, and caves). Notably, sympatric speciation scenarios cannot be disregarded throughout their freshwater invasion (Yeo et al. 2008).

The initial records on aeglid feeding habits were derived from general observations of their biology, assuming that the presence of chelae might indicate a carnivorous habit. Rodriguez and Hebling (1978) considered *Aegla perobae* to be carnivorous; this preference was subsequently confirmed in other species by observing aeglids preying on insect larvae, pupae (e.g., Simuliidae, Diptera, and others) and molluscs (e.g., *Diplodon chilensis*, Bivalvia: Hyriidae) (Magni and Py-Daniel 1989; Lara and Moreno 1995). More recent information regarding the natural diet of aeglids was obtained through stomach content analyses. Unlike what was previously indicated, plant debris and algae were the most abundant items collected in nature in the stomachs of *A. abtao* (see Lara and Moreno 1995), *A. camargoi* (see Castro and Bond-Buckup 2004), *A. cavernicola* (see Trajano 1987), *A. jujuyana* (see Williner 2007), *A. laevis* (see Bahamonde and López 1961), *A. leptodactyla* (see Castro and

Bond-Buckup 2004), *A. ligulata* (see Bueno and Bond-Buckup 2004), *A. longi-rostri* (see Santos et al. 2008), *A. neuquensis* (see Williner 2007), *A. perobae* (see Rodrigues and Hebling 1978), *A. platensis* (Bueno and Bond-Buckup 2004; Castro and Bond-Buckup 2004), *A. ringueleti* (see Manattini et al. 2017), *A. singularis* (see Williner 2007), and *A. uruguayana* (Williner 2007; Williner 2010) (Table 4.1). Insect and other animal fragments, sand grains, and unidentified materials, however, were also present in their stomachs. Moreover, the opportunistic predation over fish was observed as a natural occurrence in the feeding of *A. grisella* (see Savaris et al. 2012), *A. ligulata* (see Bueno and Bond-Buckup 2004), *A. platensis* (see Bueno 2003), *A. singularis* (see Savaris et al. 2012), and other species (Burns 1972; Bueno 2003; Savaris et al. 2012) (Table 4.1). All these studies concluded that aeglids were generalist and opportunist omnivores.

Studies on the natural diet indicated differences in the trophic web participation or use of some prey types, according to the locations in the river. In headwater streams, aeglids can act as both shredders and predators (Cogo and Santos 2013; Cogo et al. 2014). The high proportion of plant items found in the stomachs of aeglids (Williner 2007; Colpo et al. 2012) suggested that they might be primarily shredders. For shredders, the detritus together with microorganisms colonizing the surface of leaves have nutritional importance, since they enhance the food quality of the plant litter (Vannote et al. 1980; Graça 2001; Colpo et al. 2012). The effects of microbial colonization on leaf litter on the diet of these aeglids, however, are scarcely known (Colpo et al. 2012). On the other hand, when shredder insects are scarce, the aeglids can assume this role in the community (Cogo and Santos 2013). In other zones of streams and rivers, the functional trophic role of aeglids could be to serve as predators and scavengers or as grazers, scrapers, and collectors (Bueno 2003; Bueno and Buckup 2004; Williner 2007; Santos et al. 2008). These roles show the capacity of the aeglids to adjust to the habitat conditions, being opportunistic and omnivores. The variability in feeding habits also depends on the availability and behavior of the captured items (Table 4.1). For example, in tropical environments, larvae and pupae of Simuliidae insects, Ephemeroptera nymphs, protozoans, epizootics, algae, and bacteria are common stomach items (Magni and Py-Daniel 1989). Additionally, *Aegla* species that live in cave habitats (e.g., *A. cavernicola, A. perobae, A. strinatii*) probably prey on other crustaceans and insects living therein (Trajano 1987).

The more frequent animal food items in aeglid stomachs are those presenting slow movements and those that have a body with a long structure that is easy to grab and ingest, relying on easy capture with the chela or maxillipeds. Feeding on these preys require low energy costs for their consumers, and their energy content is positive in relation to the capture and handling costs (Baumart 2014).

To determine the feeding preference of *Aegla* spp. in streams and the importance of microorganisms in its detritivorous diet, researchers conducted experiments designed to evaluate food preferences, offering leaves with different levels of microbial colonization and several taxa of insect larvae (Chironomidae, Simuliidae, and Hydropsychidae). Under these conditions, the species preferred animal over plant food items; when only leaves were offered, aeglids selected preferentially leaves with higher levels of microorganism diversities (Colpo et al. 2012). Moreover, Devercelli

Table 4.1 List of Items Observed in Stomachs of Some Aeglid Species (*)

Items	Items	Items	Items
AQUATIC ITEMS	FUNGI	ANNELIDA	Trichoptera
	Ascomicete	Oligochaeta	Hidroptilidae
Heterotrophic bacteria	Mixomicete	Naididae	NI
		Allonais sp.	Hydropsychidae
ALGAE	EUKARYOTA	*Dero* sp.	NI
Cyanophyceae	Amoebidae	*Nais* sp.	Coleoptera
Aphanothece sp.	Arcellidae	*Pristina* sp.	NI
Coelosphaerium sp.	*Arcella* sp.		Hemiptera
Chroococcus sp.	Diflugiidae	CRUSTACEA	NI
Hapalosiphon sp.	*Difflugia* sp.	Cladocera	Lepidoptera
Microcystis sp.	Centropyxidae	*Ilyocryptus* sp.	NI
Nostoc sp.	*Centropyxis* sp.	Copepoda	Plecoptera
Spirulina sp.	*Ciliophora* sp.	Calanoidea	NI
Chlorophyceae	*Coleps* sp.	*Notodiaptomus* sp.	
Chaetophora sp.	Lagenophryidae	NI	PISCES
Cladophora sp.		Nauplii	Fishes scales
Coleochaete sp.	PORIFERA	Copepodit	NI
Dicranochaete sp.	NI	Ostracoda	Salmonidae
Eudorina sp.		Not identified	NI
Hydrodictyon sp.	NEMATODA	Amphipoda	
Microthamion sp.	NI	Hyalellidae	AQUATIC PLANTS
Mougeotia sp.		*Hyalella* sp	NI
Oedogonium sp.	PLATYHELMINTHES	Decapoda	Vegetal remain
Ulothrix sp.	Temnocephalidae	*Aegla* sp.	Leaves remain
Zygophyceae	Planariidae	Palaemonidae	Pollen-grains
Cosmarium sp.	Trematoda		
Euglenophyceae	Metacercaria NI	INSECTA	TERRESTRIAL ITEMS
Trachelomonas sp.		Larvae	ARTHROPOD
Phacus sp.	BRIOZOA	Pupae	NI
Dinophyceae	NI	Nymphs	
Peridinium sp.			INSECTA
Xanthophyceae	ROTIFERA	Collembola	Formicidae
Tribonema sp.	Bdelloidea	NI	NI
Not identified	NI	Diptera	
Bacillariophyceae	Lecanidae	Chironomidae	ARACHNIDA
Amphiprora sp.	*Lecane* sp.	Chironominae	ACARI
Amphora sp.	Brachionidae	*Cryptochironomus* sp.	NI
Attheya sp.	*Brachionus* sp.	*Procladius* sp.	Hydrachnidia
Cyclotella sp.	Notommatidae	*Orthocladiinae* sp.	*Arrenurus* sp.
Cymbella sp.	*Cephalodella* sp.	*Cricotopus* sp.	NI

(Continued)

Table 4.1 (Continued)   List of Items Observed in Stomachs of Some Aeglid Species (*)

Items	Items	Items	Items
Diploneis sp.	*Mytilina* sp.	*Pseudosmittia* sp.	
Eunotia sp.		NI	RIPARIAN VEGETATION
Gomphonema sp.	TARDIGRADA	Simuliidae	NI
Gyrosigma sp.	Macrobiotidae	*Simulium pertinax*	Vegetal remain
Luticola sp.	*Dactylobiotus* sp.	*Simulium* sp.	Leaves remain
Meridion sp.	NI	*Simulium riograndense*	Pollen-grains
Nistchia sp.		Ephemeroptera	
Navicula sp.	MOLLUSCA	Baetidae	INORGANIC ITEMS
Pinnularia sp.	Gastropoda	NI	Calcareous material
Pleurosigma sp.	NI	Leptophlebiidae	Sand and mud
Roicosphenia sp.	Bivalvia	Caenidae	
Rhopalodia sp.	NI	NI	
Stauroneis sp.	Hyriidae		
	Diplodon chilensis		

NI: not identified items; (*) Bahamonde and López (1961); Lara and Moreno (1995); Trajano (1987); Magni and Py-Daniel (1989); Bueno and Bond-Buckup (2004); Castro and Bond-Buckup (2004); Devercelli and Williner (2006); Williner (2007); Santos et al. (2008); Williner (2010); Colpo et al. (2012); Cogo and Santos (2013); Manattini et al. (2017).

and Williner (2006) evaluated the digestibility and viability of diatoms following gut passage, a common assembly in the streams and rivers in South America, observing that *A. uruguayana* has a high degree of consumption and digestion of these algae (Devercelli and Williner 2006).

Feeding activities and food spectrum of approximately 20% of the currently recognized 87 species of *Aegla* have been studied so far (Table 4.2). The indicated variation is probably associated to (1) differences of food resources available in the environment and (2) the use of distinct observation methodologies that permit the recognition of the consumed items with more precision and assurance.

The variations on the taxa level of the food items, however, correspond to the geographic distribution of each prey or consumed item. The presence of one or more items in a stomach shows the occurrence of some species, genera, or families in one or more geographic areas or microhabitats where aeglids live (Figure 4.4).

The presence of generalist and omnivorous organisms in an unstable or fluctuating system, such as freshwater continental systems, can be explained from an evolutionary point of view with some differences and nuances among species. The importance of omnivores in increasing food web linkages, which promote the stability and complexity of the system, should be emphasized (Woodward and Hildrew 2002; Majdi and Traunspurger 2015). With more linkages, the system has a broader range of possible responses available (Eubanks et al. 2003; Blanchette et al. 2014).

Table 4.2 **The Distribution Area of the Observed Species of the Family Aeglidae with Some Aspect in Trophic Ecology Study. The Biogeographic Province Was Assigned According to Morrone (2001)**

Region	Sub Region	Biogeographic Province	Species	Range Distribution	Reference
Neotropical	Paranaense	Paranaense Forest	*Aegla perobae*	Narrow	1
			Aegla cavernicola	Narrow	2
			Aegla singularis	Medium	3, 4
		Atlantic Forest	*Aegla ligulata*	Narrow	5
			Aegla camargoi	Narrow	6
			Aegla leptodactyla	Narrow	6
		Araucaria Forest	*Aegla schmitti*	Medium	7
			Aegla grisella	Narrow	3
			Aegla longirostri	Medium	8, 9
	Chaqueña	Chaco; Pampa; Monte	*Aegla platensis*	Wide	5, 10,11
			Aegla uruguayana	Wide	4, 12, 13
Andean	Paramo-Puneña	Puna	*Aegla jujuyana*	Narrow	4
			Aegla ringeletti	Narrow	14
	Patagonica	Central Patagonia	*Aegla neuquensis*	Wide	4
	Chilena Central	Santiago	*Aegla laevis*	Medium	15
			Aegla abtao	Medium	16

1- Rodriguez and Hebling (1978); 2- Trajano (1987); 3- Savaris et al. (2012); 4- Williner (2007); 5- Bueno and Bond-Buckup (2004); 6-Castro and Bond-Buckup (2004); 7- Trevisan et al.(2014); 8- Santos et al (2008); 9- Cogo and Santos (2013); 10- Magni and Py-Daniel (1989); 11- Colpo et al.(2012); 12- Devercelli and Williner (2006); 13- Williner (2010); 14- Manattini et al.(2017); 15- Bahamonde and López (1961); 16- Lara and Moreno (1995).

The omnivore condition is considered a key characteristic that permits the recognition of how benthic decapods regulate the energy flow and direct nutrient recycling (Buck et al. 2003; Mao et al. 2016), combining resources with different conversion efficiencies (Krivan and Diehl 2005; Mancinelli et al. 2013). These aeglids, as any other macroconsumer, modify the structure of the invertebrate community associated with leaf decomposition and energy transformation in the stream (Cogo et al. 2014).

The selection of food items fluctuates according to the cycles and movement of prey, e.g., vertical and horizontal movement of zooplankton, and availability affected by circadian, seasonal, and annual cycles occurring in decapods such as aeglids, with the hydric cycle being the most important structuring factor (Collins et al. 2006, 2007). Aeglids can be considered in several regions and seasonal periods as efficient phytoplankton grazers (Devercelli and Williner 2006; Williner 2007). In other regions or seasons, aeglids are known to be effective predators of slow-moving benthic macroinvertebrates. In some species, however, such as *A. longirostri*, no marked seasonal differences have been observed, which might be related to the availability of food items (Santos et al. 2008).

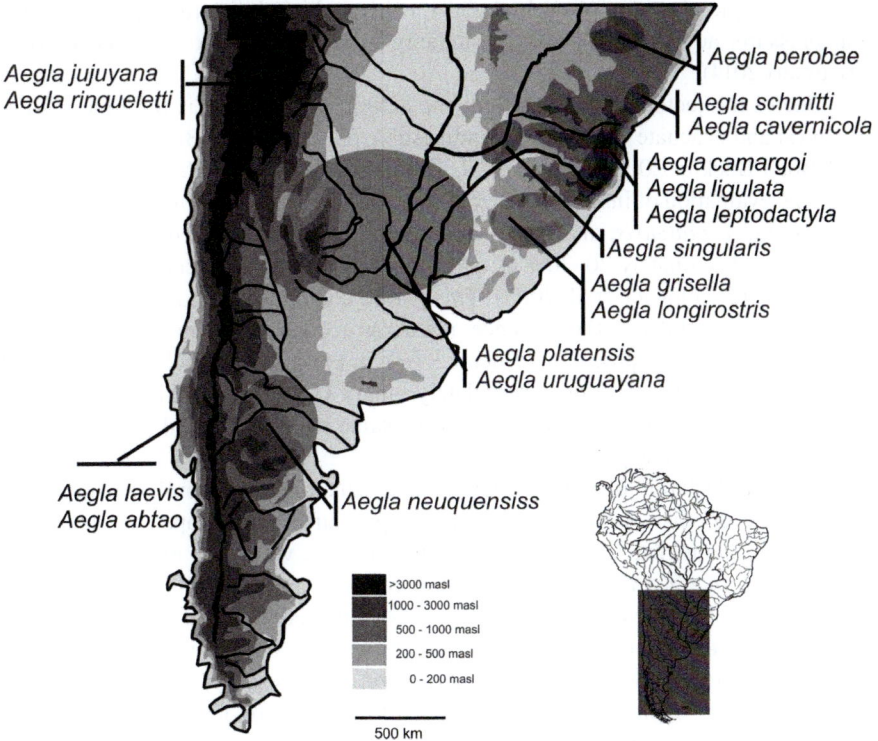

Aegla jujuyana
Aegla ringueletti

Aegla perobae

Aegla schmitti
Aegla cavernicola

Aegla camargoi
Aegla ligulata
Aegla leptodactyla

Aegla singularis

Aegla grisella
Aegla longirostris

Aegla platensis
Aegla uruguayana

Aegla laevis
Aegla abtao

Aegla neuquensiss

>3000 masl
1000 - 3000 masl
500 - 1000 masl
200 - 500 masl
0 - 200 masl

500 km

Figure 4.4   Approximate distribution of the species with information about trophic ecology in South America.

The rhythmic manifestation of any activity is influenced by exogenous and endogenous factors, which act to improve the energy equation of an activity with the lowest possible risk (Collins 2005; Collins et al. 2006). Environmental conditions, the succession of days and nights, the presence of prey and predators, together with the characteristics of the prey associated with the higher or lower quantity of hard structures constitute external agents that adjust the rhythm of trophic activities (Aréchiga and Rodríguez-Sosa 1997). Therefore, the trophic activity of freshwater anomurans occurs over a 24-hour period, but not with the same intensity. Qualitative and quantitative variations in the ingested material indicate periods of higher or lower activity (Bueno and Bond-Buckup 2004; Castro and Bond-Buckup 2004; Williner 2007). Such periods are not only observed due to the fullness of digestive tracts, but it might also happen through oscillations in enzymatic secretion, such as occur in other freshwater decapods (Sacristán et al. 2013). Activity variations occur differentially according to age or stage of development, more irregularly in juveniles than in adults, and are characteristic for species and populations (Oliveira et al. 2003; Sokolowicz et al. 2007; Sacristán et al. 2013). These cycles can vary according to the phenotypic plasticity of each aeglid species and its ability to respond to external pressure, adjusting to a new periodicity in which the risk of attack is lower (Baumart 2014).

One of the factors that determines the definition of a cycle is the time allotted to digest and evacuate the excess or indigestible prey elements (Bueno 2003; Castro and Bond-Buckup 2004; Collins et al. 2006; Carvalho et al. 2011, 2013). Results from the evaluation of insect larvae consumption (i.e., *A. camargoi, A. leptodactyla, A. ligulata, A. platensis*; and others freshwater decapods) revealed that the process occupies approximately 20–30% of one entire day and varies according to the individual size (Sokolowicz et al. 2007). This period could also be modified by the type of prey consumed due to the presence of hard structures, which may lead to a slow-down of the digestive processes (Bueno 2003). In the case of aeglids, a full stomach requires 5–6 hours to become completely empty (Bueno 2003; Sokolowicz et al. 2007; Baumart 2014). This indicates that the same individual could fill its digestive tract at least twice a day.

The manifestation of a daily cycle is determined by behavioral pressures (e.g., competence, agonistic behavior, and predator pressure) and is shaped by physiological and morphological characteristics of the species, which act also on a population level so that a certain activity rhythm can be detected (Collins 2005; Diawol et al. 2016). These behaviors tend to avoid visual predators and are coupled with the rhythm of the potential prey. This character displacement can be observed when the trophic offer is low and the population size, which uses this resource, is large. Additionally, the molting cycle is associated to a short non-feeding phase directly after molting, stopping the feeding before ecdysis, which masks the normal rhythm (Diawol and Collins 2012). When all the structures controlling the manipulation of prey hardened, aeglids start feeding again (Diawol and Collins 2012; Diawol 2018).

In general, aeglids increase their trophic activity from sunset to the hours immediately preceding and succeeding dawn (Bueno 2003; Bueno and Bond-Buckup 2004; Castro and Bond-Buckup 2004). This rhythm results from the interaction between the moment of the molting cycle and a stimulus, which, in the case of aeglids, may be the photoperiod (Collins et al. 2006), with the exception of aeglids living in caves. The most frequent consumption of certain items at a certain day period also indicates the habitat of the species. For example, animal resources, such as oligochaetes, amoebas, and rotifers are more common at night and are associated with the intake of sand and sediment particles. Other animals can be consumed together with plants, such as ostracods. The same occurs with cladocerans and copepods, which are frequent members of and associated with the littoral and benthic community. Plant remains are more prevalent in aeglid stomachs during the day (Bueno 2003; Castro and Bond-Buckup 2004; Sokolowicz et al. 2007), indicating that these individuals move within the water column to feed and find shelter, climbing from the bottom by the stems or any physical structure and/or submerge from the surface. The tidal effects are not relevant in shaping daily cycles of freshwater decapods; the importance of lunar effects, however, is yet to be studied.

Regarding the functionality of the feeding process, it has been observed that the trophic niche width is similar among aeglid species. Food acquirement depends on the relationship between the incorporation of energy and the expense of metabolic consumption, growth, reproduction, maintenance of body processes, food searches, and prey manipulation (Begon et al. 1995; Williner 2007). The predation process consists of a series of linked events likely divided into three stages (Collins et al. 2006): search, capture, and intake. During the first phase, decapods perceive the movement of potential food by means of mechanoreceptors and follow it using chemical receptors (Tudge et al. 2012). In some environments, visual recognition may be considered important and efficient because natural conditions permit this sense (Collins et al. 2007). For the recognition of dead or motionless food, only chemical receptors may be effective (Collins et al. 2006; Tudge et al. 2012).

4.2.3 Ontogenetic Variation

Ontogenetic variation in the trophic spectrum was observed in several aeglid species, e.g., *A. ligulata* (see Bueno and Bond-Buckup 2004), *A. longirostri* (see Santos et al. 2008), *A. platensis* (see Bueno and Bond-Buckup 2004), and *A. uruguayana* (see Burres et al. 2013). These differences were related to the recorded spectrum of food items but not to the stomach repletion condition. For example, several food items were recorded in both juveniles and adults, while others, such as Planariidae and Acarina, were items observed exclusively in the stomach of juveniles (Bueno and Bond-Buckup 2004). Moreover, some groups of prey or food items are differentially consumed by juveniles and adults with respect to the seasons, indicating different feeding habits according to the life stage (Bueno and Bond-Buckup 2004). In some cases, adult aeglids consumed live items, while juveniles had necrophagic habits (Lara and Moreno 1995). The presence of juvenile *A. laevis* near e.g., aquatic plants, leaf remains, and algae could indicate that vegetation and algae are used not only as physical refuges but also as feeding sites, where the associated fauna are potential prey items (including Amphipoda, Nematoda, Oligochaeta, and various orders of immature Insecta) (Bahamonde and López 1961; Burns 1972). In addition, the presence of sand in the stomach could be considered accidental and associated with the consumption of organic matter, where algae, bacteria, and other microorganisms grow (Bueno and Bond-Buckup 2004; Williner 2007). Furthermore, cannibalism can also occur among juveniles due to the high rate of ecdysis events that take place during this period (P. Collins, personal observation).

A study with *A. uruguayana* indicated changes in feeding types, from herbivorous-detritivorous to omnivorous during the life stages (Burres et al. 2013). Moreover, in this species the ossicle descriptions suggest a macrophagous pattern, and it is possible that the organisms make certain adjustments to ambient pressures (Williner 2010), in this case incorporating microorganisms and/or larger prey in the diet during the ontogenetic development (Figure 4.5). Furthermore, *A. uruguayana* assimilated a larger proportion of invertebrates in later life stages, a pattern also shared with subtropical populations of *A. platensis* and *A. ligulata* (Bueno and Bond-Buckup 2004; Burres 2012). In these two species occurred a shift in the consumption, from ~8% invertebrates (by volume) to 20% throughout ontogeny, and in

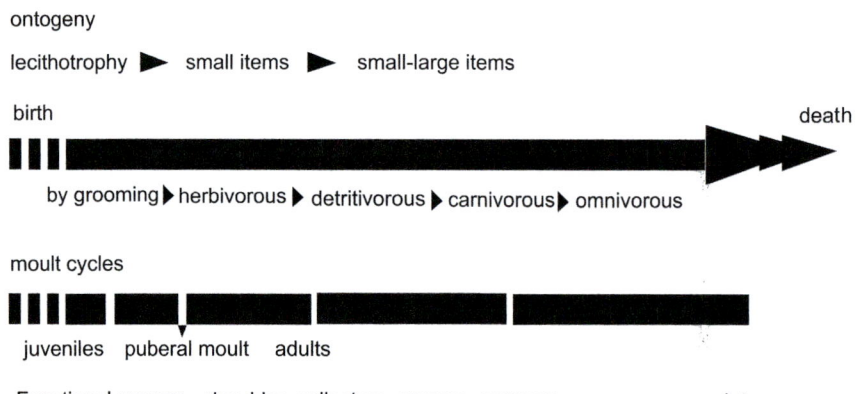

Figure 4.5 Variation of the trophic activities in the family Aeglidae according to ontogeny and molt cycle with recognition of the functional groups.

A. platensis from 8.5% to 10.5%. This is lower than the estimated 16.5% to 43% for *A. uruguayana* according to isotopic analyses (Burres 2012).

The variation observed in the diets of juvenile and adult aeglids may be related to differences in both the functional morphology of the mouthparts, locomotion system, and sensory capacities. Furthermore, the microhabitat and cycles of the preys result in the availability of distinct size classes at different times of the year, and they should keep into consideration when and where the different food items are available (Laughlin 1982). Except for the sensory capacity, differences in morphology cannot explain diet variation because aeglids exhibit direct development and have already developed the mouthparts and locomotion structures in the early juvenile stages (Bond-Buckup et al. 1996; Bueno and Bond-Buckup 1996; for more details on postembryonic development see Chapter 6). The hatchlings remain nearby the mother. Thus, there is extended parental care, and the juveniles with the grooming habit could begin to consume different items, e.g., the fouling that grows upon the exoskeleton of the mother as occurs with other decapods (Anger 1995; Glas et al. 1997; Senkman et al. 2015).

The hatchlings remain for a while in proximity to the mother before leaving and beginning the period of independent life. Juveniles generally go through an early lecithotrophic stage (feeding for a time on the reserves from the yolk they possess), thus being still independent of exogenous food (Anger 1995). Subsequently, the inclusion of items in the diet begins through the grooming behavior incorporating diatoms, rotifers, protozoans, and other epibionts (Martin and Felgenhauer 1986). Subsequently, the use of filamentous algae and zooplankton (cladocerans and copepods) has been observed, incorporating fungi and a bacterial soup into their feeding spectrum (Bueno and Bond-Buckup 2004; Colpo et al. 2012; Cogo and Santos 2013). Later on, juveniles consume a few items, including annelids and Platyhelminthes, and vegetation is added as a food resource (Bueno and Buckup 2004; Williner 2007, 2010; Santos et al. 2008). Additional types of vegetation and different degrees of plant decomposition are included, with the trophic spectrum being similar to an adult (Figure 4.5). The necessary time to attain trophic maturity varies with latitude, according to the relation with the molt cycle or growth velocity. After this period, aeglids use the full food potential of the local environment (Bueno and Buckup 2004; Collins et al. 2006; Williner 2007, 2010).

4.2.4 Temporal Variation in the Trophic Spectrum

Aeglid radiation in lotic and lentic environments requires the colonization of freshwater environments and started in different southern areas along the South American continent. The establishment of aeglid populations occurred in a variety of environments, such as rivers of different orders, lakes, mountain streams, mountain lakes, and caves (Bond-Buckup et al. 2008).

The main adaptations in crustaceans for living in freshwater environments include an adjustment to hypoosmotic conditions, which represents a tool for the more efficient use of trophic resources, avoiding the cost of migration towards estuaries for reproduction (Anger 2006; Vogt 2013). Additionally, for this adaptation, the environment must provide qualitative and quantitative food resources according to biological and population requirements (e.g., molting, reproduction). The environments inhabited by freshwater decapods pose similar pressures that are related to the conditions in continental aquatic systems and their cycles, though the response of decapods to each environment will differ according to the resource being analyzed (e.g., trophic resources, habitat, or reproduction) (Bueno and Bond-Buckup 2004; Collins et al. 2006, 2007; Bücker et al. 2008; Mancinelli et al. 2013).

Ecological stoichiometry indicates the relative proportions of chemical elements, mostly C, N, and P, in the environment and in the biota, and associates the biological transformations of these elements through the food web (Feijoó et al. 2014). Elemental imbalances exist between resources and consumers because organisms meet their internal elemental demands by consuming food sources of different stoichiometric ratios (C:N:P). These imbalances can constrain the growth and reproduction of consumers unless organisms can adjust their internal elemental composition following changes in resource stoichiometry (Frost et al. 2005). The organisms that can maintain their internal stoichiometry regardless of changes in their food C:nutrient ratios are considered homeostatic. However, an exception occurs with *Aegla* spp. when feeding on epiphyton, which exhibits weak plasticity for N:P content. This assumption is supported by the lower degree of homeostasis in *Aegla* spp. for C:N, C:P, and N:P. A possible mechanism to explain the weak homeostasis would be higher storage of P through food intake with high P content (Feijoó et al. 2014).

Aeglids present an elevated level of endemism and are currently found at sites with very different physical and chemical environmental characteristics (Bond-Buckup and Buckup 1994; Collins et al. 2011; Bueno et al. 2016; Tumini et al. 2016). The food resources can differ in the ratios of C:N, C:P, and N:P (Mancinelli et al. 2013). Therefore, one should be careful to indicate the degree of homeostasis that species present in relation to these elements. The relations C:P and N:P are lower in aeglids than in other decapods, permitting them to live in environments with low values of these components. *Agla* spp., however, exhibit a weak plasticity for N:P when feeding occurs upon epiphyton (Feijoó et al. 2014).

Abiotic factors (i.e., altitude, stability of the water body, annual temperature range (ATR), pH, and conductivity) of the environment inhabited by aeglids are highly variable in the southern region of South America. While some environments have perennial rivers, others only have water during the rainy season or the thaw (spring–summer) (Figure 4.6) (Tumini et al. 2016). The stability of the water environment depends on the soil, rain, snow, underground water, and superficial water. These factors determine flood and drought cycles or periodicity, which are reflected in the phenology of the trophic offer of the aeglids. The water level influences a series of patterns in the population dynamics (e.g., variation in trophic intensities, movements, reproduction, agonistic behavior, and intraspecific and interspecific interaction) (Collins et al. 2006, 2011; Tumini et al. 2016).

The relationship between stream flow variability and decapods must consider variations in the time intervals between flood and drought events, the availability of food, and sufficient time for the development of prey in the river or stream. The time since the last extreme event determines the biotic responses to this disturbance. In explaining benthic community responses, one study focused on floods (Blanchette et al. 2014), and another one indicated the importance of infrequent and low flows (Covich et al. 2003). In several environments in mountainous areas, such as those occurring in the Andes chain, floods are usually discrete, short events (Tumini et al. 2016).

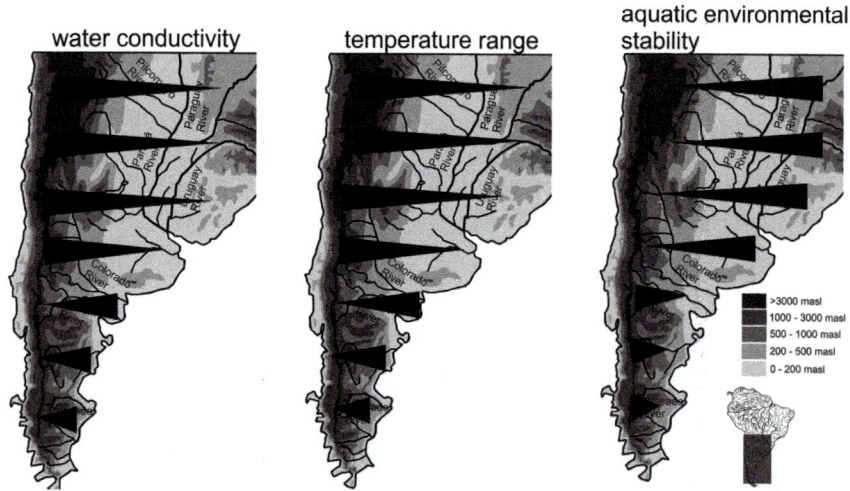

Figure 4.6 Three main factors in the south of South America that affect the family Aeglidae and the aquatic environments with their biotic assemblages. The valuation is indicated by a triangle, the base represents the higher values, and the extreme point the low values. The water conductivity is registered from high to low concentration; temperature range is shown as the difference during the day and by season, and the aquatic stability is assigned to the amount of days with water present in the environment.

In contrast, extremely low flows can persist for months or years, and it is likely that protracted scarce flows or droughts along the altitudinal gradients will occur frequently (Tumini et al. 2016). The impact of droughts on feeding patterns of decapods will likely vary regionally according to the frequencies and duration of droughts, resulting in a variety of reactions and responses.

Altitude is an important factor due to the effects of extremely low oxygen uptake, water regime, temperature, availability of prey, and refuge. The differences in the altitude, latitude, and longitude, where these aquatic systems occur, also relate to considerable temperature variations. For example, aeglids live in environments with air temperatures that range from 0 to 40°C in the same location during the entire year (e.g., shallow water wells with the rocky bottom) (Collins et al. 2011; Tumini et al. 2016). The variations can occur both across different aquatic systems and in a single environment during a daily and/or seasonal cycle (Tumini et al. 2016).

Trophic activity does not have the same intensity throughout the year. This is related to the presence and abundance of certain potential preys, which are affected by thermal and hydric cycles (Bueno 2003; Bueno and Bond-Buckup 2004; Castro and Bond-Buckup 2004; Williner 2007; Santos et al. 2008). According to the latitude, the highest level of food consumption occurs between spring (September–October) and autumn (April–May) (i.e., high latitude and/or altitude) or when hydric and thermal cycles have high values as in aquatic environments at low latitude and/or altitude regions (Bueno 2003; Bueno and Bond-Buckup 2004; Castro and Bond-Buckup 2004; Williner 2007; Santos et al. 2008; Tumini et al. 2016). During this period, growth and development are enhanced and typically include the occurrence of reproductive events. In winter, when riparian vegetation decreases in coverage and the leaves begin to decompose due to cold temperatures and low river levels, plankton is used as an important alternative resource (Williner 2007). Contrastingly, in spring, there is an increase in the consumption of algae and plants to supply and supplement diets with vitamins and other essential substances, which allows for optimal and also frequent ecdysis together with successful reproduction (Collins et al. 2006; Williner 2007).

4.2.5 Food Handling in Aeglidae and Its Importance for the Trophic Ecology

The appendages used for feeding are the mandibles, maxillules, maxillae, maxillipeds 1 to 3, and the first pereopods with its chela. These appendages have several functions ranging from food detection to ingestion (Castro et al. 2015).

Aeglid species have different types of setae on the mouthparts, each one with a specific function (mechanoeffectory, mechanosensory, or chemosensory). The functions of these appendages are capturing, transporting, transporting-aligning, sorting-aligning, current generation, cutting-crushing, and ingestion processes (Martin and Abele 1988). The unidirectional current generated by the multi-articulated flagellum of the distal portion of maxillipeds 1 and 2 transfers chemical signal information of possible food nearby. Furthermore, algae, bacteria, and microfauna are also dragged by the other appendages towards the gnathal area, where they are quickly placed into the mouth and transported to the cardiac chamber of the foregut by the esophagus (Williner 2010).

The third maxilliped is well developed, pediform, and carries out functions in feeding and grooming (Tudge et al. 2012). Maxilliped 3 and the first pair of pereopods have been frequently observed combining several movements towards the ventral body parts or entire exoskeleton, where bacteria, fungi, algae, and unicellular metazoans are loaded in a high density in the setae (Martin and Felgenhauer 1986; Williner 2007; Colpo et al. 2012). After repeated combined movements, the appendages are then moved towards the mandible and the captured food is ingested (P. Collins, personal observations). These epibiont resources are an important food source for *A. uruguayana* (see Martin and Felgenhauer 1986); additionally, maxilliped 3 and all pairs of pereopods may play a role in cleaning and grooming these epibionts (Martin and Felgenhauer 1986; P. Collins, personal observation). Extreme cases of biofouling have been observed with the fixation of freshwater bivalve *Limnoperna fortunei* on *A. platensis* and *A. uruguayana* (Lopes et al. 2009), which could suggest that these infested aeglids would have more difficulty finding shelter and/or avoiding predation with an increased rate of energy consumption (Lopes et al. 2009).

The indirect filtration process occurs using the setae present in the maxillipeds and maxillules. These appendages make rhythmic movements and capture the microparticles that are continuously placed into the mouth (P. Collins, personal observations).

The chelae of the first pereopod can crush and open the valves of molluscs or capture animals ranging from oligochaetes, insects (larvae, pupae, and adults) to small fishes that have difficulty in moving (Williner 2007; Savaris et al. 2012). Furthermore, the teeth of the chelae cut leaves of different levels of decomposition into small parts (Collins et al. 2006).

4.3 TROPHIC OFFER AND ENERGETIC REQUIREMENT DURING DIFFERENT CYCLES OF AEGLIDS

All species use the available resources of the environment, thus creating a framework that gives rise to trophic webs. The participation of all the members is inevitable since they consume or are consumed, and that is why each component has a location in a specific level of that web, but levels can change according to the members that are present in the communities (Burres 2012; Cogo et al. 2014). For this reason, seasonality and the geographic region where the species live may influence the variation. The entire food supply reflects the local environment in which it has developed and how it has evolved together with the system. The food web is characteristic for each biogeographic province where aeglids are found (Collins et al. 2011).

At the level of species, genera and in some cases families, variations occur in relation to the presence of each integrant or the degree of the participation in the trophic relations, according to the latitudinal, longitudinal, and altitudinal development of the optimal ranges of these taxa and the macrofactors that determine their distribution and phenology (Bueno and Bond-Buckup 2004; Williner 2007). The available food must provide enough energy and micro- and macromolecules to live and build the somatic structure of the aeglids (Buckup et al. 2008; Oliveira et al. 2011; Majdi and Traunspurger 2015; Musin et al. 2017; Musin 2018). In contrast, this food availability can be identified by the capacity for mobilization and/or the probability of being captured and manipulated by the aeglid feeding appendages. That is why aeglids can identify groups of food and/or prey with slow movements and other prey with no or little escape capacity or mobility (Williner 2010). This strategy is used preferentially, although the nutritional quantity and quality of prey items can be limited in some trophic items (Collins et al. 2006; Colpo et al. 2012; Burres et al. 2013; Cogo and Santos 2013). Furthermore, there are soft-bodied prey items and others that have structures of resistance and protection. The selection of one or another prey item is a question of cost versus benefit that depends on the density of each prey (Collins et al. 2006; Williner 2010). Elongated preys have a higher chance of being caught than those that are rounded, which are more difficult to capture and handle with the appendages (Collins et al. 2006).

The relationship between the cost and benefit of the catches of each of these resources must be analyzed in terms of energy and nutrients that are provided by each one of the available prey items. The densities of the prey populations and the probability of capture or encounter of this resource are factors that determine which food item will be captured by the aeglids (Collins et al. 2006).

The trophic offer used by the aeglids presents daily cycles. These cycles determine whether trophic items are available or not at a certain time of the day. The potential food items can be moving, being more accessible or staying in shelters. In addition, these food items can be more available under different conditions or in microhabitats of the aquatic environments, i.e., in pools, under rocks and stones, in areas with low current speed, or between decomposed plant leaves (Cerezer et al. 2016). The distribution and seasonal presence of potential prey is related to climatic conditions and the geographical distribution of each food item (Miserendino 2001; Bueno and Bond-Buckup 2004; Williner 2007, 2010).

One aspect that must be considered is that each prey item or member of the trophic web has certain characteristics regarding their macromolecules that comprise its body and the condition of its bioavailability according to a species capacity for digestion and assimilation (Castro and Bond-Buckup 2004; Devercelli and Williner 2006; Burres et al. 2013). Moreover, the relation with both the "Bauplan" of each trophic item, whether it converts light or chemical energy, and the complexity of the molecules that are present in the trophic items have significance for the nutritional efficiency. The energy of each intake food will be relevant, but also the elemental components, both for their basic stoichiometry (i.e., C-H-O-N-S) and in relation to proteins, lipids, and carbohydrates contents. These components will allow the establishment of the structural basis and functioning of every individual (Blanchette et al. 2014). Each of the previously mentioned elements is related to the needs and requirements of the aeglids in each period of their ontogeny or molt or reproductive cycles.

Multiple molecules need to be obtained from the environment because aeglids do not have the ability to synthesize them "de novo" from other compounds (e.g., some fatty acids and amino acids), similar to other decapod crustaceans and other animals that rely on the substratum to acquire these constituents of the organisms (D'Abramo and Conklin 1995; Wouters et al. 2001). Therefore, each food item has a certain nutritional quality in relation to the aeglid requirements, providing a certain quantity and quality of proteins, lipids, and carbohydrates that is needed for suitable development, molting, and/or the reproductive cycle. As in other freshwater decapods of South America, the lipid levels in the haemolymph are lower in winter, mainly in females, suggesting a mobilization for vitellogenesis (Oliveira et al. 2003), while in the spring, the levels of triglycerides and cholesterol increase in the haemolymph, as well as the glycogen in the muscle (Musin et al. 2017). The cholesterol level is in both sexes low in summer when the synthesis of sex hormones and growth occurs (Buckup et al. 2008; Musin et al. 2017) (Figure 4.7). These contributions are related not only to the quantity of these elements consumed but also to the development of a determined profile of amino acids and specific fatty acids required by aeglids (Table 4.3) (Collins and D'Alessandro 2017).

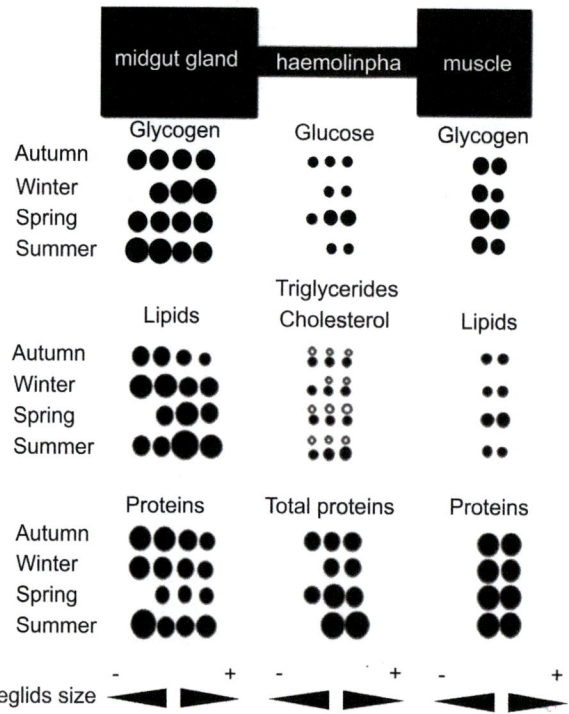

Figure 4.7 Seasonal movement of concentrations of glycogen/glucose, lipids/triglycerides/cholesterol, and proteins among different tissues and range sizes of *Aegla uruguayana*. Each circle represents the concentration of a metabolite, whose size is proportional to the quantity measured. In each tissue, the size varies from juveniles to adults (> 20 mm) going from left to right, respectively. [Modified from Musin et al. (2017).]

Table 4.3 Essential (EAA) and Non Essential Amino Acids (NEAA) Registered in Muscles of *Aegla uruguayana* (see Collins and D'Alessandro 2017)

EAA	g/100 g Prot Mean	Standard Deviation	NEAA	g/100 g Prot Mean	Standard Deviation
Histidine	3.853	0.0547	Aspartic/ glutamic acids	18.444	0.2362
Threonine	4.968	0.0302	Serine	5.550	0.0491
Valine	3.809	0.3730	Glycine	8.312	0.0929
Methionine	3.224	0.1544	Arginine	11.670	0.0509
Isoleucine	4.930	0.0114	Alanine	8.307	0.0221
Leucine	10.136	0.1144	Proline	9.642	0.7508
Phenylalanine	5.450	0.1748	Tyrosine	4.546	0.1576
Lysine	10.900	0.6224	Cysteine	0.957	0.1181

All these components must be evaluated not only in relation to quantity and quality but also in terms of their bioavailability and the capacity for digestion and for entering the cells for their disposition and use. Food, after being handled by appendages and crushed by the gastric mill, is filtered and released into the channels of the hepatopancreas. The enzymes, produced by the F cells of the hepatopancreas, fulfill their function in the lumen of the hepatopancreas and in the pyloric chamber of the foregut, thus facilitating the entry of some molecules into the R cells (Al-Mohanna and Nott 1987; Vogt 1994).

Another aspect to consider is whether energy and elemental components originate from the aquatic environment (autochthonous) or correspond to terrestrial components (allochthonous). Thus, we observe the land-water interaction and how energy moves between the different components of an ecosystem, indicating the percentages that correspond to allochthonous or autochthonous elements, and why and how energy and molecules are transported between both compartments, according to their geographical location.

The molecular component pattern in a population is an expression of the food availability in the environment and the use of the resources by the aeglids (Collins et al. 2006). Moreover, the relationship of consumed chemical elements in each food item, such as essential components (C-H-O-N-S), permits to identify a stoichiometry that can characterize the species and their life stage (ontogeny) as well as of cycles of molt (premolt–molt–postmolt–intermolt), seasonal and reproductive cycles (Burres et al. 2013; Blanchette et al. 2014; Feijoó et al. 2014).

The variations of nutritional requirement are related to changes during the ontogeny that allow individuals to provide the energy for growth and molecules needed for organ developments and internal maintenance (D'Abramo and Conklin 1995). This nutritional pattern is modified as the individual grows and develops, changing significantly during molting and after the pubertal molt (Figure 4.5). After the pubertal molt, the changes in the pattern are related mainly to reproductive events (Diawol et al. 2015).

Nutritional requirement variation is related to the availability and location of the species throughout the geographic distribution of Aeglidae, varying in relation not only to the north-south gradient (latitudinal), where temperature is an important factor, but also to the east-west gradient (longitudinal), which is mainly affected by the availability of water (Figure 4.6), and in relation to the availability of dissolved oxygen at altitude (Tumini et al. 2016, 2018).

The trophic resource and the metabolism of *A. uruguayana* allow recognition of at least 16 amino acids (AA) in the muscle, including eight essential (EAA) and eight non-essential (NEAA) (Table 4.3), with the glutamic and aspartic acids being the more abundant ones (Collins and D'Alessandro 2017). The trophic offer must contribute to this requirement, and its abundance allows having the substrate to face distinct types of stress generated naturally or induced anthropogenically. Quantitatively, of the total AA, 40% were EAA and 60% NEAA [e.g., *A. uruguayana* (Collins and D'Alessandro 2017)]. The EAA in *A. uruguayana* were recorded in the following order: lysine > leucine > phenylalanine > threonine > isoleucine > histidine > valine > methionine (Collins and D'Alessandro 2017). The lipids present are mainly in the hepatopancreas consisting predominantly of fatty acids, eicosapentaenoic acid (EPA) and docosahexaenoic acid (DHA), as well as saturated fatty acids (Musin et al. 2018; Collins and D'Alessandro unpublished data).

The hepatopancreas is the main storage organ next to the gonads. Lipids, however, can also be found in muscles and haemolymph in very low concentrations of 5 to 7% (Musin et al. 2017). Among carbohydrates, glycogen is the form of storage and immediately available for energy production disposition. They are present in a greater quantity in muscles and haemolymph (Figure 4.7). All these components must be obtained from the environment or transformed "de novo" from precursor molecules (Musin et al. 2017).

Crustaceans have limitations in having longer acid chains, such as polyunsaturated fatty acid (PUFA), so they must be obtained from the environment and then shortened to produce linolenic and linoleic acids (Collins and D'Alessandro 2017; Musin 2018). In contrast, the remaining fatty acids have an energy expenditure according to the substrate where they are found, whether they are more easily available or not for the individuals. Minerals are additional elements that are necessary for incorporation into their natural diets. These can form part of the same prey organisms, can be incorporated by trophic elements that are consumed and/or from the inorganic environmental conditions or from forming colloids (Bueno and Bond-Buckup 2004; Burres et al. 2013; Cogo and Santos 2013; Cogo et al. 2014). Carotenoids are not only responsible for the pigmentation of decapods but are also beneficial for their health (Lim et al. 2018) because they promote their growth, survival (Daly et al. 2013; Wade et al. 2015), and the gonadal maturation of females (Paibulkichakul et al. 2008; Lim et al. 2018). Decapods, however, are not able to synthesize them "de novo," so carotenoids must be incorporated into their diet (Mezzomo and Ferreira 2016). In this sense, studies have shown that feeding these animals with foods containing carotenoids improves their responses to acute and chronic stress related to hypoxia, alterations in ions and cations, temperature, and viral infections (Daly et al. 2013). These effects seem to link the antioxidant function of carotenoids with improvement of the antioxidant response in animals and improved performance under various stressful conditions (Wade et al. 2015; Lim et al. 2018).

In a hypoosmotic environment, obtaining ions and cations is an obligatory requirement, avoiding their elimination to the environment (Bozza 2016). During the molting process, a strong imbalance occurs in these components, which adjusts their capacity for osmotic regulation (Bozza 2016).

All species, including those of the family Aeglidae, require energy and molecules, which permits the performance of their activities, including maintenance as well as growth and reproduction. This energy is divided and compartmentalized, so that, in general, reproduction and somatic growth are antagonistic events in crustaceans. The distribution of energetic resources during the day or year will shape the features of the organismal life history. Consequently, these energy sources and structural molecules need to change from their incorporation at the egg stage to the reproductive adult stage; the amount and type of molecules can differ in each ontogenetic period and year due to the needs of each individual (Musin 2018).

The eggs in aeglids are rich in yolk content, and the pre-hatching stages require energy and structural molecules stored in the eggs, which are obtained at the expense of the mother. Therefore, reproductive females must mobilize lipids, which are mainly stored in the hepatopancreas, requiring a high trophic activity prior to the breeding season (Bueno and Bond-Buckup 2004; Oliveira et al. 2011). Such breeding activity occurs throughout the year in some latitudes, while at higher latitudes or at high altitudes the reproduction of aeglids is restricted to the higher temperature period and the presence of a greater trophic supply (Williner 2007; Tumini et al. 2016). In the end, aeglid populations need to speed nutrient movement and direct it quickly towards the gonads. Lipids, protein, and glycogen are required for a successful reproductive event. Moreover, in this short time period, these populations (e.g., in high latitudes or altitudes) need to grow; thus, all molecule movements must be quick.

The increase in trophic activity is related to energy needs, including the incorporation of specific molecules that provide the required nutrients. If this does not occur, a poor or deficient reproductive event may happen, in which the embryonic stages developing inside the egg are not completed, or the hatchlings do not have the aptitude to survive and develop outside the egg, or the ecdysis fails (Diawol and Collins 2012; Diawol et al. 2015; Musin et al. 2018). The reasons for such failures can vary, ranging from limitations in the supply due to a shortage in the availability or due to an increase in consumption generated by natural or anthropic factors (e.g., stress). If the animals do not capture food with abundant antioxidants (e.g., carotenes), the rapid process of molecule transformation induces the formation of free radicals. They must be immobilized with the antioxidants that are produced within the animals and/or obtained from the environment (D'Abramo and Conklin 1995).

The reproductive stage is not the only process that requires extra energy, but the molecules that are needed induce an endogenous cycle towards the gonads for the success of the reproduction (Musin et al. 2017). Another energy-consuming process is the molting cycle, which will also force the movement of minerals, protein, lipids, and glycogen from the old exoskeleton to the hepatopancreas and subsequently to the new exoskeleton (Al-Mohanna and Nott 1987; Sousa and Petriella 2001). After the hardening of the new exoskeleton, the somatic growth requires nutrients and energy obtained from the food provided by the environment. Therefore, it needs to occur at times when the environment provides the necessary trophic resource to complete all the cycles.

Other factors also generate cycles, and the trophic activity needs to be adjusted accordingly. For example, seasonality varies between populations or species living in tropical environments and in cold temperate latitudes or in high altitudes (up to 5000 meters above sea level). The variation in tropical environments is relatively small: temperature changes and different heliophany in each season are generally imperceptible or of little influence; the periods of rain are more relevant in combination with the availability of resources (Collins et al. 2006; Tumini et al. 2016). At higher latitudes, however, where marked thermal differences occur and where sufficient food supply is available only for a short period of time during the year, the trophic activity increases along with the risk to be preyed upon. During this period, individuals must obtain sufficient energy to provide an adequate quantity and quality of nutrients, which generate the necessary reserves for aeglids during the cold season and allow maintenance, growth, and reproduction (Bueno and Bond-Buckup 2004; Collins et al. 2006). This period can be brief or extended in time, according to the latitude of the location of the aeglid population. Accordingly, each species has a tolerance range to abiotic factors (e.g., temperature, dissolved oxygen, salinity) where it is active and can search for food (Baumart 2014). Therefore, aeglid populations can be found at high latitudes or altitudes exceeding 4000 meters above sea level (Tumini et al. 2018) (Figure 4.5), and they are effective, i.e., they grow and reproduce. It is not only the seasonal cycle that imprints a certain rhythm on the trophic activity in aeglids, but also the daily activity or diurnal cycle that constitutes another element, which need to be considered (Bueno 2003; Sokolowicz et al. 2007). This cycle is driven by different factors that range from abiotic to biotic conditions. Among the latter, the availability of trophic items (prey) with their internal movement cycles or the presence and activity of predators are conditioning factors. For the last condition, the daily active movement period is adjusted according to the availability of preys that can be eaten by visual or chemical-tactile (non-visual) predators (Cerezer et al. 2016). Accordingly, the trophic offer and its activity generate a force directed to higher or lower risk exposure. In addition, it is necessary to indicate the energy needs, which can vary during the day by locomotory activity (Figure 4.8). The tendency of crustaceans to be more active in the dark is considered as a mechanism to avoid visually orientated predators, such as fish and other vertebrates (Trevisan et al. 2014).

Although aeglids feed throughout the day in aquatic environments ranging from tropical to cold temperate regions, they feed preferably during twilight and nocturnal hours (Bueno 2003). The activity occurs at temperatures higher than 10°C, limited by the movements in environments with lower temperatures (P. Collins, personal observations). Aeglids inhabiting high-latitude sites or high-altitude regions display more trophic activity after midday, when water temperatures are higher (Miserendino 2001).

The daily cycle is related to the satiety and the times of digestion and evacuation that the species present in relation to the characteristics of the most abundant

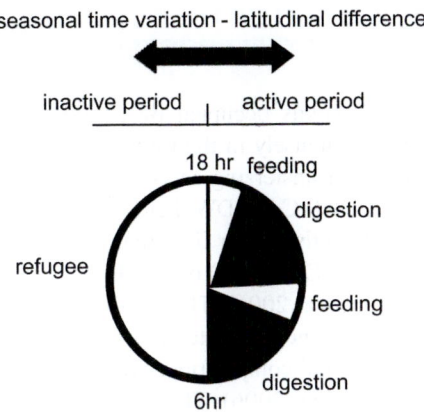

Figure 4.8 Variation of activities of species of the family Aeglidae during a daily cycle. [Modified from Sokolowicz et al. (2006).]

prey items that are consumed (Figure 4.8). The items with abundant calcareous or chitinous structures in the body delay the digestion process. Vegetal parts with more advanced stages of decomposition digest more quickly. *Aegla* selects locations with abundant old plant fragments that are usually colonized by fungi and bacteria (Bücker et al. 2008).

The time employed to obtain energy and macro- and micro-molecules is relevant for the selection of prey or food items, and this could be an element underlying the preference or selectivity of each trophic item (Colpo et al. 2012). Moreover, there must be synchronicity between the time of the highest energy requirement and the occurrence of each stage of the internal cycles (i.e., ontogeny, molt cycle, and reproductive cycle) (Sacristán et al. 2017).

The molt cycle needs to be recognized and observed, as it affects the trophic activity. The molt is a process that ends or begins with ecdysis; starvation occurs before and after this period, with a series of processes that mobilizes nutrients and minerals prior and even posterior to the exuvial elimination. During the premolt, there is a recondition and mobilization of minerals and nutrients that are scarce in the environment (Sousa and Petriella 2001). After molting, similar movements of minerals and nutrients occur towards the new exoskeleton, giving it structure and hardness. The duration of the lack of trophic activity varies among species, type of environment, or availability of minerals in the environment. Moreover, the starvation time is related to the mineral reserve stored in each animal and the capacity of the aeglids to take from the environment those minerals necessary to harden the exoskeleton. The old exocuticle is consumed by the aeglids favoring the recovery of a series of minerals, which are difficult to obtain from the environment (P. Collins, personal observations). All of the this leads to the existence of a trophic dynamic not only defined by energy needs and presence of structural elements but also by environmental dynamics inherent in the trophic supply and by the ways in which both are conditioned by the cycles of the macrofactors that regulate the environment.

4.4 CONCLUSION OF THE TROPHIC ECOLOGY, AN INTERACTIVE PROCESS IN THE WEB

In all environmental systems occupied by aeglids, the individual conforms, composes, and participates intensely in the trophic web, representing a highly valued link of carrying off and transferring energy and matter. The energy density of *Aegla* spp. is indicated by 3880 Cal/g DW, being lower than that of fishes, insects, and amphipods, but significantly higher than that of crayfish, e.g., from Patagonia (*Samastacus spinifrons*, 3364 Cal/g DW) or North America (*Orconectes* spp., 2950 Cal/g DW) (Ciancio and Pascual 2006). The quality of a particular species as prey is not given exclusively by its energy content. Other factors will also determine its intrinsic value, such as proximal composition, availability, handling time, and individual size (Ciancio and Pascual 2006). The capacity of transforming low-quality elements (e.g., cellulose) into high-value nutrients (e.g., protein) or components that are at the limit of exploitation due to their position in the physical matrix makes the aeglids an essential component of a community. This relevant participation is defined by the use of the vegetal material and the transformation of typically unavailable nitrogen to nitrogen of high nutritional value (Burres et al. 2013; Feijoó et al. 2014) (Figure 4.9).

Although the average protein content of algae or aquatic and riparian vegetation is low [approximately 5–30% of dry weight (Burtin 2003)], this is an interesting component in the natural diet of aeglids due to its nutritional properties. The vegetation could be used by aeglids as a complement incorporated together with fungi and a bacterial cocktail with high nutritional value due to the presence of exogenous enzymes. This component is available for the individuals at a low energy

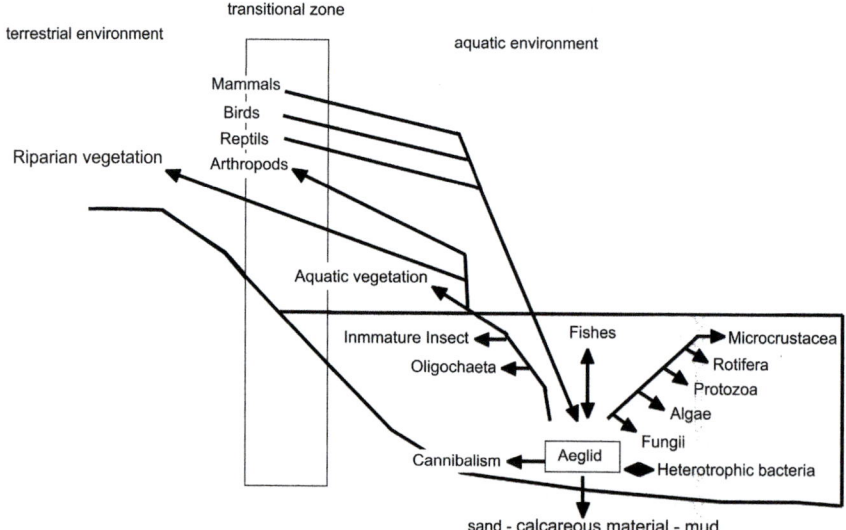

Figure 4.9 Qualitative trophic web that includes species of the family Aeglidae.

cost, providing incomparable amounts of fiber, trace minerals, vitamins, lipids, and amino acids (Burtin 2003). In addition to its nutritional components, algae and aquatic vegetation or riparian leaves that fall into the water and begin their decomposition contain bioactive compounds with high antioxidant capacity, such as carotenoids, polyphenols, and sulphated polysaccharides, which can be used to control the excessive production of free radicals and reactive oxygen species (ROS), fundamental elements used to counteract the stress in an unstable environment (Lodge et al. 1994; Negro et al. 2011).

Some community components (e.g., depredators and/or competitors) can provoke direct or indirect risks of injury to aeglids as part of trophic interrelationships. Cannibalism occurs within this group of decapods, mainly involving recently molted or small-sized individuals (Rodrigues and Hebling 1978, Bueno and Bond-Buckup 2004; Collins et al. 2006). Furthermore, aeglids take part in the matter–energy exchange between aquatic and terrestrial systems. This occurs when they are preyed upon by bird species such as the imperial shags or kingfisher (Casaux et al. 2010; Ballarini et al. 2013). In addition, aeglids are part of the trophic spectrum for some mammals and amphibians (Medina 1998; Pardini 1998). Additionally, aeglid species have been mentioned as prey of caimans (Bond Buckup and Buckup 1994), while some aeglids have been found in fish stomachs (Ferriz 1988, 1993, 2000; Moresco and Bemvenuti 2005; Kütter et al. 2009; Rodrigues et al. 2012), contributing considerably to the diet in terms of biomass eaten though not in number consumed (Klink and Ekman 1985; Juncos et al. 2014) (Figure 4.9).

BIBLIOGRAPHY

Al-Mohanna, S. Y., and J. A. Nott. 1987. R-cells and the digestive cycle in *Penaeus semisulcatus* (Crustacea: Decapoda). *Marine Biology* 95(1):129–37.

Anger, K. 1995. The conquest of freshwater and land by marine crabs: adaptations in life-history patterns and larval bioenergetics. *Journal of Experimental Marine Biology and Ecology* 193:119–45.

Anger, K. 2006. Contributions of larval biology to crustacean research: a review. *Invertebrate Reproduction and Development* 49:175–205.

Aréchiga, H., and L. Rodríguez-Sosa. 1997. Coupling of environmental and endogenous factors in the control of rhythmic behavior in decapod crustaceans. *Journal of the Marine Biological Association of the United Kingdom* 77:17–29.

Bahamonde, N., and M. T. López. 1961. Estudios biológicos en la población de *Aegla laevis laevis* (Latreille) del Monte (Crustacea, Decapoda, Anomura). *Investigaciones Zoológicas Chilenas* 7:19–58.

Ballarini, Y., M. R. Frizzas, and M. Â. Marini. 2013. Stomach contents of Brazilian non-passerine birds. *Revista Brasileira de Ornitologia* 21(4):235–42.

Baumart, J. S. 2014. Dinâmica espacial, migração e preferência de micro-habitat de *Aegla longirostri* Bond-Buckup e Buckup, 1994 (Crustacea, Anomura, Aeglidae). PhD diss., Universidade Federal de Santa Maria, RS, Brasil.

Begon, M., J. L Harper, and C. R. Townsend. 1995. *Ecología: Individuos, poblaciones y comunidades*. Ediciones Omega, Barcelona.

Blanchette, M. L., A. M. Davis, T. D. Jardine, and R. G. Pearson. 2014. Omnivory and opportunism characterize food webs in a large dry-tropics river system. *Freshwater Science* 33(1):142–58.

Bond-Buckup, G., and L. Buckup. 1994. A família Aeglidae (Crustacea, Decapoda, Anomura). *Arquivos de Zoologia* 32:159–346.

Bond-Buckup, G., A. A. P. Bueno, and K. Keunecke. 1996. O primeiro estágio juvenil de *Aegla prado*, Schmitt, 1942 (Crustacea, Anomura, Aeglidae). *Revista Brasileira de Zoologia* 13:1049–61.

Bond-Buckup, G., C. G. Jara, M. Pérez-Losada, L. Buckup, and K. A. Crandall. 2008. Global diversity of crabs (Aeglidae: Anomura: Decapoda) in freshwater. *Hydrobiologia* 595:267–73.

Bott, R. 1969. Die Süsswasserkrabben Süd-Amerikas und ihre Stammesgeschichte. Eine Revision der Trichodactylidae und der Pseudothelphusidae östlich der Anden (Crustacea, Decapoda). *Abhandlungen der Senckenbergischen Naturforschenden Gesellschaft* 518:1–94.

Bozza, D. C. 2016. Regulação osmo-iônica da espécie dulcícola *Aegla schmitti* (Crustacea, Anomura, Aeglidae) submetida a choque hiper-salino. Magister diss., Universidade Federal do Paraná, Curitiba, Brazil.

Buck, T. L., G. A. Breed, S. C. Pennings, M. E. Chase, M. Zimmer, and T. H. Carefoot. 2003. Diet choice in an omnivorous saltmarsh crab: different food types, body size, and habitat complexity. *Journal of Experimental Marine Biology and Ecology* 292:103–16.

Bücker, F., R. Goncalves, G. Bond-Buckup, and A. S. Melo. 2008. Effect of environmental variables on the distribution of two freshwater crabs (Anomura: Aeglidae). *Journal of Crustacean Biology* 28(2):248–51.

Buckup, L., B. K. Dutra, F. P. Ribarcki, F. A. Fernandes, C. K. Noro, G. T. Oliveira, and A. S. Vinagre. 2008. Seasonal variations in the biochemical composition of the crayfish *Parastacus defossus* (Crustacea, Decapoda) in its natural environment. *Comparative Biochemistry and Physiology, Part A* 149:59–67.

Bueno, A. A. 2003. Dinâmica da alimentacao em Aeglidae em ambiente natural (Crustacea, Decapoda, Anomura). PhD. diss., Universidad Federal de Rio Grande do Sul, Porto Alegre, Brazil.

Bueno, A. A. P., and G. Bond-Buckup. 1996. Os estágios iniciais de *Aegla violacea* Bond-Buckup & Buckup, 1994 (Crustacea, Decapoda, Anomura). *Nauplius* 4:39–47.

Bueno, A. A. P., and G. Bond-Buckup. 2004. Natural diet of *Aegla platensis* Schmitt and *Aegla ligulata* Bond-Buckup & Buckup (Crustacea, Decapoda, Aeglidae) from Brazil. *Acta Limnologica Brasiliensia* 16:115–27.

Bueno, S. L. S., R. M. Shimizu, and J. C. B. Moraes. 2016. A remarkable anomuran: The taxon *Aegla* Leach, 1820. Taxonomic remarks, distribution, biology, diversity and conservation. In *A global overview of the conservation of freshwater decapod crustaceans*, ed. T. Kawai, and N. Cumberlidge, 23–64. Heidelberg: Springer.

Burns, J. W. 1972. The distribution and life history of South American freshwater crabs (*Aegla*) and their role in trout streams and lakes. *Transactions of the American Fisheries Society* 4:595–607.

Burres, E. D. 2012. Food web structure of a subtropical South American stream with particular focus on the co-evolution of form and function in an endemic species flock. Master of Science diss., Appalachian State Univ., Boone, United States.

Burres, E. D., M. M. Gangloff, and L. Siefferman. 2013. Trophic analysis of two subtropical South American freshwater crabs using stable isotope ratios. *Hydrobiologia* 702:5–13.

Burtin, P. 2003. Nutritional value of seaweeds. *Electronical Journal of Environmental Agricultural and Food Chemistry* 2:198–201.

Carvalho, D. A., P. A. Collins, and C. J. De Bonis. 2011. Gut evacuation time of *Macrobrachium borellii* (Caridea: Palaemonidae) feeding on three types of prey from the littoral-benthic community *Journal of Crustacean Biology* 31(4):630–34.

Carvalho, D. A., P. A. Collins, and C. J. De Bonis. 2013. Predation ability of freshwater crabs: age and prey-specific differences in *Trichodactylus borellianus* (Brachyura: Trichodactylidae). *Journal of Freshwater Ecology* 28(4):573–84.

Casaux, R., M. L. Bertolin, M. A. Tartara, P. Alarcón, and G. Porro. 2010. The unexpected diet of breeding imperial shags (*Phalacrocorax atriceps*) at the Nahuel Huapi lake, Patagonia: implications on population trends? *Ornitologia Neotropical* 21:457–62.

Castro, T. S., and G. Bond-Buckup. 2003. The morphology of cardiac and pyloric foregut of *Aegla platensis* Schmitt (Crustacea: Anomura: Aeglidae). *Memoirs of Museum Victoria* 60(1): 53–7.

Castro T. S., and G. Bond-Buckup. 2004. O nicho trófico de duas espécies simpátricas de *Aegla* Leach (Crustacea, Aeglidae) no tributário da bacia hidrográfica do Rio Pelotas, Rio Grande do Sul, Brasil. *Revista Brasileira de Zoologia* 21(4):805–13.

Castro, P., P. J. E. Davie, D. Guinot, F. R. Schram, and J. C. von Vaupel Klein. 2015. Decapoda: Brachyura (Part 1) Vol. IX, Part C-1 Decapoda: Brachyura. In *Traité de zoologie – Anatomy, taxonomy, biology the Crustacea*, ed. Series editor P.-P. Grassé, 1–638. Leiden Boston: Brill Publishers.

Cerezer, C., C. Biasi, G. B. Cogo, and S. Santos. 2016. Avoid predation or take risks in basic activities? Predator–prey relationship in subtropical streams between decapods and caddisflies. *Marine and Freshwater Research* 67(12):1880–87.

Ciancio, J., and M. Pascual. 2006. Energy density of freshwater Patagonian organisms. *Ecología Austral* 16:1–9.

Cogo, G. B., C. Biasi, and S. Santos. 2014. The effect of the macroconsumer *Aegla longirostri* (Crustacea, Decapoda) on the invertebrate community in a subtropical stream. *Acta Limnologica Brasiliensia* 26(2):143–53.

Cogo, G. B., and S. Santos. 2013. The role of aeglids in shredding organic matter in neotropical streams. *Journal of Crustacean Biology* 34:519–26.

Collins, P. 2005. A coexistence mechanism for two freshwater prawns in the Paraná river floodplain. *Journal of Crustacean Biology* 25(2):219–25.

Collins, P., and M. E. D'Alessandro. 2017. Decapods as functional food from a view of ecosystem service at Parana River basin, South America. Paper presented at the 10th Symposium for European Freshwater Science, Olomouc, Czech Republic.

Collins, P., V. Williner, and F. Giri. 2006. Trophic relationships in Crustacea Decapoda of a river with floodplain. In *Predation in organisms: A distinct phenomenon*, ed. A. M. T. Elewa, 59–86. Heidelberg: Springer Verlag.

Collins, P., V. Williner, and F. Giri. 2007. Littoral communities. Macrocrustaceans. In *The middle Paraná River: Limnology of a subtropical wetland*, eds. M. H. Iriondo, J. C. Paggi, and M. J. Parma, 277–301. Heidelberg: Springer Verlag.

Collins, P., V. Williner, and F. Giri. 2011. Biogeography of the freshwater decapods in the La Plata Basin, South America. *Journal of Crustacean Biology* 31:179–91.

Colpo, K. D., L. Colpo Ribeiro, B. Wesz, and L. O. Oliveira Ribeiro. 2012. Feeding preference of the South American endemic anomuran *Aegla platensis* (Decapoda, Anomura, Aeglidae). *Naturwissenschaften* 99:333–36.

Covich, A. P., R. A. Crowl, and F. N. Scatena. 2003. Effects of extreme low flows on freshwater shrimps in a perennial tropical stream. *Freshwater Biology* 48:1199–206.

D'Abramo, L. R., and D. E. Conklin. 1995. New developments in the understanding of the nutrition of penaeid and caridean species of shrimp. In *Swimming through troubled waters*, ed. C. L. Browdy, and J. S. Hopkings, 95–107. Proceedings of special session on shrimp farming.

Dall, W., and D. J. W. Mortiary. 1983. Functional aspects of nutrition and digestion. In *The biology of Crustacea: internal anatomy and physiological regulation*, Vol. 5, ed. L. H. Mantel, 215–61. New York: Academic Press.

Daly, B., J. Swingle, and G. Eckert. 2013. Dietary astaxanthin supplementation for hatchery-cultured red king crab, *Paralithodes camtschaticus*, juveniles. *Aquaculture Nutrition* 19:312–20.

Dalziel, I. W. D. 1991. Pacific margins of *Laurentia* and East Antarctica-Australia as a conjugate rift pair: Evidence and implications for an Eocambrian supercontinent. *Geology* 19:598–601.

del Río, C. J., and S. A. Martínez Chiappara. 1998. El Mioceno marino en la Argentina y en el Uruguay. *Monografías de la Academia Nacional de Ciencias Exactas, Físicas y Naturales* 1:6–25.

Devercelli, M., and V. Williner. 2006. Diatom grazing by *Aegla uruguayana* (Decapoda: Anomura: Aeglidae): digestibility and cell viability after gut passage. *Annales of Limnology - International Journal of Limnology* 42(2):73–77.

Diawol, V. 2018. Ajustes biológicos de crustáceos de la familia Aeglidae (Decapoda, Anomura) en distintos ambientes de la Argentina. PhD diss., Universidad Nacional del Litoral, Santa Fe, Argentina.

Diawol, V., and P. Collins. 2012. Caracterización de los estadios del ciclo de muda del pseudocangrejo dulceacuícola *Aegla uruguayana* Schmitt, 1942 (Decapoda, Anomura). *Natura Neotropicalis* 43(1y2):21–29.

Diawol, V., F. Giri, and P. Collins. 2015. Shape and size variations of *Aegla uruguayana* (Anomura, Aeglidae) under laboratory conditions: A geometric morphometric approach to the growth. *Iheringia Série Zoologia* 105(1):76–83.

Diawol, V., V. Torres, and P. Collins. 2016. Field evaluation of oxygen consumption by two freshwater decapod morphotypes (Trichodactylidae and Aeglidae); the effect of different times of the day, body weight and sex. *Marine and Freshwater Behaviour and Physiology* 49(4):251–63.

Dutra, B. K., C. Zank, K. M. da Silva, M. R. Conter, and G. T. Oliveira. 2008. Seasonal variations in the intermediate metabolism of the crayfish *Parastacus brasiliensis* (Crustacea, Decapoda, Parastacidae) in the natural environment and experimental culture. *Iheringia Série Zoologia* 98(3):355–61.

Eubanks, M. D., J. D. Styrsky, and R. E. Denno. 2003. The evolution of omnivory in heteropteran insects. *Ecology* 84: 2549–56.

Feijoó, C., L. Leggieri, C. Ocón, et al. 2014. Stoichiometric homeostasis in the food web of a chronically nutrient-rich stream. *Freshwater Science* 33(3):820–31.

Feldmann, R. M. 1984. *Haumuriaegla glaessneri* n. gen. and sp. (Decapoda; Anomura; Aeglidae) from Haumurian (Late Cretaceous) rocks near Cheviot, New Zealand. New Zealand *Journal of Geology, and Geophysics* 27:379–85.

Feldmann, R. M., F. J. Vega, S. P. Applegate, and G. A. Bishop. 1998. Early Cretaceous arthropods from the Tlayúa formation at Tepexi de Rodríguez, Puebla, México. *Journal of Paleontology* 72:79–90.

Felgenhauer, B. E., and L. G. Abele. 1989. Evolution of the foregut in the lower Decapoda. In *Functional morphology of feeding and grooming in Crustacea*, Crustacean Issues 10, eds. Felgenhauer, B. E., L. Watling, and A. B. Thistle, 205–19. Rotterdam: A. A. Balkema.

Ferriz, R. 1988. Relaciones tróficas de Trucha Marrón, *Salmo fario* Linné, y Trucha Arco Iris, *Salmo gairdneri* Richardson, (Osteichthyes, Salmoniformes) en un embalse norpatagónico. *Studies on Neotropical Fauna and Environment* 23(3):123–31.

Ferriz, R. 1993. Algunos aspectos de la dieta de cuatro especies ícticas del río Limay (Argentina). *Revista de Ictiología* 2/3(1/2):1–7.

Ferriz, R. 2000. Feeding of *Percichthys colhuapiensis* Mac Donagh 1955 (Pisces: Percichthyidae) in the river Negro upper basin, Argentina. *Bioikos* 14(1):44–48.

Francis, J. E., D. Pirrie, and J. A. Crame. 2006. *Cretaceous–Tertiary high-latitude palaeoenvironments: James Ross Basin Antarctica.* London: Geological Society Special Publications N° 258.

Frost, P. C., W. F. Cross, and J. P. Benstead. 2005. Ecological stoichiometry in freshwater benthic ecosystems: an introduction. *Freshwater Biology* 50:1781–85.

Gibbs, S. J., T. J. Bralower, P. R. Bown, J. C. Zachos, and L. M. Bybell. 2006. Shelf and open-ocean calcareous phytoplankton assemblages across the Paleocene-Eocene Thermal Maximum: Implications for global productivity gradients. *Geology* 34(4):233–36.

Glas, P. S., L. A. Courtney, J. R. Rayburn, and W. S. Fisher. 1997. Embryonic coat of the grass shrimp *Palaemonetes pugio. Biological Bulletin* 192(2):231–42.

Graça, M. A. S. 2001. The role of invertebrates on leaf litter decomposition in streams – a review. *International Review of Hydrobiology* 86: 383–93.

Greenaway, P. 2003. Terrestrial adaptations in the Anomura (Crustacea: Decapoda). *Memoirs of Museum Victoria* 60(1):13–26.

Icely, J. D., and J. A. Nott. 1992. Digestion and absorption: digestive system and associated organs. In *Microscopic anatomy of invertebrates: decapod, Crustacea, Vol. 10*, eds. Harrison, F. W., and A. G. Humes, 147–201. New York: Wily-Liss Inc.

Jackson, M. C., J. Grey, K. Miller, J. R. Britton, and I. Donohue. 2016. Dietary niche constriction when invaders meet natives: evidence from freshwater decapods. *Journal of Animal Ecology* 85:1098–107.

Juncos, R., D. Milano, P. J. Macchi, and P. H. Vigliano. 2014. Niche segregation facilitates coexistence between native and introduced fishes in a deep Patagonian lake. *Hydrobiologia* 747:53–67.

Kelly, D. C. 2002. Response of Antarctic (ODP Site 690) planktonic foraminifera to the Paleocene–Eocene thermal maximum: Faunal evidence for ocean/climate change. *Paleoceanography* 17(4):1071.

Kennet, J. P. 1977. Cenozoic evolution of Antarctic glaciation, the Circum-Antarctic Ocean, and their impact on global paleoceanography. *Journal of Geophysical Research* 82(27):3843–60.

Klink, A., and R. Ekman. 1985. Age and growth, feeding habits, and reproduction of *Cauque mauleanum* Steindachner 1896) (Pisces: Atherinidae) in Southern Chile. *Studies on Neotropical Fauna and Environment* 20(4):239–49.

Krivan, V., and S. Diehl. 2005. Adaptive omnivory and species coexistence in tri-trophic food webs. *Theoretical Population Biology* 67:85–99.

Kütter, M. T., M. de Azevedo Bemvenuti, and A. Moresco. 2009. Feeding strategy of the jundiá *Rhamdia quelen* (Siluriformes, Heptapteridae) in costal lagoons of southern Brazil. *Acta Scientiarum-Biological Sciences* 31(1):41–47.

Lara, G. P., and C. A. Moreno. 1995. Efectos de la depredación de *Aegla abtao* (Crustacea, Aeglidae) sobre la distribución espacial y abundancia de *Diplodon chilensis* (Bivalvia, Hyriidae) en el Lago Panguipulli, Chile. *Revista Chilena de Historia Natural* 68:12–129.

Laughlin, R. A. 1982. Feeding habits of the crab, *Callinectes sapidus* Rathbun, in the Apalachiola estuary, Florida. *Bulletin of Marine Science* 32:807–22.

Lehmann, U., and G. Hillner. 1983. *Fossil invertebrates*. Cambridge: Cambridge Univ. Press.

Lim, K. C., F. M. Yusoff, M. Shariff, et al. 2018. Astaxanthin as feed supplement in aquatic animals. *Reviews in Aquaculture* 10(3):738–73.

Lodge, D. M., M. W. Kershner, J. E. Aloi, and A. P. Covich. 1994. Effects of an omnivorous crayfish (*Orconectes rusticus*) on a freshwater littoral food web. *Ecology* 75(5):1265–81.

Lopes, M. N., J. P. Vieira, and M. D. M. Burns. 2009. Biofouling of the golden mussel *Limnoperna fortunei* (Dunker, 1857) over the Anomura crab *Aegla platensis* Schmitt, 1942. *Pan-American Journal of Aquatic Sciences* 4(2):222–25.

Lovrich, G., and M. Thiel. 2011. Ecology, physiology, feeding and trophic role of squat lobsters. In *The biology of squat lobsters*, ed. G. C. B. Poore, S. T. Ahyong, and J. Taylor, 183–222. Boca Raton USA: CRC Press.

Magni, S. T., and V. Py-Daniel. 1989. *Aegla platensis* Schmitt, 1942 (Decapoda: Anomura) um predador de imaturos de Simuliidae (Diptera: Culicomorpha). *Revista de Saúde Pública* 23:258–59.

Majdi, N., and W. Traunspurger. 2015. Free-living nematodes in the freshwater food web: A review. *Journal of Nematology* 47(1):28–44.

Manattini, M. C., V. Williner, and D. Carvalho. 2017. Dieta natural de *Aegla ringueleti* (Decapoda, Aeglidae) del rio Calchaquí, Cachi, Salta. Paper presented at the 80° Reunión de Comunicaciones Científicas de la Asociación de Ciencias Naturales del Litoral, Santa Fe, Argentina.

Mancinelli, G., F. Sangiorgio, and A. Scalzo. 2013. The effects of decapod crustacean macroconsumers on leaf detritus processing and colonization by invertebrates in stream habitats: A meta-analysis. *International Review of Hydrobiology* 98:206–16.

Mao, Z., X. Gu, and Q. Zeng. 2016. Food sources and trophic relationships of three decapod crustaceans: insights from gut contents and stable isotope analyses. *Aquaculture Research* 47(9):2888–98.

Martin, J. W., and L. G. Abele. 1988. External morphology of the genus *Aegla* (Crustacea: Anomura: Aeglidae). *Smithsonian Contributions to Zoology* 453:1–46.

Martin, J. W., and B. E. Felgenhauer. 1986. Grooming behaviour and the morphology of grooming appendages in the endemic South American crab genus *Aegla* (Decapoda, Anomura, Aeglidae). *Journal of Zoology* 209:213–24.

McLaughlin, P. A., R. Lemaitre, and U. Sorhannus. 2007. Hermit crab phylogeny: A reappraisal and its "fall-out". *Journal of Crustacean Biology* 27:97–115.

Medina, G. 1998. Seasonal variations and changes in the diet of southern river otter in different freshwater habitats in Chile. *Acta Theriologica* 43(3):285–92.

Mezzomo, N., and S. R. S. Ferreira. 2016. Carotenoids functionality, sources, and processing by supercritical technology: A review. *Journal of Chemistry* 2016:1–16.

Miserendino, M. L. 2001. Macroinvertebrate assemblages in Andean Patagonian rivers and streams: environmental relationships. *Hydrobiologia* 444:147–58.

Moresco, A., and M. A. Bemvenuti. 2005. Morphologic features and feeding analysis of the black catfish *Trachelyopterus lucenai* Bertoletti, Pezzi da Silva & Pereira (Siluriformes, Auchenipteridae). *Acta Limnologica Brasiliensia* 17(1):37–44.

Morrone, J. J. 2001. *Biogeografía de América Latina y el Caribe*. M&T–Manuales & Tesis SEA, Zaragoza.

Musin, G. E. 2018. Rol trófico de los crustáceos decápodos dulciacuícolas: Buscando respuestas desde una perspectiva fisiológica. PhD diss., Universidad Nacional del Litoral, Santa Fe, Argentina.

Musin, G. E., A. Rossi, V. P. Diawol, P. A. Collins, and V. Williner. 2017. Dynamic metabolic pattern of *Aegla uruguayana* (Schmitt, 1942) (Decapoda: Anomura: Aeglidae):

responses to seasonality and ontogeny in a temperate freshwater environment. *Journal of Crustacean Biology* 37(4): 436–44.

Musin, G. E., A. Rossi, V. P. Diawol, P. A. Collins, and V. Williner. 2018. Development of enzymes during ontogeny of two freshwater Decapoda: *Aegla uruguayana* (Aeglidae) and *Macrobrachium borellii* (Palaemonidae). *Aquaculture Research* 49(12):3889–97.

Negro, L., E. Senkman, M. Montagna, and P. Collins. 2011. Freshwater decapods and pesticides: an unavoidable relation in the modern world. In *Pesticides in the Modern World/ Book 3*, ed. M. Stoytcheva, 197–226. London: Intech Open Access Publishers.

Oliveira, G. T., F. A. Fernandes, G. Bond-Buckup, A. A. Bueno, and R. S. M. Silva. 2003. Circadian and seasonal variations in the metabolism of carbohydrates in *Aegla ligulata* (Crustacea: Anomura: Aeglidae). *Memoirs Museum Victoria* 60:59–62.

Oliveira, G. T., F. A. Fernandes, A. A. P. Bueno, and G. Bond-Buckup. 2007. Seasonal variations in the intermediate metabolism of *Aegla platensis* (Crustacea, Aeglidae). *Comparative Biochemical and Physiology* A 147:600–06.

Oliveira G. T., C. Hack, M. Almerão, G. Bond-Buckup, and B. K. Dutra. 2011. Tissue composition and haemolymphatic metabolites during gonadal development in *Aegla platensis* (Crustacea, Decapoda) maintained in experimental culture. *Revista Brasileira de Biociências* 9(1):64–71.

Oyanedel Pérez A. B. 2015. Refugios glaciales pleistocénicos de invertebrados bentónicos en Patagonia: Áreas prioritarias para la conservación de biodiversidad dulceacuícola. PhD. Diss., Universidad de Concepción, Concepción, Chile.

Paibulkichakul, C., S. Piyatiratitivorakul, P. Sorgeloos, and P. Menasveta. 2008. Improved maturation of pond-reared, black tiger shrimp (*Penaeus monodon*) using fish oil and astaxanthin feed supplements. *Aquaculture* 282:83–89.

Pardini, R. 1998. Feeding ecology of the neotropical river otter *Lontra longicaudis* in an Atlantic Forest stream, south-eastern Brazil. *Journal of Zoology* 245:385–91.

Pérez-Losada, M., G. Bond-Buckup, C. G. Jara, and K. A. Crandall. 2004. Molecular systematics and biogeography of the southern South American freshwater crabs *Aegla* (Decapoda: Anomura: Aeglidae) using multiple heuristic tree search approaches. *Systematic Biology* 53:1–14.

Pérez-Losada, M., C. G. Jara, G. Bond-Buckup, M. L. Porter, and K. A. Crandall. 2002. Phylogenetic position of the freshwater anomuran family Aeglidae. *Journal of Crustacean Biology* 22:670–76.

Potter, P. E. 1997. The Mesozoic and Cenozoic paleodrainage of South America: a natural history. *Journal of South American Earth Sciences* 10:331–44.

Poulsen, C. J., A. S. Gendaszek, and R. L. Jacob. 2003. Did the rifting of the Atlantic Ocean cause the Cretaceous thermal maximum? *Geology* 31(2):115–18.

Ribeiro, A. C. 2006. Tectonic history and the biogeography of the freshwater fishes from the coastal drainages of eastern Brazil: an example of faunal evolution associated with a divergent continental margin. *Neotropical Ichthyology* 4(2):225–46.

Rodrigues, L. R., N. F. Fontoura, and D. M. Marques. 2012. Feeding dynamics of *Oligosarcus jenynsii* (Günther, 1864) in a subtropical coastal lake assessed by gut-content analysis and stable isotopes. *International Journal of Plant, Animal and Environment Sciences* 2(2):126–34.

Rodrigues, W., and N. J. Hebling. 1978. Estudos biologicos em *Aegla perobae* Hebling & Rodrigues, 1977 (Decapoda, Anomura). *Revista Brasileira de Biologia* 38(2): 383–90.

Rosas, C., O. Carrillo, R. Wilson, and E. R. Andreatta. 2006. *Estado actual y perspectivas de la nutrición de los camarones peneidos cultivados en Iberoamérica.* CYTED México.

Sacristán, H. J., L. M. Franco-Tadic, and L. S. López-Greco. 2013. Influence of feeding on the circadian rhythm of digestive enzymes in cultivated juveniles of the freshwater crayfish *Cherax quadricarinatus* (Parastacidae). *Latin American Journal of Aquatic Research* 41(4):753–61.

Sacristán, H. J., Y. E. Rodriguez, A. N. Pereira, L. S. López Greco, G. A. Lovrich, and A. V. Fernández Gimenez. 2017. Energy reserves mobilization: Strategies of three decapod species. *PLoS ONE* 12(9):e0184060.

Santos, S., L. Ayres-Peres, R. C. F. Cardoso, and C. C. Sokolowicz. 2008. Natural diet of the freshwater anomuran *Aegla longirostri* (Crustacea, Anomura, Aeglidae). *Journal of Natural History* 42:13–14.

Savaris, M., S. Lampert, A. Trevisan, and S. Masunari. 2012. Opportunistic predation of fish by anomuran crabs (Crustacea, Anomura, Aeglidae) in rivers of southern Brazil. *Biota Neotropical* 12(4):248–51.

Schnabel, K. E., and S. T. Ahyong. 2010. A new classification of the Chirostyloidea (Crustacea: Decapoda: Anomura). *Zootaxa* 2687:56–64.

Schnabel, K. E., S. T. Ahyong, and E. W. Maas. 2011. Galatheoidea are not monophyletic—molecular and morphological phylogeny of the squat lobsters (Decapoda: Anomura) with recognition of a new superfamily. *Molecular Phylogenetics and Evolution* 58:157–68.

Schweitzer, C. E., and R. M. Feldmann. 2005. Decapod crustaceans. The K/P event and Paleocene recovery. In *Crustacean and arthropod relationships*, ed. S. Koenemann, and R. Jenner, 17–53. Rotterdam: Balkema/CRC Press/Taylor & Francis.

Senkman, E., L. Negro, E. Lopretto, and P. Collins. 2015. Reproductive behavior of three species of freshwater crabs of the family Trichodactylidae (Crustacea: Decapoda) including forced copulation by males. *Marine and Freshwater Behaviour and Physiology* 48(2):77–78.

Simpson, G. G. 1953. *The major features of evolution.* Columbia University Press, New York.

Sokolowicz, C. C., L. Ayres-Peres, and S. Santos. 2007. Atividade nictimeral e tempo de digestão de *Aegla longirostri* (Crustacea, Decapoda, Anomura). *Iheringia Série Zoologia* 97(3):235–38.

Sousa, L. G., and A. M. Petriella. 2001. Changes in the hepatopancreas histology of *Palaemonetes argentinus* (Crustacea, Caridea) during moult. *Biocell* 25(3): 275–81.

Trajano E. 1987. Fauna cavernicola brasileira: Composição e caracterização preliminar. *Revista brasileira de Zoologia* 3(8):533–61.

Trevisan, A., Z. M. Murilo, and S. Masunaria. 2014. Circadian rhythm in males of *Aegla schmitti* (Decapoda, Anomura, Aeglidae) under laboratory conditions. *Biological Rhythm Research* 45:803–16.

Tsang, L. M., T. -Y. Chan, S. T. Ahyong, and K. H. Chu. 2011. Hermit to king, or hermit to all: Multiple transitions to crab-like forms from hermit crab ancestors. *Systematic Biology* 60:616–29.

Tudge, C. C., A. Asakura, and S. T. Ahyong. 2012. Infraorder Anomura MacLeay, 1838. In *Traité de Zoologie – Anatomy, taxonomy, biology the Crustacea. Vol IX Part B Eucarida Decapoda: Astacidea (Enoplometopoidea, Nephropoidea), Glypheidea, Axiidea, Gebiidea, and Anomura*, ed. F. R. Schram and J. C. von Vaupel Klein, Series editor P.-P. Grassé, 221–333. Leiden Boston: Brill Publishers.

Tudge, C. C., and D. M. Scheltinga. 2002. Spermatozoal morphology of the freshwater anomuran *Aegla longirostri* Bond-Buckup & Buckup, 1994 (Crustacea: Decapoda: Aeglidae) from South America. *Proceedings of the Biological Society of Washington* 115:118–28.

Tumini, G., F. Giri, V. Williner, and P. Collins. 2016. The importance of biogeographical history and extant environmental conditions as drivers of freshwater decapod distribution
in southern South America. *Freshwater Biology* 61:715–28.

Tumini, G., F. Giri, V. Williner, P. Collins, and J. J. Morrone. 2018. Distributional patterns of endemic southern South American freshwater aeglids (Crustacea: Decapoda:
Anomura: Aeglidae). *Zoologischer Anzeiger* 277:55–64.

Vannote, R. L., G. W. Minshall, K. W. Cummins, J. R. Sedell, and C. E. Cushing. 1980.
The river continuum concept. *Canadian Journal of Fisheries and Aquatic Sciences*
37:130–37.

Vogt, G. 1994. Life-cycle and functional cytology of the hepatopancreatic cells of *Astacus
astacus* (Crsutacea: Decapoda). *Zoomorphology* 114(2): 83–101.

Vogt, G. 2013. Abbreviation of larval development and extension of brood care as key features of the evolution of freshwater Decapoda. *Biological Review* 88:81–116.

Wade, N. M., S. Cheers, N. Bourne, et al. 2015. Dietary astaxanthin levels affect colour,
growth, carotenoid digestibility and the accumulation of specific carotenoid esters in
the Giant Tiger Shrimp, *Penaeus monodon. Aquaculture Research* 48:395–406.

Williner, V. 2007. Ecología trófica de poblaciones de especies de la familia Aeglidae en la
Argentina. PhD diss., Universidad Nacional de La Plata, La Plata, Argentina.

Williner, V. 2010. Foregut ossicles morphology and feeding of the freshwater anomuran crab
Aegla uruguayana (Decapoda, Aeglidae). *Acta Zoologica* 91:408–15.

Woodward, G., and A. G. Hildrew. 2002. Food web structure in riverine landscapes.
Freshwater Biology 47:777–98.

Wouters, R., P. Lavens, J. Nieto, and P. Sorgeloos. 2001. Penaeid shrimp broodstock nutrition:
an updated review on research and development. *Aquaculture* 202:1–21.

Yeo, D. C. J., P. K. L. Ng, N. Cumberlidge, C. Magalhães, S. R. Daniels, and M. R. Campos.
2008. Global diversity of crabs (Crustacea: Decapoda: Brachyura) in freshwater.
Hydrobiologia 595:275–86.

Zwirglmaier, K., W. D. K. Reid, J. Heywood, et al. 2015. Linking regional variation of epibiotic bacterial diversity and trophic ecology in a new species of Kiwaidae (Decapoda,
Anomura) from East Scotia Ridge (Antarctica) hydrothermal vents. *MicrobiologyOpen*
4(1):136–50.

CHAPTER 5

Reproductive Biology and Gonadal Development in Aeglidae

Carolina Sokolowicz

CONTENTS

5.1 INTRODUCTION

The early studies on the genus *Aegla* focused mostly on general morphology, systematics, and taxonomy (Schmitt 1942; Ringuelet 1949; Lopretto 1978a, b, 1980; Buckup and Rossi 1979; Martin and Abele 1988; Bond-Buckup and Buckup 1994). Reproductive aspects were always part of biological surveys, which generally included information about the reproductive period, fecundity, and other baseline biological data (Bahamonde and López 1961; López 1965; Rodrigues and Hebling 1978). The evaluation of specific reproductive features of *Aegla* started in the 1990s, focusing on the description of juveniles, population structure, and reproductive biology of species from southern and southeastern Brazil (Bond-Buckup et al. 1996, 1999). Since the 2000s, many studies on reproductive aspects of the genus have been done (Bueno and Bond-Buckup 2000; Swiech-Ayoub and Masunari 2001; Noro and Buckup 2002; Fransozo et al. 2003; Sokolowicz et al. 2006, 2007; Viau et al. 2006; Bueno and Shimizu 2008, 2009; Collins et al. 2008; Almerão et al. 2010;

Rocha et al. 2010; Oliveira et al. 2011; Oliveira and Santos 2011; Trevisan et al. 2012; Trevisan and Santos 2012; Grabowski et al. 2013; Siqueira et al. 2013; Copatti et al. 2015, 2016; Da Silva et al. 2016, 2017; Takano et al. 2016).

In this chapter I provide a broad review of studies on different reproductive aspects of *Aegla*. Later, a more detailed description of the reproductive morphology and physiological aspects of Aeglidae will be given.

5.2 GENERAL REPRODUCTIVE ASPECTS

Aeglids are dioecious, the sperm transfer is achieved through the copula, and females incubate the eggs in a brood chamber (Martin and Abele 1988; Almerão et al. 2010). Female gonopores are located on the coxae of the third pereopods. Male gonopores are located at the tip of a tube-like extension (sexual tube) on the coxal segment of the reduced fifth pair of pereopods (Lopretto 1978a; Viau et al. 2006).

The sexual dimorphism in pleocyemate decapods is a conspicuous feature. Males are often larger than females since they grow continuously, while females allocate a large amount of energy to gonadal development and egg incubation, and do not molt during these processes (Hartnoll 1974). Both sexes often have other morphological differences that are closely related to reproductive functions. This morphological dimorphism becomes evident during growth, and the identification of traits that have ontogenetic allometric growth provides valuable information about which traits are related to reproductive advantages (Da Silva et al. 2017). Studies on the allometric growth of aeglids have increased during recent years and some patterns can be identified (Swiech-Ayoub and Masunari 2001; Viau et al. 2006; Bueno and Shimizu 2008, 2009; Rocha et al. 2010; Trevisan et al. 2012; Trevisan and Santos 2012; Takano et al. 2016; Da Silva et al. 2017).

Male aeglids often have larger bodies and chelipeds, which reflect their different life history strategies: males with larger chelipeds are more successful in courtship rituals and post-copula behaviors (Almerão et al. 2010). The larger abdomen of females increases the space available for egg incubation (Viau et al. 2006). These dimorphic features are not present in males and females during all life stages: during ontogeny some body structures grow faster than others. Although there are references to differential morphotypes related to the maturity molt in male aeglids (Bueno and Shimizu 2009), a single pre-mating molt defining the allometric change in both males and females has not been recorded in Aeglidae.

Many studies evaluated the functional maturity, i.e., the actual capacity of each sex of performing reproductive events, which includes the morphometric and gonadal maturity (Viau et al. 2006). Although allometric growth evaluations indicate the average size at the onset of sexual morphometric maturity, the functional maturity occurs later than the morphometric maturity (Viau et al. 2006; Bueno and Shimizu 2008). The presence of eggs attached to the pleopods of females has been used to indicate the minimum size at the onset of functional maturity along with other information about the reproduction of several species: *A. castro* (Swiech-Ayoub and Masunari 2001; Takano et al. 2016), *A. franca* (see Bueno and Shimizu

2008), *A. grisella*, *A. ludwigi*, *A. platensis* (see Copatti et al. 2015), *A. parana* (see Grabowski et al. 2013), *A. strinatii* (see Rocha et al. 2010), and *A. uruguayana* (see Viau et al. 2006).

The size of the first pair of chelipeds and abdomen have been used as the main traits to characterize the differential growth of immature and mature stages, confirming the importance of these structures in reproductive events (Viau et al. 2006; Bueno and Shimizu 2008, 2009; Da Silva et al. 2017). The development of secondary sexual traits, used to define the morphometric maturity, occurs prior to the gonadal and functional maturity (Viau et al. 2006). This feature has been determined for several species from different geographical zones: *A. castro* (see Takano et al. 2016), *A. franca* (see Bueno and Shimizu 2008, 2009), *A. manuinflata* (see Trevisan and Santos 2012), *A. platensis*, *A. grisella*, *A. ludwigi* (see Copatti et al. 2015), and *A. uruguayana* (see Viau et al. 2006). The cephalothorax length (CL) at the onset of the morphometric maturity ranges from 10.91 mm in males and 10.03 mm in females of *A. castro* (see Takano et al. 2016) to 19.15 mm in males and 16.5 mm in females of *A. platensis* (see Oliveira and Santos 2011).

Besides confirming the maturity status of females, the ovigerous condition is used to determine the reproductive period of aeglid populations. Most species have a seasonal reproduction, often concentrated in colder months in different latitudes, such as *A. lepctoctyla* (Noro and Buckup 2002), *A. castro* (Fransozo et al. 2003), *A. franca* (Bueno and Shimizu 2008), *A. strinatii* (Rocha et al. 2010), *A. paulensis* (Cohen et al. 2011), *A. parana* (Grabowski et al. 2013), *A. perobae* (Bueno et al. 2014), and *A. marginata* (Da Silva et al. 2016). A continuous reproduction has been recorded for fewer species: *A. grisella* (see Copatti et al. 2015), *A. ludwigi* (see Copatti et al. 2015), *A. platensis* (see Bueno and Bond-Buckup 2000; Copatti et al. 2015), and *A. uruguayana* (see Viau et al. 2006). Several authors have linked the reproductive pattern of *Aegla* to a latitudinal variation in the distribution of these species (Rocha et al. 2010; Cohen et al. 2011; Grabowski et al. 2013; Bueno et al. 2014; Chiquetto-Machado et al. 2016). The opposite latitudinal paradigm has been proposed for Aeglidae, where the length of the breeding period increases with latitude (Bueno and Shimizu 2008). This differs from most marine benthic decapods in which the reproduction tends to be continuous in tropical species and seasonal in species from higher latitudes (Bauer 1992; Castilho et al. 2007). This trend is probably linked to environmental variables such as water temperature, food availability, and photoperiod (see Bueno and Shimizu 2008 for further information).

Female decapods, such as marine brachyurans, generally produce numerous but small eggs, which go through a relatively high number of planktonic stages. In aeglids, however, females incubate large, but few eggs; the development is direct with no planktonic stages (Mouchet 1932; Bueno and Bond-Buckup 1996). Maximum fecundity ranges from 113 eggs in *A. paulensis* (mean egg size 1.3 mm±0.006 mm) (López 1965) to 1043 eggs (mean egg size 1.2 mm±0.003 mm) in *A. rostrata* (see Jara 1977). A positive relationship between egg number and female size has been shown for *A. leptodactyla* (Noro and Buckup 2002), *A. franciscana* (Gonçalves et al. 2006), and *A. franca* (Bueno and Shimizu 2008).

Eggs of aeglids contain a large amount of yolk, and larval development takes place inside the egg and the hatchlings have a morphology similar to the adult (Bond-Buckup et al. 1996, 1999; Bueno and Bond-Buckup 1996). Aeglids have an extended brood care, which includes egg incubation and grooming as well as carrying hatchlings under the female abdomen (López-Greco et al. 2004; Vogt 2014). Along with the direct development, this type of maternal care allowed the colonization of freshwater environments and it is a strategy to increase reproductive efficiency. Thus, aeglid species can be considered as k-strategists since fewer offspring have improved survival rates (Vogt 2013, 2014). During egg incubation, a grooming behavior is performed by females, using their fifth pereopods to remove solid particles and organisms from the egg mass (Da Silva et al. 2016). After hatching, the offspring remain confined to the ventral abdominal surface of females for a few days. This period ranges from four days in *A. uruguayana* (see López-Greco et al. 2004) to up to 15 days in *A. castro* (see Swiech-Ayoub and Masunari 2001). During this period, juveniles move around the female but always return to the abdomen (López-Greco et al. 2004).

Studying the reproductive biology of a population requires the analysis of several aspects; each one will give separate answers that, when combined, provide a general description of a species reproductive profile. Reproductive period, maturity, fecundity, and juvenile development are features that have been widely studied in several species of Aeglidae: *A. castro* (Swiech-Ayoub and Masunari 2001; Takano et al. 2016), *A. franca* (see Bueno and Shimizu 2008, 2009), *A. grisella*, *A. ludwigi*, *A. georginae* (see Copatti et al. 2015, 2016), *A. leptodactyla* (see Noro and Buckup 2002), *A. manuinflata* (see Trevisan and Santos 2012), *A. marginata* (see Da Silva et al. 2016, 2017), *A. prado* (see Bond-Buckup et al. 1996), *A. parana* (see Grabowski et al. 2013), *A. paulensis* (see Moraes and Bueno 2013), *A. platensis* (Bond-Buckup et al. 1999; Bueno and Bond-Buckup 2000; Oliveira and Santos 2011; Copatti et al. 2015), *A. strinatii* (see Rocha et al. 2010), *A. uruguayana* (see Collins et al. 2008), and *A. violacea* (see Bueno and Bond-Buckup 1996). The available information about the gonadal morphology and histology and the physiological processes involved in gonadal maturation are limited to two species: *A. platensis* (Sokolowicz et al. 2006, 2007; Oliveira et al. 2011) and *A. uruguayana* (Viau et al. 2006; Castiglioni et al. 2009). The pattern, however, is probably similar in the other species of the genus, following environmental conditions faced by different species such as water temperature and food availability.

5.2.1 Gonadal Morphology and Histology

The paired ovary of decapod females has two oviducts, each one leading to one of the gonopores on the sternite or coxa of the third pereopod (López-Greco 2013). In Decapoda, structures used as seminal receptacles are diverse, varying from modifications of the ventral surface of females to internal storage structures (López-Greco 2013). Such structures have not been reported in Aeglidae. Males have paired testes and vas deferens leading to paired gonopores at the coxae of the fifth pereopods (López-Greco 2013).

In aeglids, the histologic maturity of a population can be inferred based on the gonadal maturation status. The gonadal maturity is reached when there are viable spermatids or spermatozoa in the vas deferens of males and vitellogenic oocytes in females (Sastry 1983; López-Greco and Rodríguez 1999). The maturity can also be evaluated based on the gonadal morphology, mainly the ovaries. The size, shape, and color of ovaries indicate different stages of gonadal development (Sokolowicz et al. 2007). This knowledge is an important criterion to evaluate the maturity of individuals in the field or for experiments in which mature females are needed since mature ovaries are visible through the ventral part of the abdomen (Sokolowicz et al. 2007; Bueno and Shimizu 2008; Almerão et al. 2010) (Figure 5.1).

Color and size changes of ovaries and testicles correspond to cellular changes; as the gonadal development proceeds, morphology also changes. The most complete information about the gonadal morphology and histology of Aeglidae refers to *A. uruguayana* (see Viau et al. 2006) and *A. platensis* (see Sokolowicz et al. 2007). The following description is based on these two species.

5.2.1.1 Females

The H-shaped ovaries are located in the cephalothorax and, during the most developed ovarian stage, the posterior lobes can reach the third abdominal somite (Sokolowicz et al. 2007). The anterior and posterior lobes are connected by an ovary portion located in the mid-dorsal cephalothorax region (Viau et al. 2006). The anterior portion of each posterior lobe is connected by a straight and narrow oviduct with many secretory cells (Figure 5.2A) ending in the genital pore on the coxa of the third pereopod (Viau et al. 2006; Sokolowicz et al. 2007). The anterior lobes are shorter and narrower than the posterior ones.

The ovary has three stages of germinative cells: oogonia, primary oocytes, and secondary oocytes. The oogonia have a scarce basophilic cytoplasm, and size ranges

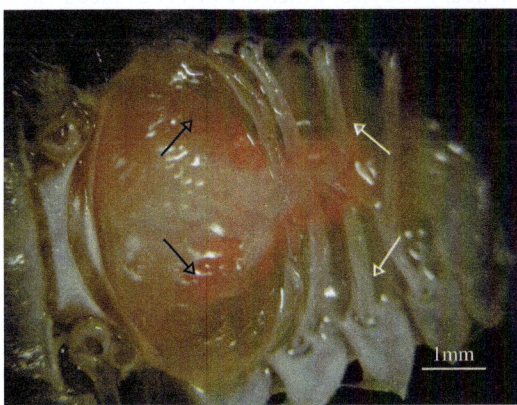

Figure 5.1   Ventral view of a female *Aegla platensis* abdomen; Stage III ovary (orange) is seen by transparency (black arrows). White arrows indicate the pleopods. (Adapted from Almerão [2005].)

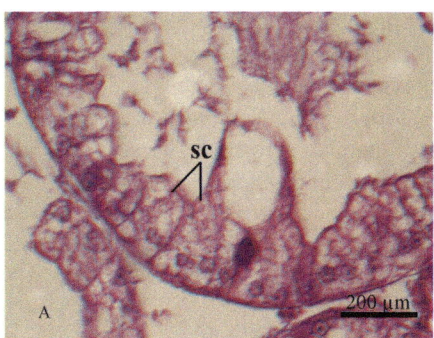
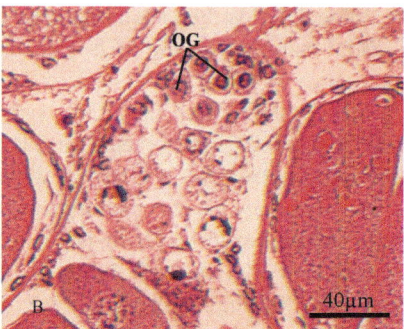

Figure 5.2 Histological sections of female *Aegla platensis*. A. Transverse section of oviduct. SC: secretory cells. B. Oogonia cluster in Stage III ovary. OG: oogonia. (Adapted from Sokolowicz et al. [2007].)

between 8.75 and 31.25 µm in *A. platensis*, including all ovary developmental stages (Sokolowicz et al. 2007). The high nucleus/cytoplasm rate is related to the small amount of yolk concentrated in these cells (Viau et al. 2006). The oogonia are often arranged in clusters located either centrally or peripherally in the ovary and surrounded by follicular cells (Figure 5.2B) (Viau et al. 2006; Sokolowicz et al. 2007).

As vitellogenesis proceeds, the yolk deposition increases in the germinative cells. The primary oocytes, also known as previtellogenic oocytes, have homogeneous basophilic cytoplasm and are often surrounded by oogonia; smaller oocytes may appear inside the oogonia clusters (Sokolowicz et al. 2007). The diameter varies from 17.5 to 172.5 µm in *A. platensis*, including all ovary developmental stages (Sokolowicz et al. 2007). Larger primary oocytes are surrounded by follicle cells, characterizing the beginning of folliculogenesis (Sokolowicz et al. 2007) (Figure 5.3B). Secondary oocytes, which are in an advanced vitellogenic stage, have an eosinophilic cytoplasm full of yolk and are larger than the primary oocytes. Their diameters vary from 95 to 770 µm in *A. platensis*, including all developmental stages (Figure 5.3C and 5.3D) (Sokolowicz et al. 2007).

During the folliculogenesis, follicle cells completely surround the secondary oocytes to assist the process of exogenous uptake of vitellogenin (vitellogenesis), characterizing the beginning of gonadal maturation (Viau et al. 2006). As this process proceeds, cells become larger and the nucleus/cytoplasm rate decreases; yolk droplets can be observed in the cytoplasm resultant from the yolk uptake from the haemolymph (Viau et al. 2006). A rematuration process has been described for the ovary of *A. platensis*, characterized by a disorganized tissue and cellular resorption in which the oogonia are scattered throughout the ovary and vitellogenic oocytes migrate to the periphery (Sokolowicz et al. 2007).

All these changes in the histology of the ovaries are reflected by simultaneous color and size changes of the gonad. Immature ovaries are colorless and posterior lobes are short and thin, while ripe ovaries are bright orange/red with posterior lobes occupying the entire abdominal extension (Viau et al. 2006; Sokolowicz et al. 2007).

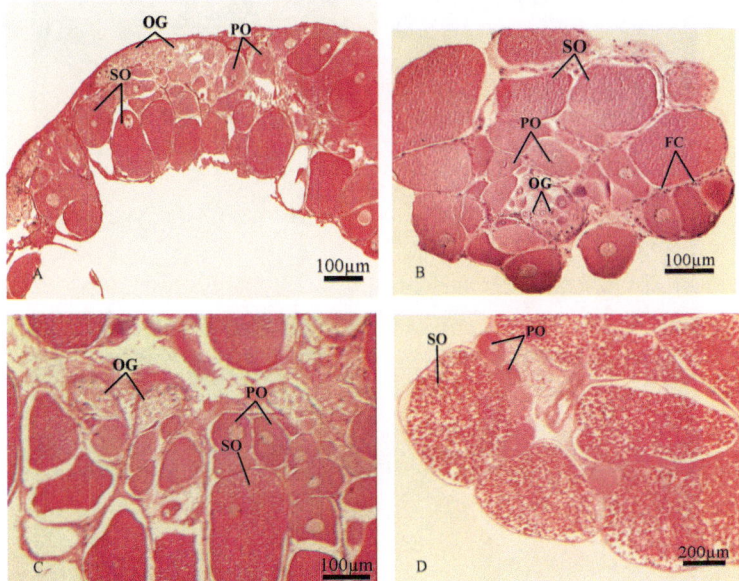

Figure 5.3 Histological sections of *Aegla platensis* ovaries in different development stages. A. Stage I; B. Stage II; C. Stage III; D. Stage IV. OG: oogonia; PO: primary oocytes; SO: secondary oocytes; FC: follicle cells. (Adapted from Sokolowicz et al. [2007].)

The histological examination confirmed this morphological development pattern of ovaries. Although gonads in the earlier developmental stages have secondary oocytes, these are few and little yolk has accumulated, which can be seen by comparing the diameter of the secondary oocytes in the different stages of development (Sokolowicz et al. 2007).

The following macroscopic color-based classification of the ovarian maturation was established for *A. platensis* and can be used as a parameter to verify the sexual maturity in other species of the genus (Figure 5.4A–D) (Sokolowicz et al. 2007).

Stage I. The ovary is small and white. The anterior lobes are in the center of the cephalothorax and the posterior lobes reach the first abdominal somite (Figure 5.4A). Histologically, the ovary contains many oogonia and primary oocytes, but few secondary oocytes (Figure 5.3A). According to Sokolowicz et al. (2007), the smallest *A. platensis* female with these characteristics measured 11.52 mm CL, and the largest one 16.14 mm CL. Although this stage may have secondary oocytes, they are few and have a small yolk accumulation in comparison with the subsequent stages. As maturation proceeds, both the accumulation of yolk in the oocyte and its size increase.

Stage II. The ovary is small and yellow. The anterior lobes are located behind the stomach and the posterior lobes reach the first abdominal somite (Figure 5.4B). The characteristics of the germinative cells are similar to those of Stage I; however, Stage II has more secondary oocytes (Figure 5.3B). Female CL in this stage ranged from 11.68 to 17.01 mm CL (Sokolowicz et al. 2007).

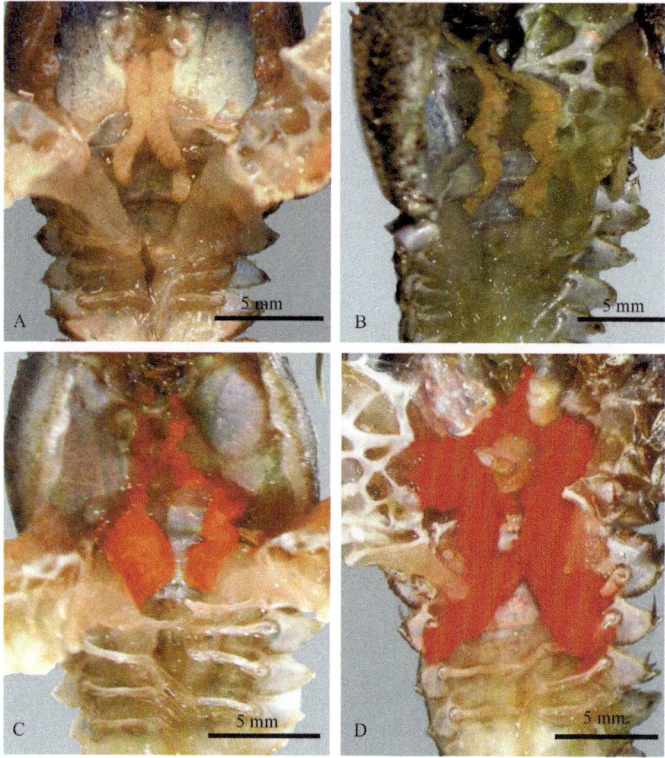

Figure 5.4 Macroscopic development stages of *Aegla platensis* ovaries. A. Stage I; B. Stage II; C. Stage III; D. Stage IV. (Reprinted from Sokolowicz et al. [2007].)

Stage III. Orange ovaries with anterior lobes emerging from behind the stomach. Posterior lobes are wider than in the previous stages and reach the first abdominal somite (Figure 5.4C). There are several oogonia clusters and primary oocytes of different sizes; secondary oocytes are more abundant and have a more advanced vitellogenesis (Figure 5.2B and 5.3C). The size of females with Stage III gonads varied between 11.63 and 18.68 mm CL (Sokolowicz et al. 2007). At this stage, ovaries in rematuration were often observed in *A. platensis*, suggesting that the female might have previously spawned, and maturing oocytes and oocytes are undergoing resorption. According to Charniaux-Cotton (1985), oogenesis from oogonia to the end of primary vitellogenesis is a continuous process. The resorption process explains why large females, which probably had already gone through ovulation, sometimes have white ovaries (Sokolowicz et al. 2007), corroborating the presence of females of variable sizes in different gonadal development stages in the population (Sokolowicz et al. 2006).

Stage IV. The ovary is bright red, with secondary oocytes in abundance with large yolk content, comprising the most developed ovary stage. The lobes are wider than the previous stages, and the posterior lobes can reach the third abdominal somite (Figure 5.4D). Histologically it contains many secondary oocytes in advanced vitellogenesis (Figure 5.3D). The oocytes are larger than in the previous stages and have a cytoplasm full of yolk granules, resulting in an intense coloration. There are few or no oogonia or primary oocytes. The size of females with Stage IV gonads ranged from 12.85 to 19.25 mm CL (Sokolowicz et al. 2007).

A similar classification was established for *A. uruguayana* (see Viau et al. 2006); although the description of gonadal stages was less detailed, the results are basically the same as those of *A. platensis*. Immature ovaries of female *A. uruguayana* are colorless with thin and short posterior lobes. As gonadal maturation starts, the ovary changes color from pale white to pale orange, and the posterior lobes become macroscopically visible from a ventral view of the abdomen. In mature females, ovaries have an intense orange color and when oocytes are ready for ovulation, gonads are bright orange and the posterior lobes full of oocytes occupy the entire abdominal length.

The functional maturity is generally confirmed by the presence of eggs attached to the female abdomen. Histologically, although ovaries with secondary oocytes are present in females at the beginning of their maturation, fully developed ovaries are attained after reaching the morphometric maturity. Most ovigerous females of the population are comprised of large-sized individuals, when both secondary reproductive structures and ovaries attain full reproductive capacity (Sokolowicz 2005; Viau et al. 2006; Bueno and Shimizu 2008, 2009; Takano et al. 2016). At an individual level, functional maturity is attained when the animal is fully mature (gonadal and morphometric maturity) and when it is able to transfer genetic information to the next generation (Bueno and Shimizu 2009). At a population level, the functional maturity is reached when 50% of the female population of a certain size class carries eggs or has gonads in the last developmental stage (often mentioned as CL_{50}, in reference to the average CL in which 50% of females are mature) (Bueno and Shimizu 2008). The size of the smallest ovigerous female of a population can be considered as the minimum functional size, and the CL_{50} corresponds to the population functional maturity size (López-Greco and Rodríguez 1999).

The gonadal maturation starts early in female adulthood, since small females ($\approx$9.00 mm CL in *A. platensis*) already have germinative cells in the early stages of maturation (Sokolowicz 2005), and there are records of ovigerous females of similar sizes (Bueno and Bond-Buckup 2000). The morphometric maturity is also attained earlier when body structures that play an important role in reproduction (such as chelipeds in males and pleon in females) reach a certain size to function adequately for reproductive purposes (Viau et al. 2006). Males with fully developed chelipeds (i.e., larger) would have an advantage in courtship behavior and sexual confrontations (Almerão et al. 2010), and females with a larger pleon would increase the area available for egg incubation (Viau et al. 2006).

Most of the females in a population of *Aegla platensis* have Stage III-ovaries throughout the year, demonstrating that they are constantly producing mature oocytes (Sokolowicz et al. 2006). This feature is reflected by the year-round presence of ovigerous females (Bueno and Bond-Buckup 2000; Sokolowicz et al. 2006).

Ongoing data analysis with the same population studied by Sokolowicz et al. (2006, 2007) revealed that at the CL_{50} size of 12.03 mm the female population would reach functional maturity (Sokolowicz 2005). This size is larger than the minimum maturity and morphologic maturity, corroborating data for other species such as *A. castro* (see Takano et al. 2016), *A. franca* (see Bueno and Shimizu 2008, 2009) and *A. uruguayna* (see Viau et al. 2006).

5.2.1.2 *Males*

In males, the gonadal maturation is not as visible morphologically as it is in females. Although there is a notable increase in the size of the testis and vas deferens, a conspicuous change in size and color does not occur since the physiology of gonadal maturation is slightly different. Similarly to the description of female gonads, the information presented here is based on the analyses of the male reproductive system of *A. platensis* (see Sokolowicz et al. 2007) and *A. uruguayana* (see Viau et al. 2006).

The paired genital ducts of males are located dorsoventrally next to the digestive system; these ducts are not interconnected. The anterior portion of each duct is the testis with multiple lobes. Each testis extends into a vas deferens (VD), which has no lobes; however, as gonadal development proceeds, parts of the VD become more dilated.

According to Lopretto (1978a), male gonads have small capsules full of cellular elements and intermediary connective threads. Later, these structures were referred to as being the spermatozoa with microtubular arms wrapped up by lobes in the testis and VD (Tudge and Scheltinga 2002; Tudge 2003).

Up to now, it has been concluded that members of the genus *Aegla* do not have spermatophores (Mouchet 1932; Tudge 2003; Viau et al. 2006). In *A. platensis*, the testes are polymorphic and there are different stages of spermatogenesis within a single lobe (Figure 5.5A) (Sokolowicz et al. 2007). Light micrographs of the testicular lobes of *A. rostrata* demonstrate the spermatozoa scattered amongst different cell types (Tudge 2003). The spermatozoa have long and filamentous arms radiating from the central cell mass (Tudge 2003).

In *A. platensis*, the anterior part of each vas deferens (AVD) has a simple cubic or round epithelium with secretory cells; below the epithelium lies a thin muscle wall (Figure 5.5B). Few spermatozoa can be seen within an eosinophilic matrix at this part of the VD. The epithelium can participate in the transport of gametes down the ducts, producing a protective covering (Krol et al. 1992). The epithelium of the middle part of each vas deferens (MVD) has cells of different sizes and a thinner muscle wall; many spermatozoa can be seen in the eosinophilic matrix (Figure 5.5C). The posterior vas deferens part (PVD) is narrower and the heterogeneously organized epithelium has cells of different sizes; a thick muscle wall can be seen in this region (Figure 5.5D).

The AVD of *A. platensis* has many secretory cells, indicating a probable sperm transportation and protection function (Sokolowicz et al. 2007). The concentration of spermatozoa in the AVD and MVD suggests that the gametes are stored in this region until mating occurs. Upon mating, spermatozoa are moved to the PVD where

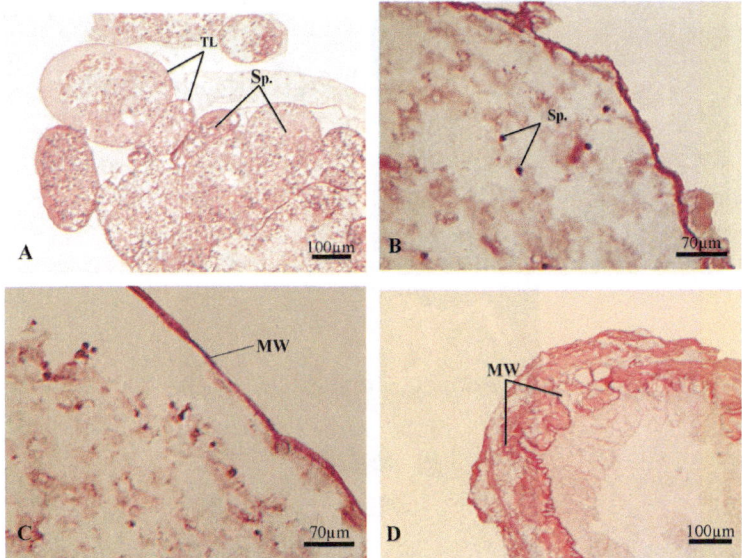

Figure 5.5 Histological sections of *Aegla platensis* male gonad. A. Testes; B. Anterior vas deferens; C. Middle vas deferens; D. Posterior vas deferens. TL: testicular lobes; Sp.: spermatozoa; MW: muscle wall. (Adapted from Sokolowicz et al. [2007].)

the thick muscle wall helps the sperm expulsion. Since aeglids have no spermatophore, a mass of unbound spermatozoa is transferred to the female during mating (Sokolowicz et al. 2007; Almerão et al. 2010).

This description of the *A. platensis* male reproductive system is based on males with large gonads. A macroscopic examination, however, provides a morphological differentiation of the male reproductive system (Sokolowicz et al. 2007).

Type I. Testes, and especially the different VD portions, are slightly distinguishable since both structures have a similar width (Figure 5.6A). Testicular lobes are poorly developed. The same pattern was established for immature males of *A. uruguayana*, where only spermatogonia are seen in the first morphological stage (Viau et al. 2006).

Type II. Testis can be distinguished from the VD, which is poorly developed and folded strictly in the cephalothorax (Figure 5.6B).

Type III. There is a clear differentiation between the testicular portion and the VD parts (Figure 5.6C). The testis has several lobes and the VD is wider than the testis and coils around itself, extending over the third or fourth abdominal somite in large males (CL above 18 mm). The three VD portions are visible at this gonadal stage. These features are also seen in *A. uruguayana*, including the color differentiation between testis and VD (Viau et al. 2006). In both species, the testis has whitish lobes and VD parts are colorless and translucid.

Males also attain morphometric maturity before functional maturity (Sokolowicz 2005; Viau et al. 2006; Bueno and Shimizu 2009). This feature implies that the

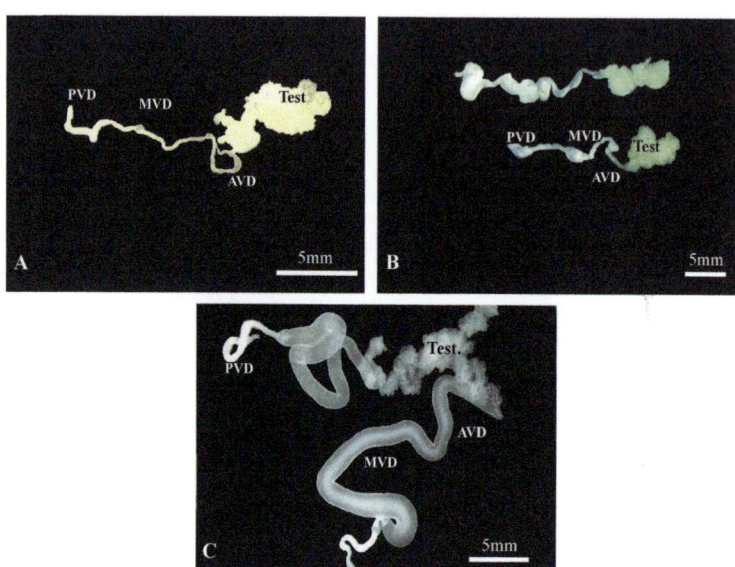

Figure 5.6   Different types of *Aegla platensis* male gonads. A. Type I; B. Type II; C: Type III. Test.: testes; AVD: anterior vas deferens; MVD: middle vas deferens; PVD: posterior vas deferens. (Adapted from Sokolowicz et al. [2007].)

effective development of body structures—such as chelae in males and abdomen in females—is crucial for reproduction and occurs before gonadal development to ensure an effective mating (Viau et al. 2006).

5.3 GONADAL DEVELOPMENT

In most decapods, the gonadal development in females can be followed through ovarian modifications (color and size) throughout the reproductive cycle. These modifications include multiplication of gonadal cells, growth of gametes to maturation, ovulation, and oviposition (Grassé 1994).

In males, the gonadal modifications are less conspicuous morphologically. The spermatogenesis is the development of spermatogonia into primary and secondary spermatocytes until mature spermatids and then into mature spermatozoa (Krol et al. 1992). The gonadal maturation requires a considerable amount of energy. The lipid deposition during gonadal and embryo development is crucial for their maturation and development, respectively (Krol et al. 1992; Cavalli et al. 2001). The yolk is composed of water, proteins, and lipids and is an essential component in embryonic development, tissue formation, and an important energy source (Krol et al. 1992). During the vitellogenesis, there is a rapid deposition of yolk and other proteins in the oocyte, which rapidly increases the cell size (Eastman-Reks and Fingerman 1985). The main storage organ of organic and inorganic reserves in decapods is the hepatopancreas (Pillay and Nair 1973). It is involved in food digestion and absorption, and in the storage of carbohydrates and lipids (Gibson and Barker 1979). During the gametogenesis, organic and mineral reserves are transferred from storage tissues to the gonad (Lawrence 1976).

This relationship between gonadal development and energy storage organs has been studied in *A. platensis* (Sokolowicz et al. 2006; Oliveira et al. 2011) and might represent the way that this process is conducted in aeglids. Since *A. platensis* has a year-round reproduction, the dynamics of gonadal development can be more easily followed (Bueno and Bond-Buckup 2000; Sokolowicz et al. 2006).

Lipid and triglycerides are possibly related to the synthesis of vitellogen and egg transfer during gonadal maturation in females and act as an energetic source for behavioral demands in males (López-Greco et al. 2004; Oliveira et al. 2011). In females of *A. platensis*, the highest levels of these metabolites in gonadal tissue are associated with an increase of the gonadosomatic index (GI) and a decrease of the hepatosomatic index (HI), which reflect their use during gonadal development (Oliveira et al. 2011). The GI and HI indexes express how much the weight of the gonad and the hepatopancreas represent in the total animal weight, respectively. The use of these indexes is a qualitative method to estimate maturity in decapods as their values fluctuate throughout the reproductive cycle (López-Greco and Rodríguez 1999; Sokolowicz et al. 2006; Oliveira et al. 2011). Sokolowicz et al. (2006) stated that in the natural environment the HI decreased significantly in months when the GI was higher. This suggests that the animals are using the hepatopancreatic reserves for gonadal maturation. Oliveira et al. (2011) found similar results for *A. platensis* maintained under laboratory conditions; the amount of lipids stored in the ovaries was higher than that in the hepatopancreas. The hepatopancreatic reserves, however, were not completely absorbed, because the HI was always higher than the GI. Although this pattern occurs year-round, a peak in gonadal development has been detected at the beginning of autumn (April), when 68% of females had ovaries in Stage III in the natural environment (Sokolowicz et al. 2006). At this stage, the ovaries already contain many oocytes in secondary vitellogenesis with enough yolk for reproduction (Sokolowicz et al. 2007), which is corroborated by the high GI during these months. At this stage, most of the population is preparing the gonads for reproduction. The growth of oocytes occurs through endocytotic uptake of vitellogenin (Charniaux-Cotton 1985). This accumulation of yolk in the ovary during preparation for reproduction can be observed as an increase of the GI in females as the ovary develops.

In females, the triglyceride levels in the haemolymph decrease during winter–spring (June–November) and increase in summer–autumn (December–May) (Oliveira et al. 2007). Although the level of this metabolite is high in males during the entire year, it increases in summer (Oliveira et al. 2007). The increased triglyceride levels during summer in both males and females suggests an intensification of the energetic demand in summer probably due to gamete production, incubation, and egg laying in autumn and winter, and parental care in spring and summer (Oliveira et al. 2007).

The use of hepatopancreatic reserves for gonadal maturation was observed for *A. platensis* under experimental conditions (Oliveira et al. 2011). After 150 days of experimental culture, the levels of lipids and triglycerides decreased in the gonads, which indicated the use of these substrates for vitellogen synthesis and egg transfer in females and reproductive behaviors in males (Oliveira et al. 2011).

Oliveira et al. (2007) analyzed the seasonal variation of several metabolites in different tissues of *A. platensis* in the natural environment and demonstrated that lipid levels decreased in muscle tissue of males during the end of summer and beginning of autumn. This is a period of high activity of this species, as well as the peak of reproductive activity (Sokolowicz et al. 2006), which would explain the use of lipidic reserves from the muscles. In both males and females, the triglyceride levels increase during summer in muscles, hepatopancreas, and gills, and are considerably depleted during autumn (Oliveira et al. 2007).

Carbohydrates are also an important energy source during the early stages of gonadal development, corresponding to the primary vitellogenic stage of oocytes in aeglids as suggested by Oliveira et al. (2011) for *A. platensis*. As the process of primary vitellogenesis decreases, there is a reduction of carbohydrate levels in the gonads, suggesting that it is not crucial for the subsequent stages such as the embryonic development (Oliveira et al. 2011).

Although there are changes in the carbohydrate content during gonadal development of males, the largest amount of energy is directly used for behavioral demands. During the pre-copulatory, copulation, and post-copulatory behaviors, much energy is required. Oliveira et al. (2011) demonstrated that after 150 days under experimental conditions the hyperglycemic levels of males decreased markedly, and by the end of 180 days the glycogen reserves from the hepatopancreas were completely used. During a 150-day observation period, the authors found intense gonadal development in both sexes, which explained the high glycogen demand in males for reproductive activities. In the field, the glycogen levels in the hepatopancreas of females vary throughout the year: it is elevated in autumn and gradually diminished in summer (Oliveira et al. 2007). This suggests that the glycogen stored in the hepatopancreas during the autumn is used for the increased reproductive demands for gonadal development and the subsequent egg laying and parental care (Bueno and Bond-Buckup 2004; Sokolowicz et al. 2006, 2007).

Apart from lipid and triglycerides, protein is another important component of the yolk during the vitellogenesis. It is a component of hormones and lipoproteins involved in gonadal maturation (Rosa and Nunes 2003). Since the reproductive peak of *A. platensis* occurs in the autumn, during spring and summer proteins are directed toward the vitellogenic process. Oliveira et al. (2011) observed that after 150 days under experimental conditions (which was equivalent to spring in the natural environment), both sexes had a peak of protein levels in the haemolymph. In the field, there is a decrease in protein levels in the haemolymph during autumn, which seems to indicate the use of this metabolite for reproductive demands (Oliveira et al. 2007). The protein mobilized from the tissues is used during the synthesis of vitellogen in the ovaries and for reproductive behavioral demands in males (Oliveira et al. 2007).

The aforementioned features make clear that *A. platensis* does not completely use its energetic reserves from the main storage organ (Sokolowicz et al. 2006; Oliveira et al. 2011). Although females with fully developed ovaries and males with VD in the highest dilatation degree are found year-round, the hepatosomatic reserves are not completely depleted in any month of the year (Sokolowicz et al. 2006; Oliveira et al. 2011). These aeglids may mobilize metabolites from other tissues such as muscle, gills, and haemolymph to the gonads (Oliveira et al. 2007, 2011) or obtain their energetic input by food intake. An intense feeding activity is probably responsible for the maintenance of nutrients and metabolites in organs and tissues of aeglids. Studies with *A. platensis*, *A. ligulata*, and *A. longirostri* have demonstrated that most individuals have some degree of food content in their stomachs (Oliveira et al. 2003, 2007; Santos et al. 2008). Oliveira et al. (2003, 2007) showed that there are no representative changes in metabolic parameters during different periods of the day. These findings indicate that these individuals are constantly feeding, which corroborates their adequate nutritional status.

5.4 MATING BEHAVIOR

The process of sperm transfer and copulation behavior in Aeglidae has been uncertain for a long time. The first studies on the morphology and histology of gonads highlighted the lack of information about the fertilization process and the pre- and post-copulatory behaviors (Tudge 2003; Viau et al. 2006). In most decapods, the copulation occurs after the female molt, while the exoskeleton is still soft, enabling the insertion of male gonopods into the female genital pore (Hartnoll 1969). This process requires a pre- and post-copulatory behavior to ensure male's paternity. In Aeglidae, the mating behavior was described for *A. platensis* by Almerão et al. (2010) and some peculiarities became apparent when compared to the general mating behavior in decapods.

The mating behavior of *A. platensis* is divided into three phases (Almerão et al. 2010): pre-copulatory, copulatory, and post-copulatory phases. In the pre-copulatory phase, males exhibit an agonist behavior toward both immature and mature females; the success depends on the female gonadal development status (fully mature ovary). During this phase, male and female are close to one another and the courtship takes place. The courtship is comprised by a series of body movements of both sexes. The copulatory phase begins a few seconds after finishing the pre-copulatory phase, when the couple assumes the supine position (male under the female with his back on the substrate) (Figure 5.7). This position is rare and occurs in some crayfish species (Barki and Karplus 1999; Kawai and Saito 2001). The supine position might favor the oocyte transfer to the abdominal chamber and also increases the copulation success (Hartnoll 1969). In the post-copulatory phase, the male stays near the female as she begins grooming the abdominal chamber full of eggs with the fifth pair of pereopods. After this process, a white mass becomes visible in the female brood mass, which is supposedly the spermatozoa mass (Figure 5.8) (Almerão et al. 2010). To date, the direct observation of sperm transfer in Aeglidae has not been recorded. Male aeglids have no copulatory appendages and there are two hypotheses about the way the sperm transfer is accomplished in these decapods (Almerão et al. 2010): the genital papilla (small papilla at the end of

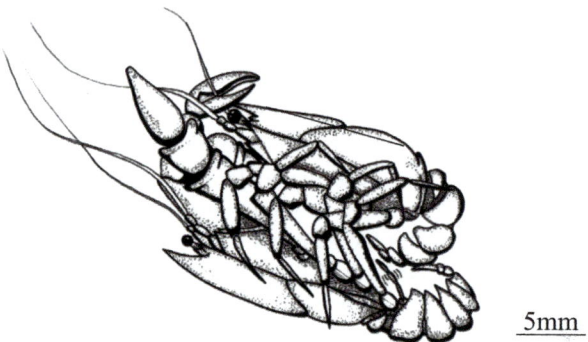

Figure 5.7 Schematic drawing of *Aegla platensis* copula in supine position (male under the female). (Adapted from Almerão [2005].)

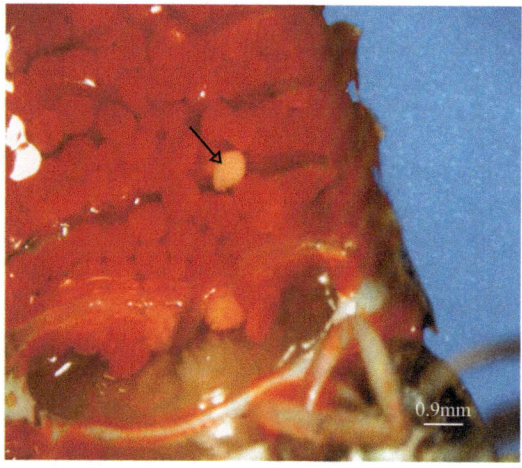

Figure 5.8 A spermatozoa mass (arrow) among the oocytes of *Aegla platensis*. (Adapted from Almerão [2005].)

male genital pore) hypothesis is less probable. The genital papilla is vestigial in *A. platensis* (species in which the mating behavior was studied) (Lopretto 1978a) and was not observed in the brood chamber during experiments conducted to observe the copulation (Almerão et al. 2010). The second hypothesis is that the fifth pair of pereopods plays a role in sperm transfer. Since females lack any structure of sperm storage and sperm is not confined into a spermatophore structure (Viau et al. 2006; Sokolowicz et al. 2007), the sperm transfer probably takes place during the final copulatory stage. Additional studies, however, are needed to clarify the details regarding the sperm transfer in aeglids.

Knowledge of the reproductive biology of aeglids has gradually been built up since the studies on the genus started. Reproductive period and size at the onset of sexual maturity are the most well-studied features and are known in a large number of species from different geographical regions. Although the morphology and histology of the gonadal development are known only for two species of *Aegla*, the respective studies are very detailed, and the general development pattern can be extended to other species. However, new studies would increase the knowledge about this subject as the gonadal description serves as baseline information in investigations regarding populational status and reproductive behavior. To date, little is known about the process of sperm maturation in Aeglidae, and studies on this subject may also contribute to a better understanding of mating behavior and life-history strategies.

ACKNOWLEDGMENTS

The author thanks Dr. Mauricio Almerão for providing the images for a better understanding of the descriptions made in this chapter. Thanks to Rodrigo Fontana for the image editing. Thanks to John Wiley and Sons for the license to reuse previously published figures [in: Sokolowicz et al. 2007. The gonads of *Aegla platensis* Schmitt (Decapoda, Anomura, Aeglidae): a macroscopic and histological perspective. *Acta Zoologica* 81: 71–79].

REFERENCES

Almerão, M. 2005. Aspectos do comportamento reprodutivo de *Aegla platensis* Schmitt em condições de cultivo (Crustacea, Anomura, Aeglidae). MS diss., Universidade Federal do Rio Grande do Sul.

Almerão, M., G. Bond-Buckup, and M. S. de Mendonça Jr. 2010. Mating behavior of *Aegla platensis* (Crustacea, Anomura, Aeglidae) under laboratory conditions. *Journal of Ethology* 28:87–94.

Bahamonde, N., and M. T. López. 1961. Estudios biológicos en la población de *Aegla laevis* (Latreille) de El Monte (Crustacea, Decapoda, Anomura). *Investigaciones Zoológicas Chilenas* 7:19–58.

Barki, A., and I. Karplus. 1999. Mating behavior and a behavioral assay for female receptivity in the red-claw crayfish *Cherax quadricarinatus*. *Journal of Crustacean Biology* 19:493–97.

Bauer, R. T. 1992. Testing generalizations about latitudinal variation in reproduction and recruitment patterns with sicyoniid and caridean shrimp species. *Invertebrate Reproduction and Development* 22:193–202.

Bond-Buckup, G., and L. Buckup. 1994. A família Aeglidae (Crustacea, Decapoda, Anomura). *Arquivos de Zoologia* 32(4):1–346.

Bond-Buckup, G., A. P. Bueno, and K. A. Keunecke. 1996. Primeiro estágio juvenil de *Aegla prado* Schmitt (Crustacea, Decapoda, Anomura, Aeglidae). *Revista Brasileira Zoologia* 13(14):1049–61.

Bond-Buckup, G., A. P. Bueno, and K. A. Keunecke. 1999. Morphological characteristics of juvenile specimens of *Aegla* (Decapoda, Anomura: Aeglidae). In *Crustaceans and the Biodiversity Crisis*, eds. F. R. Schram, and J. C. von Vaupel Klein, pp. 371–81. Leiden: Brill.

Buckup, L., and A. Rossi. 1979. O gênero *Aegla* no Rio Grande do Sul, Brasil (Crustacea, Decapoda, Anomura, Aeglidae). *Revista Brasileira de Zoologia* 37(4):39–47.

Bueno, A. A. P., and G. Bond-Buckup. 1996. Os estágios juvenis iniciais de *Aegla violacea* Bond-Buckup & Buckup (Crustacea, Anomura, Aeglidae). *Nauplius* 4:39–47.

Bueno, A. A. P., and G. Bond-Buckup. 2000. Dinâmica populacional de *Aegla platensis* Schmitt (Crustacea, Decapoda, Aeglidae). *Revista Brasileira de Zoologia* 17:43–49.

Bueno, A. A. P., and G. Bond-Buckup. 2004. Natural diet of *Aegla platensis* Schmitt and *Aegla ligulata* Bond-Buckup and Buckup (Crustacea, Decapoda, Aeglidae). *Acta Limnologica Brasiliensia* 16:115–27.

Bueno, S. L. S., and R. M. Shimizu. 2008. Reproductive biology and functional maturity in females of *Aegla franca* (Decapoda: Anomura: Aeglidae). *Journal of Crustacean Biology* 28:652–62.

Bueno, S. L. S., and R. M. Shimizu. 2009. Allometric growth, sexual maturity, and adult male chelae dimorphism in *Aegla franca* (Decapoda: Anomura: Aeglidae). *Journal of Crustacean Biology* 29(3):317–28.

Bueno, S. L. S., B. F. Takano, F. P. A. Cohen, et al. 2014. Fluctuations in the population size of the highly endemic *Aegla perobae* (Decapoda, Anomura, Aeglidae) caused by disturbance event. *Journal of Crustacean Biology* 34(2):165–73.

Castiglioni, D. S., A. V. Cahansky, E. Rodríguez, B. K. Dutra, Oliveira, G. T., and G. Bond-Buckup. 2009. Indução do crescimento ovariano em *Aegla uruguayana* (Crustacea, Anomura, Aeglidae) mediante a incorporação de neuroreguladores ao alimento. *Iheringia, Série Zoologia* 99(3):286–90.

Castilho, A. L., M. A. Gavio, R. C. Costa, E. E. Boschi, R. T. Bauer, and A. Fransozo. 2007. Latitudinal variation in population structure and reproductive pattern of the endemic South American shrimp *Artemesia longinaris* (Decapoda: Penaeoidea). *Journal of Crustacean Biology* 27:548–52.

Cavalli, R. O., M. Tamtin, P. Lavens, and P. Sorgeloos. 2001. Variations in lipid classes and fatty acid content in tissues of wild *Macrobrachium rosenbergii* (de Man) females during maturation. *Aquaculture* 193:311–24.

Charniaux-Cotton, H. 1985. Vitellogenesis and its control in malacostracan Crustacea. *American Zoologist* 25:197–206.

Chiquetto-Machado, P. I., L. C. M. Vieira, R. M. Shimizu, and S. L. S. Bueno. 2016. Life cycle of the freshwater anomuran *Aegla schmitti* Hobbs, 1978 (Decapoda: Anomura: Aeglidae) from southern Brazil. *Journal of Crustacean Biology* 36(1):39–45.

Cohen, F. P. A., B. F. Takano, R. M. Shimizu, and S. L. S. Bueno. 2011. Life cycle and population structure of *Aegla paulensis* (Decapoda: Anomura: Aeglidae). *Journal of Crustacean Biology* 31:389–95.

Collins, P. A., F. Giri, and V. Williner. 2008. Sexual maturity and ontogenetic shape variations in the freshwater anomuran crab, *Aegla uruguayana* (Decapoda, Aeglidae). *Invertebrate Reproduction and Development* 52(1–2):113–19.

Copatti, C. E., J. V. V. Machado, and A. Trevisan. 2015. Morphological variation in the sexual maturity of three sympatric aeglids in a river in Southern Brazil. *Journal of Crustacean Biology* 35(1):59–67.

Copatti, C. E., R. P. Legramanti, A. Trevisan, and S. Santos. 2016. Growth, sexual maturity and sexual dimorphism of *Aegla georginae* (Decapoda: Anomura: Aeglidae) in a tributary of the Ibicuí River in southern Brazil. *Zoologia* 33(3):e20160010.

Da Silva, A. R., G. P. Paciencia, P. C. Bispo, and A. L. Castilho. 2017. Allometry and sexual dimorphism of the Neotropical freshwater anomuran *Aegla marginata* Bond-Buckup & Buckup, 1994 (Crustacea, Anomura, Aeglidae). *Nauplius* 25:1–7.

Da Silva, A. R., M. R. Wolf, and A. L. Castilho. 2016. Reproduction, growth and longevity of the South American crab *Aegla marginata* (Decapoda: Anomura: Aeglidae). *Invertebrate Reproduction & Development* 60(1):59–72.

Eastman-Reks, S., and M. Fingerman. 1985. *In vitro* synthesis of vitelline by the ovary of the fiddler crab, *Uca pugilator*. *Jornal of Experimental Zoology* 233:111–16.

Fransozo, A., R. C. Costa, A. L. D. Reigada, and J. M. Nakagaki. 2003. Population structure of *Aegla castro* Schmitt, 1942 (Crustacea: Anomura: Aeglidae) from Itatinga (SP), Brazil. *Acta Limnologica Brasiliensia* 15(2):13–20.

Gibson, R., and P. L. Barker. 1979. The decapod hepatopancreas. *Oceanography and Marine Biology, an Annual Review* 17:285–346.

Gonçalvez, R. S., D. S. Castiglioni, and G. Bond-Buckup. 2006. Ecologia populacional de *Aegla franciscana* (Crustacea, Decapoda, Anomura) em São Francisco de Paula, RS, Brasil. *Iheringia, Série Zoologia* 96:109–14.

Grabowski, R. C., S. Santos, and A. L. Castilho. 2013. Reproductive ecology and size of sexual maturity in the anomuran crab *Aegla parana* (Decapoda: Aeglidae). *Journal of Crustacean Biology* 33(3):332–38.

Grassé, P. P. 1994. *Traité de zoologie: Anatomie, systématique, biologie*. Paris: Masson.

Hartnoll, R. G. 1969. Mating in the Brachyura. *Crustaceana* 16:161–81.

Hartnoll, R. G. 1974. Variation in growth pattern between some secondary sexual characters in crabs (Decapoda, Brachyura). *Crustaceana* 27(2):131–36.

Jara, C. 1977. *Aegla rostrata* n. sp., (Decapoda, Aeglidae), nuevo crustáceo dulceacuícola del sur de Chile. *Studies on Neotropical Fauna Environment* 12:165–76.

Kawai, T., and K. Saito. 2001. Observations on the mating behavior and season, with no form alteration, of the Japanese crayfish, *Cambaroides japonicus* (Decapoda, Cambaridae), in Lake Komadome, Japan. *Journal of Crustacean Biology* 21:885–90.

Krol, R. M., W. E. Hawkins, and R. M. Overstreet. 1992. Reproductive components. In *Microscopic Anatomy of Invertebrates*, ed. F. W. Harrison, pp. 295–343. New York: Willy-Lizz.

Lawrence, J. M. 1976. Patterns of lipid storage in post-metamorphic marine invertebrates. *American Zoologist* 16:747–62.

López, M. T. 1965. Estudios biológicos en *Aegla odebrechtti* paulensis, Schmitt (Crustacea, Decapoda, Anomura). *Boletim de Zoologia da Faculdade de Ciências* 25:301–14.

López-Greco, L. S. 2013. Functional anatomy of the reproductive system. In *Functional Morphology & Diversity – The Natural History of the Crustacea*, eds. L. Watling, and M. Thiel, vol. 1, pp. 413–50. Oxford: Oxford University Press.

López-Greco, L. S., and E. M. Rodríguez. 1999. Annual reproduction and growth of adult crabs *Chasmagnathus granulata* (Crustacea, Brachyura, Grapsidae). *Cahiers de Biologie Marine* 40:155–64.

López-Greco, L. S., V. Viau, M. Lavolpe, G. Bond-Buckup, and E. M. Rodríguez. 2004. Juvenile hatching and maternal care in *Aegla uruguayana* (Anomura, Aeglidae). *Journal of Crustacean Biology* 24(2):309–13.

Lopretto, E. C. 1978a. Estructura exoesqueletaria y miología del quinto par de pereiópodos del macho de la familia Aeglidae (Crustacea, Anomura). *Limnobios* 1(8):284–98.

Lopretto, E. C. 1978b. Las especies de *Aegla* del centro-oeste Argentino en base a la morfología comparada del quinto par de pereiópodos (Crustacea, Anomura, Aeglidae). *Neotropica* 24(71):57–68.

Lopretto, E. C. 1980. Clave para la determinación de las especies del género *Aegla* de la República Argentina en base al estudio comparativo del quinto par de pereiópodos masculinos (Crustacea, Anomura, Aeglidae). *Limnobios* 1(10):431–36.

Martin, J. W., and L. G. Abele. 1988. External morphology of the genus *Aegla* (Crustacea: Anomura: Aeglidae). *Smithsonian Contributions to Zoology* 453:1–46.

Moraes, J. C. B., and S. L. S. Bueno. 2013. Description of the newly-hatched juvenile of *Aegla paulensis* (Decapoda, Anomura. Aeglidae). *Zootaxa* 3635(5):501–19.

Mouchet, S. 1932. Notes sur la biologie du galathéide *Aeglea laevis* (Latr.). *Bulletin de la Société Zoologique de France* 57:316–40.

Noro, C. K., and L. Buckup. 2002. Biologia reprodutiva e ecologia de *Aegla leptodactyla* Buckup & Rossi (Crustacea, Anomura, Aeglidae). *Revista Brasileria de Zoologia* 19(4):1063–74.

Oliveira, D., and S. Santos. 2011. Maturidade sexual morfológica de *Aegla platensis* (Crustacea, Decapoda, Anomura) no Lajeado Bonito, norte do estado do Rio Grande do Sul, Brasil. *Iheringia, Série Zoologia* 101(1–2):127–30.

Oliveira, G. T., F. A. Fernandes, G. Bond-Buckup, A. A. Bueno, and R. S. M. Da Silva. 2003. Circadian and seasonal variations in the metabolism of carbohydrates in *Aegla ligulata* (Crustacea: Anomura: Aeglidae). *Memoirs of Museum Victoria* 60:59–62.

Oliveira, G. T., F. A. Fernandes, A. A. P. Bueno, and G. Bond-Buckup. 2007. Seasonal variation in the intermediate metabolism of *Aegla platensis* (Crustacea, Aeglidae). *Comparative Biochemistry and Physiology, Part A* 147:600–06.

Oliveira, G. T., C. Hack, M. Almerão, G. Bond-Buckup, and B. K. Dutra. 2011. Tissue composition haemolymphatic metabolites during gonadal development in *Aegla platensis* (Crustacea, Decapoda) maintained in experimental culture. *Brazilian Journal of Biosciences* 9(1):64–71.

Pillay, K. K., and N. B. Nair. 1973. Observations on the biochemical changes in gonads and other organisms of *Uca annulipes, Portunus pelagicus* and *Metapenaeus affinis* (Decapoda: Crustacea) during the reproductive cycle. *Marine Biology* 18:167–98.

Ringuelet, R. A. 1949. Consideraciones sobre las relaciones filogenéticas entre las espécies del género *Aegla* Leach. *Notas del Museo de La Plata* 14:111–18.

Rocha, S. S., R. M. Shimizu, and S. L. S. Bueno. 2010. Reproductive biology in females of *Aegla strinatii* (Decapoda: Anomura: Aeglidae). *Journal of Crustacean Biology* 30(4):589–96.

Rodrigues, W., and N. J. Hebling. 1978. Estudos biológicos em *Aegla perobae* Hebling & Rodrigues, 1977 (Decapoda, Anomura). *Revista Brasileira de Biologia* 38:383–90.

Rosa, R. A., and M. L. Nunes. 2003. Changes in organ indices and lipid dynamics during the reproductive cycle of *Aristeus antennatus, Parapenaeus longirostris* and *Nephrops norvegicus* (Crustacea: Decapoda) females from the south Portuguese coast. *Crustaceana* 75:1095–105.

Santos, S., L. Ayres-Peres, R. C. F. Cardoso, and C. C. Sokolowicz. 2008. Natural diet of the freshwater anomuran *Aegla longirostri* (Crustacea, Anomura, Aeglidae). *Journal of Natural History* 42(13–14):1027–37.

Sastry, A. N. 1983. Ecological aspects of reproduction. In *Environmental Adaptation. The Biology of Crustacea*, eds. F. J. Vernberg, and W. B. Vernberg, pp. 179–270. New York: Academic Press.

Schmitt, W. 1942. The species of *Aegla*, endemic South American freshwater crustaceans. *Proceedings of the United States National Museum* 91:431–520.

Siqueira, A. F., A. V. Palaoro, and S. Santos. 2013. Mate preference in the neotropical freshwater crab *Aegla longirostri* (Decapoda: Anomura): does the size matter? *Marine Freshwater Behaviour and Physiology* 46(4):219–27.

Sokolowicz, C. C. 2005. Aspectos da biologia reprodutiva de *Aegla platensis* Schmitt, 1942 (Crustacea, Anomura, Aeglidae). MS diss., Universidade Federal do Rio Grande do Sul.

Sokolowicz, C. C., G. Bond-Buckup, and L. Buckup. 2006. Dynamics of gonadal development of *Aegla platensis* Schmitt (Decapoda, Anomura, Aeglidae). *Revista Brasileira de Zoologia* 23(4):1153–58.

Sokolowicz, C. C., L. S. López-Greco, R. Gonçalves, and G. Bond-Buckup. 2007. The gonads of *Aegla platensis* Schmitt (Decapoda, Anomura, Aeglidae): a macroscopic and histological perspective. *Acta Zoologica* 81:71–79.

Swiech-Ayoub, B. P., and S. Masunari. 2001. Biologia reprodutiva de *Aegla castro* Schmitt (Crustacea, Anomura, Aeglidae) no Buraco do Padre, Ponta Grossa, Paraná, Brasil. *Revista Brasileira de Zoologia* 18:1019–30.

Takano, B. F., F. P. A. Cohen, A. Fransozo, R. M. Shimizu, and S. L. S. Bueno. 2016. Allometric growth, sexual maturity and reproductive cycle of *Aegla castro* (Decapoda: Anomura: Aeglidae) from Itatinga, state of São Paulo, southeastern Brazil. *Nauplius* 24:1–15.

Trevisan, A., and S. Santos. 2012. Morphological sexual maturity, sexual dimorphism and heterochely in *Aegla manuinflata* (Anomura). *Journal of Crustacean Biology* 32(4):517–27.

Trevisan, A., M. Z. Marochi, M. Costa, S. Santos, and S. Masunari. 2012. Sexual dimorphism in *Aegla marginata* (Decapoda: Anomura). *Nauplius* 20(1):75–86.

Tudge, C. 2003. Endemic and enigmatic: the reproductive biology of *Aegla* (Crustacea: Anomura: Aeglidae) with observations on sperm structure. *Memoirs of Museum Victoria* 60(1):63–70.

Tudge, C., and D. M. Scheltinga. 2002. Spermatozoal morphology of the freshwater anomuran *Aegla longirostri* Bond-Buckup & Buckup, 1994 (Crustacea: Decapoda: Aeglidae) from South America. *Proceedings of the Biological Society of Washington* 115:118–28.

Viau, V. E., L. S. López-Greco, G. Bond-Buckup, and E. M. Rodriguez. 2006. Size at the onset of sexual maturity in the anomuran crab, *Aegla uruguayana* (Aeglidae). *Acta Zoologica* 7:253–64.

Vogt, G. 2013. Abbreviation of larval development and extension of brood care as key features of the evolution of freshwater Decapoda. *Biological Reviews of the Cambridge Philosophical Society* 88:81–116.

Vogt, G. 2014. Life span, early life stage protection, mortality, and senescence in freshwater Decapoda. In *Advances in Freshwater Decapod Systematics and Biology*, eds. D. C. J. Yeo, S. Klaus, and N. Cumberlidge, pp. 17–52. New York: Crustaceana Monographs.

CHAPTER 6

Postembryonic Development, Parental Care, and Recruitment

**Sérgio Luiz de Siqueira Bueno, Roberto Munehisa Shimizu,
and Juliana Cristina Bertacini Moraes**

CONTENTS

6.1 INTRODUCTORY REMARKS ON THE
PATTERNS OF POSTEMBRYONIC DEVELOPMENT
IN THE ANOMURA MACLEAY, 1838

The pattern of postembryonic development in marine anomurans is metamorphic and is similar to that of most pleocyemate decapods, meaning that there is a larval phase in which the hatching form is typically a zoea, followed by one megalopal (= decapodid) stage; the larval phase is completed subsequently with the appearance of the first juvenile form (Anger 2006). Postembryonic development in the zoeal phase may be either regular or abbreviated. In general terms, the regular and the abbreviated types of larval development can be distinguished from each other by the size and number of brooding eggs, feeding mode of the newly hatched larvae, total number of zoeal stages before reaching the megalopal stage, and degree of development of some larval structures at hatching or at specific larval instars (Anger 2001).

In regular larval development of anomurans, for instance, free-swimming planktotrophic zoeae emerge from numerous small-sized eggs, and go through 4–5 instars before metamorphosing into the megalopa [see Baba et al. (2001) for detailed information on postembryonic development in squat lobsters]. They initially exhibit rudiments in some of the appendages, such as the pereopods, which will develop gradually as the larval phase proceeds through a series of molts, and then become well developed and functional in the megalopal stage (Baba et al. 2001). On the other hand, several gnathal appendages are already functional in newly hatched zoeae, such as the grinding processes of the mandibles and the setose endites of the first and second pairs of maxillae, although none of these pairs of appendages show their complete and definitive morphology until the magalopal stage is reached (Baba 2001). Species from the superfamilies Galatheoidea Samouelle, 1819, Paguroidea Latreille, (1802) and Hippoidea Latreille, (1825) are representative examples of anomurans with larval development of the regular type (Knight 1967; Pike and Wear 1969; Fagetti and Campodonico 1971; Stuck and Truesdale 1986; Gherardi and McLaughlin 1995; Brodie and Harvey 2001; Fujita et al. 2002; Fujita and Osawa 2005; Fujita and Shokita 2005; Barria et al. 2006; Fujita 2010).

Larval development of the abbreviated type has been reported for several species of anomurans belonging to the superfamilies Lithodoidea Samouelle, 1819, Lomisoidea Bouvier, (1895) and Chirostyloidea Ortmann, 1892 (Campodonico 1971; Cormie 1993; McLaughlin et al. 2001, 2003; Anger et al. 2004; Clark and Ng 2008; Fujita and Clark 2010). Species with this type of development produce usually few but large eggs. The lecithotrophic larvae hatch with a large amount of yolk reserve in the thoracic region, which is gradually consumed during the larval phase composed of 2–3 zoeal instars (Baba et al. 2001). Gnathal appendages, therefore, are initially not functional and typically show rudimentary morphology, such as a pair of mandibles with incipient grinding processes, and endites of the first and second pairs of maxillae devoid of setae. Conversely, the newly hatched larvae already exhibit an advanced degree of development in other paired appendages as compared to the regular type of larval development. Worthy of mention are the biramous buds of pleopods and the pereopods, though the degree of development of the latter group of appendages may vary among genera (Baba et al. 2001).

As already mentioned for larvae that go through the regular type of postembryonic development, larvae with abbreviated larval development also show well-developed and functional pairs of appendages by the time the megalopal stage is reached. In both cases, the immediate instar after the megalopa is the first juvenile, also known as first crab. Megalopae still have functional pleopods for swimming purposes, whilst juveniles typically show epibenthic habits and use their functional locomotory pereopods to move on the substrate (Baba et al. 2001).

Aegla Leach, 1820 is the only taxon of Anomura whose life cycle is entirely confined to freshwater habitats (Schmitt 1942; Bond-Buckup and Buckup 1994; Bueno et al. 2016). One remarkable evolutionary adaptive feature shared by all *Aegla* species—and this feature finds no similar parallel within the Anomura—is that the postembryonic development is of the direct (= epimorphic) type. This means that hatchlings emerging from eggs are epibenthic juveniles that resemble adults in general morphology and habits (with the exception of adult traits associated with reproduction). Therefore, the life stage of the hatching form would be equivalent to the first juvenile of other marine anomurans.

There is also a brief period of paternal care in freshwater aeglids, during which the newly hatched juveniles remain for a few days in the brood chamber of their mother before venturing out in the surroundings on their own. Direct postembryonic development and parental care have evolved independently in different groups of freshwater decapods, and these traits are considered as key features of the successful adaptation and colonization of freshwater environments by marine ancestors [see review on this topic by Vogt (2016, and references therein)].

Detailed morphological descriptions of newly hatched juveniles are limited to eight aeglid species from Brazil: *Aegla castro* Schmitt, 1942 (Silva et al. 2017); *A. franca* Schmitt, 1942 (Francisco et al. 2007); *A. jaragua* Moraes, Tavares and Bueno, 2016 [as *A. paulensis* Schmitt 1942 s. lat.] (Moraes and Bueno 2013); *A. perobae* Hebling and Rodrigues, 1977 (Moraes and Bueno 2015); *A. platensis* Schmitt, 1942 (Bond-Buckup et al. 1999); *A. prado* Schmitt, 1942 (Bond-Buckup et al. 1996); *A. schmitti* Hobbs III, 1978 (Teodósio and Masunari 2007); and *A. violacea* Bond-Buckup and Buckup, 1994 (Bueno and Bond-Buckup 1996). Brief morphological descriptions are also available for two additional Brazilian species: *Aegla ligulata* Bond-Buckup and Buckup, 1994 and *A. longirostri* Bond-Buckup and Buckup, 1994 (Bond-Buckup et al. 1999). Brief remarks regarding the second juvenile instar have been provided for *A. violacea* and *A. schmitti* (Bueno and Bond-Buckup 1996; Teodósio and Masunari 2007): no significant morphological differences from the preceding instar were encountered, except for the general increase in the number of setae in certain appendages.

Besides *Aegla*, two additional fossil genera of marine origin (*Protaegla* Feldmann et al., 1998 and *Haumuriaegla* Feldmann, 1984) are recognized as belonging to the family Aeglidae Dana, 1852. These findings support the marine origin of the aeglids, probably in the Pacific region, from where these fossil specimens were reported (Feldmann 1984; Feldmann et al. 1998). No available information exists about the type of postembryonic development these two extinct aeglid genera might have had. It would be very reasonable, however, to infer that postembryonic development most probably was characterized by the presence of zoeal larval stages. In any case, dispersal of ancient aeglid larvae towards the South American Continental Plate could well have benefited from a counter-clockwise gyre of water circulation in the Pacific Ocean during the early Cenozoic (Feldmann 1986). Current knowledge on the origin and dispersal routes of freshwater aeglids through the Cenozoic Paleobasins of Continental South America is presented in Chapter 1 of this book.

6.2 EMBRYONIC AND POSTEMBRYONIC DEVELOPMENT IN *AEGLA* SPP.

6.2.1 Eggs

The earliest report speculating about the type of postembryonic development *Aegla* might have was provided by the German naturalist Fritz Müller in the late nineteenth century. In a brief commentary remark in his paper on the abbreviated larval development of the palaemonid freshwater prawn, *Macrobrachium potiuna* (Müller 1880) and written in Portuguese, Müller (1892, p. 204) became intrigued by the large size of the eggs he had observed in the endemic Brazilian aeglid species, *Aegla odebrechtii* Müller, 1876. That observation led him to infer that the postembryonic development of that species would not be of the regular type, but, rather, "...muito abreviada ou quasi nulla" [...very abbreviated or almost null].

Half a century later, Mouchet (1932) also noticed that eggs are large in ovigerous females of *Aegla laevis laevis* (Latreille, 1818) from Uruguay (a possible case of misidentification since the actual distribution of this species is limited to Chile) and described hatching forms as fully developed juveniles that very much resembled the general morphology of the adults, thus establishing that the postembryonic development in *Aegla* was actually of the direct type.

Indeed, aeglid eggs are large and few. They are brooded externally on the ventral side of the pleon and attached to the pleopods, just like in all other pleocyemate decapods. The pleon is wider in adult females than it is in adult males, a secondary dimorphic trait for egg incubation and whose definitive shape is achieved at the puberty molt (Colpo et al. 2005; Viau et al. 2006). The brood chamber is formed by the folded pleon of females as two opposite halves with their ventral surfaces facing each other.

The number of incubating eggs may vary from small numbers in the upper tens (Jara 1980; Verdi 1985; Lizardo-Daudt and Bond-Buckup 2003) to as many as the lower hundreds (Bahamonde and López 1961; Noro and Buckup 2002; Gonçalves et al. 2006; Bueno and Shimizu 2008; Bueno et al. 2014; da Silva et al. 2016), although figures reaching over 1000 have also been documented for the Chilean species, *Aegla rostrata* Jara, 1977 (Jara 1977). The number of eggs is positively correlated with the size of adult females (López 1965; Jara 1980; Bueno and Shimizu 2008; da Silva et al. 2016). The egg shape may be elliptical or round (López 1965; Rodrigues and Hebling 1978; Lizardo-Daudt and Bond-Buckup 2003; Bueno and Shimizu 2008) and it increases in size during embryogenesis (Bahamonde and López 1961; López 1965; Bueno and Shimizu 2008).

Embryonic development may extend for over a month, as in *Aegla prado* (32 days) (Verdi 1985) and *A. platensis* (35 days) (Lizardo-Daudt and Bond-Buckup 2003) to 45–50 days as observed in *A. uruguayana* Schmitt, 1942 (López-Greco et al. 2004). Eggs have a large amount of yolk (Mouchet 1932; López 1965), which is gradually consumed as the embryo develops. A quick evaluation of egg developmental stages under field conditions has been proposed by Bueno and Shimizu (2008). Three sequential embryonic stages are recognized: 1) Early eggs (Figure 6.1A) are observed soon after oviposition and during the initial days of development; the color of eggs may vary from bright orange to red soon after oviposition, due to the homogenous distribution of the yolk reserve within; no sign of embryo development. 2) Intermediate eggs (Figure 6.1B), the yolk mass occupies 50% to 80% of the egg volume; a pair of narrow, darkly pigmented compound eyes are noticeable in the developing embryo. 3) Late eggs (Figure 6.1C), several morphological details of the embryo are discernible, including large stalked compound eyes, pleonal somites and developing pereopods; the yolk mass is reduced to less than 50% of the egg volume and concentrated in the cephalothoracic region.

Details regarding the embryonic development in *Aegla* have been addressed by Verdi (1985) and Lizardo-Daudt and Bond-Buckup (2003). Although there is no free-living larval phase in *Aegla* whatsoever, zoeal and megalopal morphological traits can still be recognized within the egg during embryonic development (Lizardo-Daudt and Bond-Buckup 2003). In Figure 6.1D, for instance, one can still see the narrow shape of the pleon reminiscent of the early zoeal condition, as well as the absence of paired uropods and presence of nonfunctional developing pereopods in the still developing embryo. The well-segmented pereopods, the wider and flattened second, third, and fourth pleonites, and a well-defined separation between the sixth pleonal somite and the telson can be clearly observed later on when hatching is imminent and late embryos are fully developed (Figure 6.1E).

Hatching of all juveniles from a single brood is not synchronous, meaning that it may take 3 to 5 days before the last juvenile hatches (López-Greco et al. 2004; Moraes and Bueno 2013, 2015). Although newly hatched juveniles remain in the brood pouch for a while (see Section 6.3, "Parental Care") in association with any unhatched eggs, these are not subject to cannibalism by juveniles during this period (López-Greco et al. 2004).

A most rare and intriguing case of conjoined twins among newly hatched juveniles has been reported in *Aegla abtao* Schmitt, 1942 by Jara and Palacios (2001). Similar cases (Figure 6.1G) were observed twice in *Aegla franca* (SLS Bueno personal observation). In both species, the twins were joined together by the dorsal region of their carapace. The occurrence of conjoined twins in protostome metazoans with determinate cleavage, such as crustaceans, remains as yet unexplained (Jara and Palacios 2001).

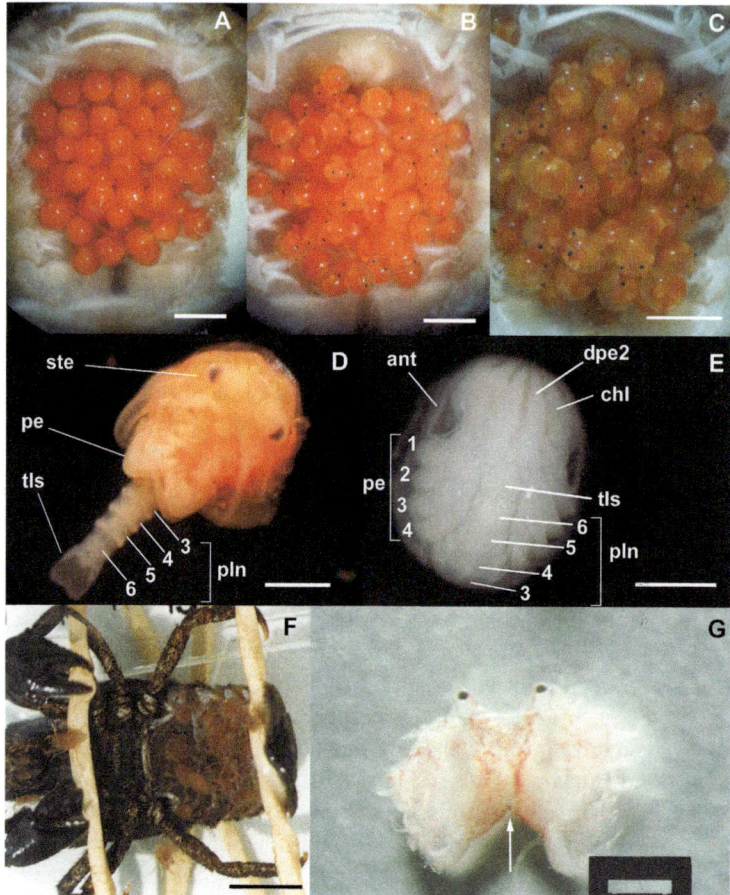

Figure 6.1   A, B, and C: *Aegla jaragua*. Embryonic stages of eggs attached to ovigerous female pleopods: (A) early eggs; (B) intermediate eggs; and (C) late eggs. Scale bars A, B, and C: 2 mm. (D) *Aegla franca*. Developing late embryo. Scale bar: 500 µm. (E) Fully developed late embryo within preserved egg obtained from an ovigerous female of a presumed new species of *Aegla* from subterranean habitat. Scale bar 500 µm. (F) Parental care: *Aegla franca*. Exposed live specimens of newly hatched juveniles in the brood chamber after the female pleon has been kept open. Scale bar: 5 mm. (G) *Aegla franca*. Conjoined twins joined by the dorsal surface of the carapace of each individual (arrow). Scale bar: 0.8 mm. Abbreviations: ant (antenna); chl (chela); dpe2 (distal portion of second pereopods); pe (pereopods); pln (pleonites); ste (stalked eyes); tls (telson).

6.2.2 Morphology of the Newly Hatched Juvenile

Newly hatched juveniles of *Aegla* display several advanced morphological characteristics that correspond to features seen in the megalopa and the first juvenile crab instar of marine anomurans. In a way, the very same advanced morphological condition is also shared by the adult specimens, except for the adult traits associated with reproduction. Hatchlings have the same tagmosis as the adults, and, except for the pairs of pleopods, all other pairs of appendages are well developed and fully functional, including a functional tail fan (Bueno and Bond-Buckup 1996; Teodósio and Masunari 2007; Moraes and Bueno 2013; Silva et al. 2017).

6.2.2.1 Carapace

The general shape of the carapace is similar to that of the adult form, with the anterior and posterior regions clearly separated by a distinct cervical groove. The posterior region of the carapace has been described as having a rougher surface texture than the anterior one (Moraes and Bueno 2013; Silva et al. 2017). Unique lineae (sutures) that distinguish adult aeglids from other anomurans are already present in newly hatched juveniles. These lineae (indicated on Figure 6.2B,C) are clearly discernible in scanning electron microscopy (SEM) preparations, as previously reported in the species *Aegla jaragua* (as *A. paulensis* s. lat.), *A. perobae*, *A. castro* (Moraes and Bueno 2013, 2015; Silva et al. 2017), and now also for *A. franca* (Figure 6.2B). Moraes and Bueno (2013), however, mentioned that the transverse dorsal linea was not discernible in newly hatched juveniles of *Aegla perobae*.

On the dorsal surface of the posterior region of the carapace, specific lineae contribute to demarcate the well-defined branchial and cardiac areas. The shape of the cardiac area—delimited laterally by a pair of non-parallel dorsal longitudinal lineae (dll), by the transverse dorsal linea (tdl) anteriorly and by the mid-section of the posterior border of the carapace—is strongly trapezoidal in early juveniles, whereas in adult specimens, this area is usually more weakly trapezoidal or subrectangular. Nonetheless, the shape of the areola, which is located in the central region of the cardiac area, may vary considerably [for comparison, see Figure 6 in Moraes and Bueno (2015), Figure 6A in Silva et al. (2017), and Figure 6.2B, in this chapter], and differs greatly from the definite shape observed in adult specimens (Martin and Abele 1988; see also Moraes et al. 2016 for details on shapes of the cardiac area and of the areola in adults).

6.2.2.2 Compound Eyes

The pedunculate (stalked) condition of the pair of compound eyes is clearly recognizable in the developing embryo (Figure 6.1D) and in juveniles soon after hatching (Figure 6.2A). Moraes and Bueno (2013) suggested that the number of small simple setae on the eye peduncle near the cornea (Figure 6.3A) might be of taxonomic value in comparative studies regarding juveniles of different aeglid species.

6.2.2.3 Gills

Developed gills (Figure 6.2A,C) are a clearly advanced trait in newly hatched juveniles of *Aegla* spp. Gill lamellae are located within the well-developed branchial chamber on each side of the posterior region of the carapace. Additionally, the

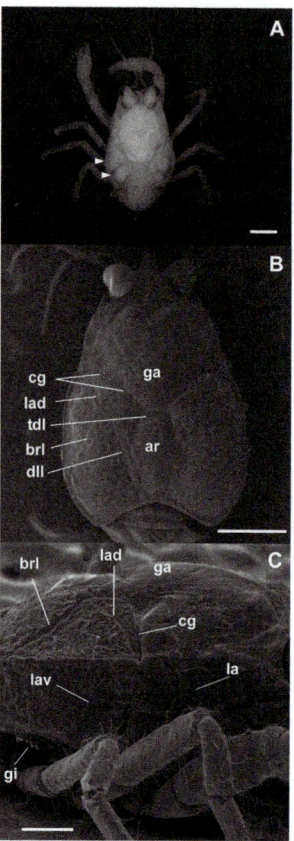

Figure 6.2   Newly hatched juveniles. (A) Preserved specimen of a newly hatched juvenile of a presumed new species of *Aegla* from subterranean habitat. Arrowheads indicate gills located within the branchial chamber. Scale bar: 0.5 mm. SEM. (B) *Aegla franca*. Carapace (dorsal view). Scale bar 500 μm. (C) *Aegla jaragua*. Carapace (lateral view). Scale bar 200 μm. Abbreviations: ar (areola); brl (branchial linea); cg (cervical groove); dll (dorsal longitudinal linea); ga (gastric area); gi (gills); la (linea aeglica); lad (linea aeglica dorsalis); lav (linea aeglica ventralis); tdl (transverse dorsal linea). Abbreviations and terminology of lineae follow those of Martin and Abele (1988) for adult aeglids.

presence of a fully developed and functional gill bailer (scaphognathite, or maxillary exopod) constitutes strong evidence that gas exchange in the newly hatched juveniles is already carried out in the same way as it is in adults.

6.2.2.4 *Antennules*

The general morphology of this first pair of uniramous head appendages is very similar between hatchlings and adult specimens, except for the higher number of segments in the two distal flagella in adults (see Martin and Abele 1988 for adult morphology). In

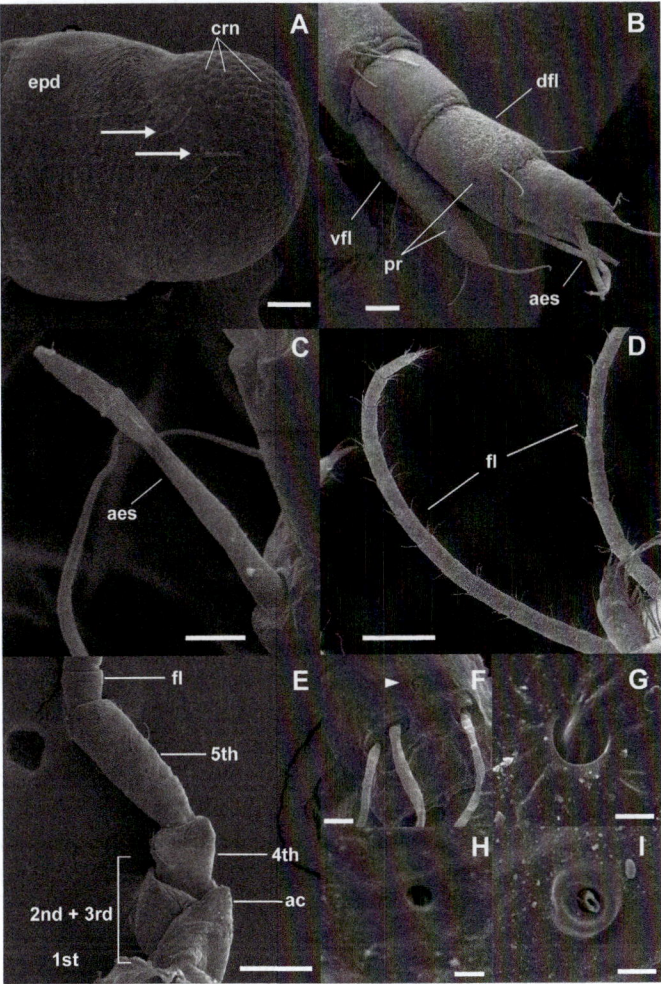

Figure 6.3 Newly hatched juvenile. SEM. (A) *Aegla schmitti*. Compound eye: small simple setae on the eye peduncle (arrows). Scale bar: 50 µm. (B) *Aegla franca*: antennule; three-segmented dorsal and one-segmented ventral rami (flagella). Scale bar: 20 µm. (C) *Aegla perobae*: aesthetasc on distolateral angle of second antennular segment (dorsal ramus). Scale bar: 10 µm. (D) *Aegla franca*: Antenna; pair of flagella (endopod). Scale bar: 200 µm. (E) *Aegla schmitti*. Antennal peduncle showing acicle (rudimentary exopod) on fused second and third segments. Scale bar: 100 µm. F, G, H, I: *Aegla jaragua*. Pores on antennal segments; (F) pore type 1 near distal setae. Scale bar: 5 µm. (G) Closer view of yet another pore type 1. Scale bar 2 µm. H) Pore type 2. Scale bar 0.5 µm. I) Pore type 3. Scale bar: 2 µm. Abbreviations: ac (acicle); aes (aesthetasc); crn (cornea with ommatidia); dfl (antennular dorsal flagellum); epd (eye peduncle); fl (antennal flagellum); pr (pore); vfl (antennular ventral flagellum). Sequential numbers indicate segments of antennal peduncle, from proximal to distal.

both juveniles and adults, the peduncle is three-segmented, from which the two flagella (or rami) originate apically. In newly hatched juveniles, the dorsal (or upper) flagellum is consistently three-segmented whilst the ventral (or lower) one is one-segmented in all *Aegla* juveniles described so far (Bond-Buckup et al. 1996, 1999; Bueno and Bond-Buckup 1996; Francisco et al. 2007; Teodósio and Masunari 2007; Moraes and Bueno 2013, 2015; Silva et al. 2017). The chemosensory function attributed to antennules is evident by the presence of olfactory aesthetascs on the distal and mesial segments of the dorsal flagellum (Figure 6.3B,C) and the presence of pore sensilla of a presumed chemoreceptor function on the surface of both flagella (Figure 6.3F,G) (Francisco et al. 2007; Moraes and Bueno 2013, 2015; Silva et al. 2017).

6.2.2.5 Antennae

The second pair of head appendages may also be regarded as uniramous; in adult aeglids only the inner ramus (endopod) is well-developed as a long, multi-articulated flagellum, whereas the outer ramus (exopod) is vestigial and reduced to a small triangular-shaped projection, known as the acicle, and located on the fused second and third segments (presumably corresponding to a fused basi-ischium) of the peduncle (Martin and Abele 1988). The acicle can be identified in the newly hatched juvenile as well (Figure 6.3E). The multi-articulated flagellum originates distally from the fifth peduncular article. The number of articles that comprise the flagellum may vary by species, from 13–14 in juveniles of *Aegla perobae* to 19–21 in juveniles of *Aegla schmitti* (Teodósio and Masunari 2007; Moraes and Bueno 2015). Flagellar articles are subequal in size and each one of them bears simple setae distally, except for the most proximal one, which is the shortest in size and is devoid of setae (Bond-Buckup et al. 1996; Bueno and Bond-Buckup 1996; Francisco et al. 2007; Teodósio and Masunari 2007; Moraes and Bueno 2013, 2015; Silva et al. 2017). Moraes and Bueno (2013) recognized and described three different morphological types of pore sensilla on the antennal segments. Pores type 1 are observed on the flagellum of the antenna and on both flagella of the antennule (Figure 6.3F,G). Pores type 2 and 3 on segments of the antennal flagellum are shown in Figure 6.3H,I.

6.2.2.6 Mandibles

Each uniramous mandible consists of a large calcified protopodal endite and a two-segmented palp (endopod). The distal border of the endite is a heavily sclerotized dentate incisor process, whereas the molar process is rudimentary in juveniles (Moraes and Bueno 2013, 2015; Silva et al. 2017) as well as in adult specimens (Martin and Abele 1988). Teeth and indentations from opposing mandibular incisor processes are asymmetrical and complementary as regards each other. The distal segment of the two-segmented mandibular palp bears a fringe of composite serrulate setae along the margins of the distal half (Figure 6.4C,D). The labrum and the paired paragnaths are already well developed in the first juvenile (Figure 6.4A). Although these structures are not true appendages, the labrum and the paragnaths mark the anterior and posterior limits of the pre-oral cavity, respectively.

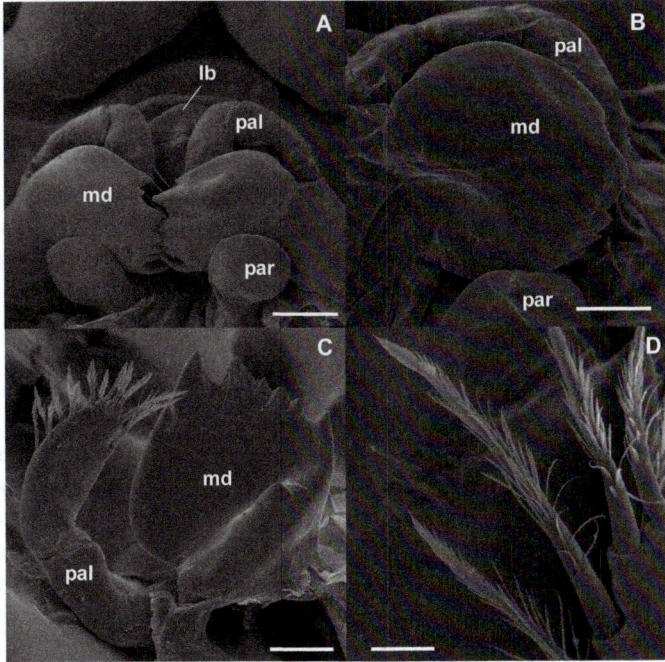

Figure 6.4 Newly hatched juvenile. SEM. (A) *Aegla franca*. Pair of well-developed mandibles. Scale bar: 100 µm. (B) *Aegla franca*. Underdeveloped mandible (right side). Scale bar: 50 µm. (C) *Aegla schmitti*. Inner surface of left mandible with two-segmented palp. Scale bar: 50 µm. (D) *Aegla schmitti*. Pappo-serrulate setae from the distal segment of the mandibular palp. Scale bar: 10 µm. Abbreviations: lb (labrum); md (mandible); pal (palp); par (paragnath).

The advanced asymmetrical morphological characteristics of the incisor processes suggest that this pair of appendages is already fully functional by the time juveniles hatch. No evidence exists, however, as to whether they actually feed on exogenous food while they are still under maternal care for a few days. Francisco et al. (2007) described the pair of mandibles from newly hatched juveniles of *A. franca* as having a rudimentary incisor process, which suggested that the mandibles would not be functional during the period of maternal care. A recent examination of additional first-juvenile specimens confirmed that some individuals of this species do show a rudimentary incisor process (see Figure 6.4B); results also showed that a well-developed incisor process with asymmetrical teeth and indentation can be observed in other individuals (Figure 6.4A). In either case, however, the mandibular palp is well developed, so perhaps the "rudimentary" condition of the mandible, as described by Francisco et al. (2007), should be more properly characterized as "underdeveloped" instead.

The last two pairs of head appendages—the maxillules and the maxillae—are located in the gnathal region and possess well-developed and setose endites on the protopod. Together, these endites are oriented mesially, covering most of the ganathobasic endites of the mandibles ventrally. They also exhibit a plethora of setal types showing a specialized morphology for food handling and the gustatory chemoreceptive function.

6.2.2.7 Maxillules

The maxillule is uniramous, flat, and well developed (Figure 6.5A). In juveniles, the maxillulary protopod already has well-developed and heavily setose coxal and basial endites. Types of setae on the basial endite include serrate setae and spine-like cuspidate setae (Figure 6.5B). Setal types on the coxal endite include pappose and serrate setae (Figure 6.5C). The endopod is short and not segmented (Figure 6.5 A).

6.2.2.8 Maxillae

Besides the important role in food handling mentioned above, the biramous maxilla is also the head appendage associated with promoting water movement inside the gill chamber of juveniles (Francisco et al. 2007). As in adult specimens, water movement is induced by the regular and rhythmic beating of the broad and flattened exopod (the scaphognathite or gill bailer, a decapod trait) bearing a fringe of long plumose setae along its free margin. In juveniles, the number of these plumose setae may vary from 42 in *Aegla violacea* to 78 in *A. schmitti* (Bueno and Bond-Buckup 1996; Teodósio and Masunari 2007). The gill bailer is oriented outwardly and well into the gill chamber, while the bilobed coxal and basial endites are oriented inwardly and ventrally covering the maxillule on the gnathal region (Figure 6.5D). In SEM preparations, Francisco et al. (2007) described groups of long setae bearing a subterminal pore of presumed chemoreceptive (gustatory) function on the distal margin of the bilobed basial endite from juveniles of *Aegla franca*. Setae showing similar morphology at the very same relative position were subsequently also reported for juveniles of other aeglid species (Moraes and Bueno 2013, 2015; Silva et al. 2017) (Figure 6.5E). Groups of serrate setae and a few pappose setae are found on the bilobed coxal endite distally (Figure 6.5F).

6.2.2.9 Maxillipeds

The first three pairs of thoracic appendages—the maxillipeds—are modified limbs used for food handling and, as in the case of the second and third pairs, also used for grooming (Martin and Felgenhauer 1986).

The first maxilliped (Figure 6.6A) is biramous, flat, and bears well-developed lobulate and setose coxal and basial endites on the protopod. Moraes and Bueno (2013, 2015) reported the presence of a rudimentary epipod [sensu (Schnabel and Ahyong 2010)] on the outer margin of the protopod in juveniles of *Aegla jaragua* (as *A. paulensis* s. lat.) and *A. perobae*. These authors mentioned that this rudimentary structure is homologous to the conspicuous lamellar lobe found in adult specimens and which has been identified as belonging to the basal portion of the exopod (Martin and Abele 1988).

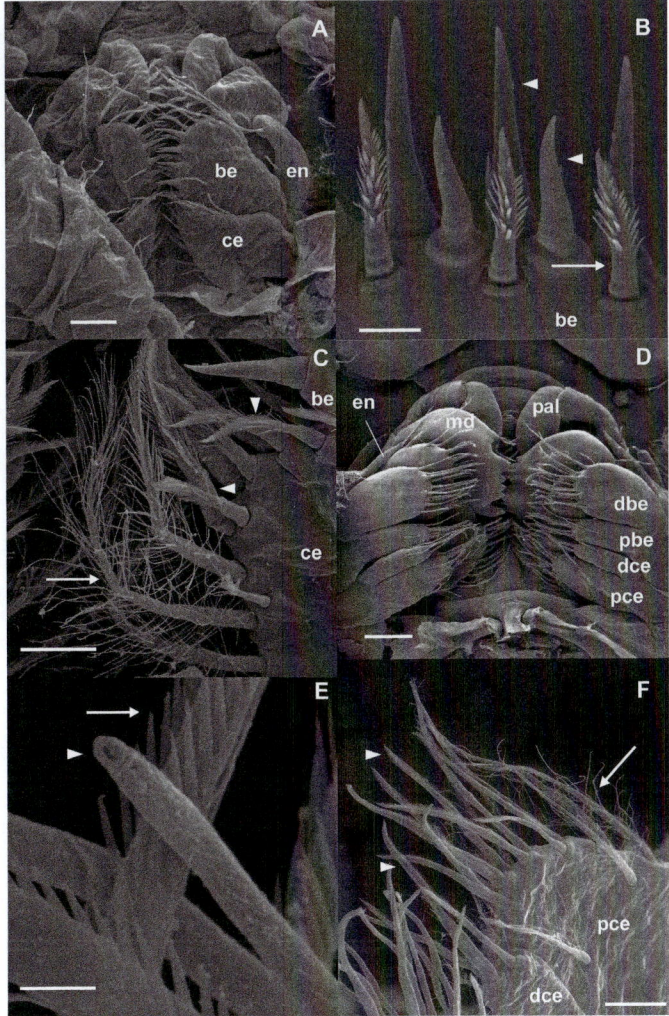

Figure 6.5 Newly hatched juvenile. SEM. (A) *Aegla jaragua*. Pair of maxillules (the maxillae and maxillipeds have been removed in this preparation). Scale bar: 50 µm. (B) *Aegla schmitti*. Maxillule. Basial endite bearing cuspidate (arrowhead) and serrate setae (arrow). Scale bar: 10 µm. (C) *Aegla jaragua*. Maxillule. Coxal endite bearing pappose (arrow) and serrate (arrowhead) setae. Scale bar: 20 µm. (D) *Aegla franca*. Pair of maxillae showing bilobed coxal and basial endites, and endopod. Exopod (scaphognathite) not shown (maxillipeds have been removed in this preparation). Scale bar: 50 µm. (E) *Aegla jaragua*. Maxilla. Gustatory setae with terminal pore (arrowhead) and serrate setae (arrow) from bilobed basial endite. Scale bar: 2 µm. (F) *Aegla jaragua*. Maxilla. Pappose (arrow) and serrate setae (arrowhead) from the bilobed coxal endite. Scale bar 20 µm. Abbreviations: be (basial endite); ce (coxal endite); dbe (distal lobe of basial endite); dce (distal lobe of coxal endite); en (endopod); md (mandible); pal (mandibular palp); pbe (proximal lobe of basial endite); pce (proximal lobe of coxal endite).

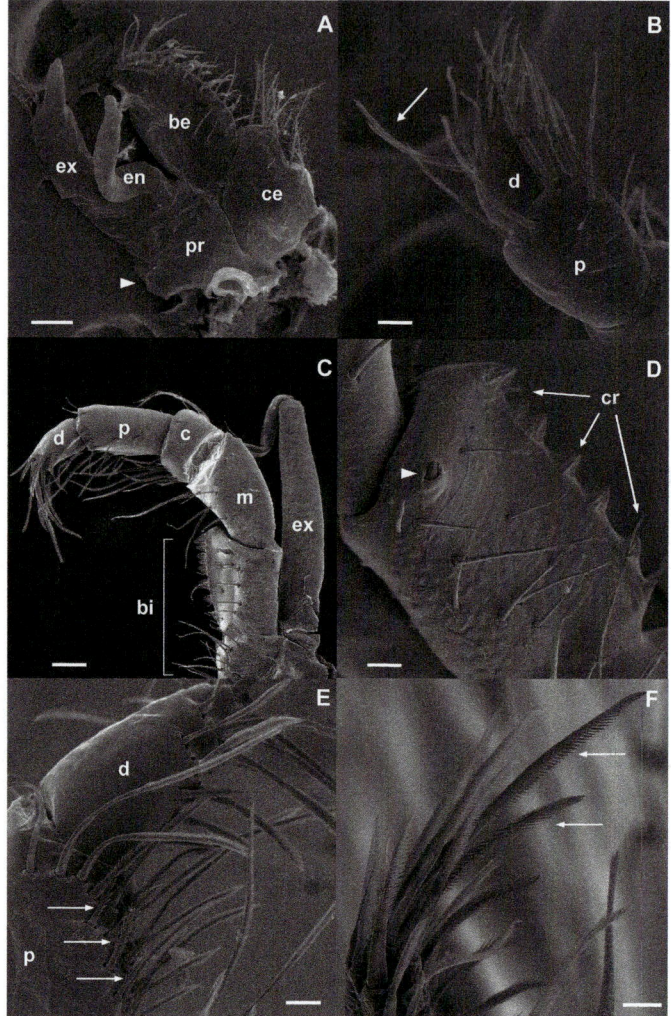

Figure 6.6 Newly hatched juvenile. SEM. (A) *Aegla franca*. First maxilliped. Arrowhead indicates vestigial epipod. Scale bar 50 μm. (B) *Aegla schmitti*. Second maxilliped. Distal portion of endopod showing group of serrate setae (arrows) on dactylus and propodus. Scale bar: 20 μm. (C) *Aegla franca*. Third maxilliped. Scale bar: 100 μm. (D) *Aegla schmitti*. Third maxilliped. Ischium portion of the fused basi-ischium, showing crista dentata on ventral border and one corneous tooth subdistally (arrowhead). Scale bar: 20 μm. (E) *Aegla perobae*. Third maxilliped. Short stout serrate setae (arrows) subterminally on ventral surface of propodus. Scale bar: 20 μm. (F) *Aegla jaragua*. Third maxilliped. Long serrate setae concentrated on distal portion of dactylus. Scale bar: 20 μm. Abbreviations: be (basial endite); bi (basi-ischium); c (carpus); ce (coxal endite); cr (crista dentata); d (dactylus); en (endopod); ex (exopod); m (merus); p (propodus); pr (protopod).

The second pair of maxillipeds (Figure 6.6B) is pediform and biramous. The endopod is strongly oriented at a right angle inwardly. The concentration of long serrulate and denticulate setae on the propodus and dactylus of the endopod clearly shows the grooming role of this pair of appendages. In juveniles, the long flattened proximal segment of the exopod bears very few pappose setae on the outer margin (Bueno and Bond-Buckup 1996; Francisco et al. 2007; Teodósio and Masunari 2007; Moraes and Bueno 2013, 2015; Silva et al. 2017). This condition strongly contrasts with that observed in adults, whose outer margin has a fringe of pappose and plumose setae along its entire length (Martin and Abele 1988).

The third pair of maxillipeds is also pediform and biramous (Figure 6.6C). In adult aeglids, it is one of the two most important pairs of appendages (the other is the reduced fifth pair of pereopods) used for grooming (Martin and Felgenhauer 1986). The third maxilliped in newly hatched juveniles bears striking resemblance to that of the adults, and, therefore, most probably participates in both grooming and food handling. Perhaps grooming presents itself first in juveniles still under parental care since exogenous food apparently is not utilized until the yolk reserve is used up. Juveniles and adults share the same groups of grooming setal types on the dactylus, propodus, and carpus, including several long serrate and denticulate setae distally on the dactylus and a group of small stout sword-like or serrulate setae on the distoventral surface of propodus and carpus (Figure 6.6E,F). These endopodal segments are used for the removal of fouling organisms and particulate matter from the antennule, antenna, and tips of ambulatory pereopods in adults (Martin and Felgenhauer 1986). These very same segments are most probably employed in the same way by newly hatched juveniles.

Besides the grooming function of the third maxillipeds, this pair of appendages in both juvenile and adult *Aegla* spp. also has a conspicuous crista dentata, a row of large teeth directed mesially on the ventral margin of the ischium (of the fused basiischium) (Figure 6.6C,D). Since the crista dentata facing each other on the right and left endopods may oppose and move against each other, this pair of structures most probably participates in food handling and maceration in juveniles as it does in adult aeglids (Martin and Felgenhauer 1986).

6.2.2.10 Pereopods

All pairs of pereopods are uniramous, well developed, and functional. Although the morphology of the first pair, the chelipeds, is similar in juvenile and adult specimens, a few distinct morphological features between them are worth mentioning. Firstly, juveniles lack the characteristic palmar crest on the dorsal margin of the manus of the propodus, an important taxonomic feature in adults of the species. Secondly, the cutting margins of both the dactylus (the moving finger) and the propodal pollex (the fixed finger) in juveniles lack a lobular basal tooth (present in adults, though it may be poorly developed in some species). They are instead ornamented by a row of long corneous scales (Figure 6.7A,B) that are morphologically different from the usual row of wider and flattened corneous scales found in adult specimens.

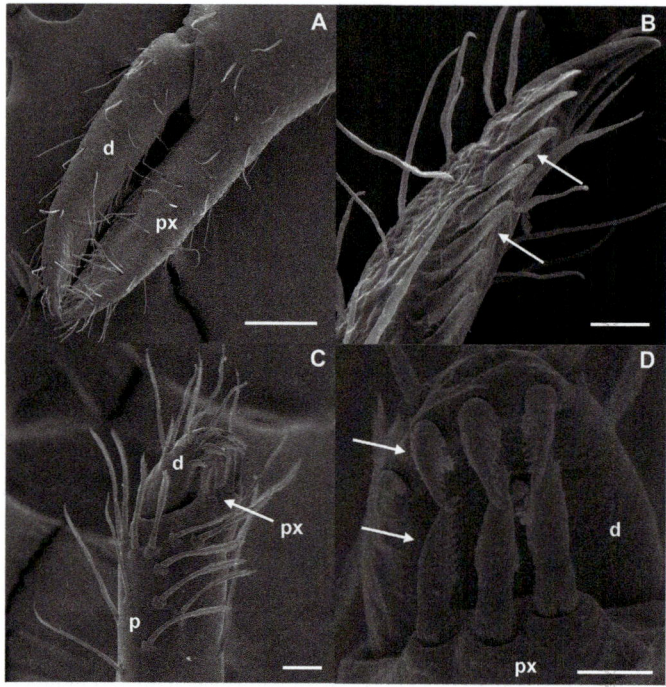

Figure 6.7 Newly hatched juvenile. SEM. (A) *Aegla schmitti*. First pereopod. Chela. Scale
bar 100 µm. (B) *Aegla jaragua*. First pereopod. Row of denticles (arrows) along
cutting margin of dactylus. Scale bar: 20 µm. (C) *Aegla schmitti*. Fifth pereopod.
Minute chela. Scale bar: 20 µm. (D) *Aegla schmitti*. Fifth pereopod. Robust stout
serrate setae (arrows) along distal margin of dactylus and pollex. Scale bar: 10
µm. Abbreviations: d (dactylus); p (propodus); px (pollex).

The second, third, and fourth pairs of pereopods are fully functional locomotory appendages and morphologically similar to those of adult specimens (Francisco
et al. 2007; Moraes and Bueno 2013; Silva et al. 2017).

In both juvenile and adult specimens, the fifth pair of pereopods is reduced,
which is an anomuran trait. Differently from the four preceding pairs of pereopods,
the basis is not fused with the ischium (Martin and Abele 1988). The fifth pair of
pereopods is highly mobile and capable of moving in multiple directions.

The fifth pereopods have a tiny chela distally (Martin and Abele 1988)
(Figure 6.7C,D; Figure 6.8A). In juveniles, the distal margins of the dactylus and
the pollex of the minute chela usually have three (variation: 2 to 4) stout curved
serrate setae; each interdigitates in alternate positions when the chela is closed
(Figure 6.7C,D). In adult specimens, the minute chelae have several small corneous
scales along the margin and show limited articulating action of the dactylus against
the pollex, raising the question as to how functional the chela really is (Martin and
Abele 1988). Nevertheless, their function as an important grooming appendage in
adult specimens is well established (Martin and Felgenhauer 1986). Notwithstanding

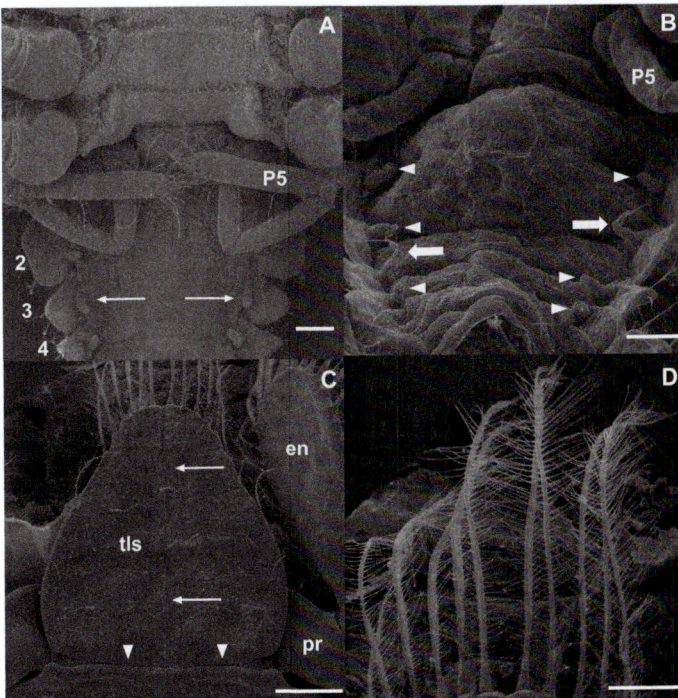

Figure 6.8 Newly hatched juvenile. SEM. (A) *Aegla franca*. Ventral surface of the pleon with pairs of rudimentary pleopod buds (arrows indicate the third pair). Scale bar: 100 μm. (B) Ventral surface of the pleon of the juvenile from a presumed new species of *Aegla* from subterranean habitat showing incomplete two-segmented uniramous pleopod buds with apical setae (arrows) and one-segmented pleopod buds (arrowheads). Scale bar: 100 μm. (C) *Aegla schmitti*. Dorsal surface of telson showing proximal transverse articulating margin (arrowheads) with sixth pleonal somite and mid-longitudinal suture (arrows). Scale bar: 100 μm. (D) *Aegla schmitti*. Fringe of plumose setae on distal margin of telson. Scale bar: 50 μm. Numbers in A, B, and C denote sequence of pleonal somites. Abbreviations: en (endopod from the left uropod); pr (protopod from left uropod); P5 (reduced fifth pereopods); tsl (telson).

its short length when compared to all other pereopods, the fifth pair of pereopods serves for grooming purposes in areas of the body that can be reached by their distal portion, such as the ventral surface of the pleon and the posterior region of the carapace (Martin and Felgenhauer 1986; Martin and Abele 1988). In adult specimens, they also play an important role in grooming the eggs of ovigerous females and cleaning gills by inserting, in the latter case, the distal portion of the appendage into the branchial chamber (Martin and Felgenhauer 1986).

In adult specimens of *Aegla platensis* and *A. uruguayana*, Martin and Felgenhauer (1986) observed a high density of composite setae of a unique setal morphology (in their own words: "with a distally-directed double row of sharp setulae and a proximally-directed border of minute papillae") distributed over large exposed surface areas of the dactylus and propodus, suggesting that the distal portion of the fifth pereopods functions as a cleaning brush for the gills. In juveniles, there are several serrate setae distributed over a similar relative position on the dactylus and the propodus (Figure 6.7C) (Moraes and Bueno 2013, 2015); these setae, however, are present in significantly fewer numbers than in adult specimens. Still, they may already play an important role in the grooming of gills, since these gas exchange structures are present and functional inside the branchial chambers of newly hatched juveniles while they are still under maternal care.

6.2.2.11 Pleon

Newly hatched juveniles have all six pleonal somites (the sixth pleonal somite is not fused with the telson) (Teodósio and Masunari 2007; Moraes and Bueno 2013, 2015). The second somite is the widest one; with the width of pleonal somites 3 to 6 decreasing progressively (Figure 6.8A,B). The width of the posterior margin of the sixth pleonal somite is similar to the width of the anterior margin of the telson (Figure 6.8C).

6.2.2.12 Pleopods

Adult males and females lack the first pair of pleopods. The second through fifth pairs of pleonal appendages (pleopods 2–5) are easily distinguishable sexual dimorphic traits in adult *Aegla* spp. because they are present and well developed only in females (Martin and Abele 1988). In this sex, pleopods are two-segmented, uniramous appendages employed for egg incubation. In adult males, pleopods 2–5 are vestigial and represented by small and calcified knobs (Martin and Abele 1988). An intriguing exception has been observed in adult male specimens of the cave-dwelling species *A. charon* Bueno and Moraes, 2017, for which short, one- and two-segmented pleopods have been reported (Bueno et al. 2017).

Pleopods 2–5 are absent in newly hatched juveniles of *A. platensis, A. prado, A. schmitti,* and *A. violacea* (Bond-Buckup et al. 1996; Bueno and Bond-Buckup 1996; Lizardo-Daudt and Bond-Buckup 2003; Teodósio and Masunari, 2007). Rudimentary buds of pleopods 2–5, however, were observed in hatchlings of *A. jaragua* (as *A. paulensis* s. lat.), *A. perobae,* and *A. castro* under SEM preparations (Moraes and Bueno 2013, 2015; Silva et al. 2017). Buds of pleopods 2–5 are also present in *A. franca* (Figure 6.8A). Moraes and Bueno (2013) speculated that the growth of pleopods 2–5 might proceed to full development in adult females, whereas they might regress to calcified knobs in adult males during ontogenetic development.

Recently, newly hatched juveniles of a presumed new species of cave-dwelling freshwater aeglid were obtained under laboratory conditions (S. Bueno unpublished data) (Figure 6.2A). All hatchlings examined had pleopod buds showing similar morphology to those previously observed in the four species from epigean habitats mentioned previously. Surprisingly, some hatchling specimens exhibited incompletely articulated two-segmented buds with one or two setae apically in some of the pleopods (Figure 6.8B). This clearly represents the most advanced developmental stage as yet observed regarding these abdominal appendages in newly hatched juveniles.

6.2.2.13 Telson

As in adult forms, the tail fan is already well developed and functional in newly hatched juveniles (Bueno and Bond-Buckup 1996; Bond-Buckup et al. 1999; Teodósio and Masunari 2007; Moraes and Bueno 2013; Silva et al. 2017). This structure consists of a broad flattened telson mesially (Figure 6.8C,D), flanked by a pair of well-developed biramous uropods (modified sixth pair of pleonal appendages) laterally. The tail fan is employed in escape-reaction behavior, which is characterized by the sudden and rapid swimming backwards caused by the quick (and sometimes repeated) flexing action of the muscular pleon while keeping the tail fan spread.

6.3 PARENTAL CARE

Parental care is the rule in freshwater aeglids (Vogt 2016). Indeed, this behavior has been observed in the following aeglid species: *Aegla castro* (Swiech-Ayoub and Masunari 2001a; Silva et al. 2017), *A. franca* (see Francisco et al. 2007), *A. jaragua* (as *A. paulensis* s. lat.) (see Moraes and Bueno 2013), *A. perobae* (Rodrigues and Hebling 1978; Moraes and Bueno 2015), *A. prado* (Verdi 1985; Bond-Buckup et al. 1996), *A. schmitti* (see Teodósio and Masunari 2007), and *A. uruguayana* (see López-Greco et al. 2004).

After hatching, juveniles remain under maternal care in the brood chamber formed by the flexed pleon (Figure 6.1F). The time juveniles remain under maternal care may vary from as few as 24–48 hours in *A. prado* (Verdi 1985) to 15 days in *A. castro* (Swiech-Ayoub and Masunari 2001a). López-Greco et al. (2004) reported that three distinct phases may be recognized during the 3- to 4-day period of maternal care in *A. uruguayana*: 1-day dependent phase, when juveniles remain protected in the brood chamber after they hatch; followed by a 2-day exploration phase, during which juveniles may leave the brood chamber for quick and occasional exploratory reconnaissance of the surroundings; then they return to the brood chamber or cling to other parts of their mother's body; and finally an independent phase on the fourth day, when juveniles leave definitively their mother, thus marking the end of the maternal-care period in this species.

Apparently, juveniles do not molt nor feed on exogenous food while they are still under maternal care. During this period, juveniles probably rely exclusively on the remaining yolk reserves that are still visible in the cephalothorax by the time juveniles hatch (Lizardo-Daudt and Bond-Buckup 2003). Indeed, Francisco et al. (2007) and Moraes and Bueno (2013, 2015) considered the hatchlings of *Aegla* as lecithotrophic juveniles. The complete depletion of yolk reserves might serve as a driving stimulus for the juveniles to gradually leave their mother, to begin exploring the surroundings and to start feeding on exogenous food (Francisco et al. 2007). By the time juveniles become independent, however, they are still in the first juvenile instar (López-Greco et al. 2004). Indeed, Teodósio and Masunari (2007) reported that the first juveniles of *A. schmitti* remained in this very same instar for 45 days before molting to the second juvenile instar long after they have left the brood chamber.

The parent-offspring association also improves the chances of survival of juveniles (Thiel 2003; Vogt 2016). During parental care, the mother provides the hatchlings protection from predators by keeping them secure in the brood chamber. Except for the presence of functional chelipeds, juveniles have no other special structures that could serve as holding devices to firmly attach themselves to the female's pleopods during parental care (Vogt 2016). The mother, however, avoids premature loss of juveniles by keeping the pleon flexed and by choosing areas of reduced water flow (López-Greco et al. 2004).

6.4 RECRUITMENT

Because postembryonic development in *Aegla* species is direct, recruitment occurs as soon as juveniles leave the brood chamber after a short period under parental care. Juveniles show a reduced dispersal capacity and recruits tend to remain in the same distributional area of the parental population (Bueno et al. 2016). They frequently hide under pebbles and rocks and show a preference for remaining in areas where hydrodynamic water flow is less intense, such as protected areas near margins and pools with accumulated plant debris on the streambed (López 1965). The reduced capacity for dispersal of epibenthic juveniles favors high endemism (Thiel 2003; Vogt 2016), as evidenced by the distributional pattern of several critically endangered species of aeglids (Bueno et al. 2016; Santos et al. 2017). Additionally, highly endemic populations may be subject to sudden disturbance events in stream ecosystems that could potentially interfere with recruitment success and, therefore, could strongly affect the population structure and reproductive input for subsequent years (Bueno et al. 2014).

Recruitment can be deemed to occur at the moment when the proportion of ovigerous females decreases considerably after reaching its peak as revealed by the analysis of the reproductive cycle (see Section 10.3.3 in Chapter 10). Peaks of females carrying eggs are discernible in species of *Aegla* with a markedly seasonal reproductive period (4 to 7 months) as well as in those species that exhibit extended reproductive periods (8 to 12 months) [see review on this topic by Bueno et al. (2016 and references therein)].

When analysing a temporal sequence of frequency distributions in size classes, the inclusion of recently recruited juveniles can be detected as an increase in the frequency of the smallest size classes, constituting a distinguishable cohort (see Section 10.3.1 in Chapter 10). The lag between the actual recruitment time and the time when the new cohort can be detected varies in accordance with the selectivity of the sampling method: it can range from 1 to 11 months, when aeglids are actively collected with the aid of nets (Bahamonde and López 1961; López 1965; Bueno et al. 2000; Bueno and Bond-Buckup 2000; Swiech-Ayoub and Masunari 2001a, b; Noro and

Buckup 2002, 2003; Fransozo et al. 2003; Colpo et al. 2005; Gonçalves et al. 2006; Trevisan and Santos 2014; Marçal et al. 2018) and from 9 to 16 months when traps (a device that is more selective against small individuals) are employed (Teodósio and Masunari 2009; Cohen et al. 2011; Grabowski et al. 2013; Bueno et al. 2014; Chiquetto-Machado et al. 2016; Copatti et al. 2016).

ACKNOWLEDGMENTS

We are very grateful to Angela Christine Charity for reviewing the English text. Special thanks to Dr. Christopher Tudge for critically reviewing both the manuscript and the English text.

REFERENCES

Anger, K. 2001. *The Biology of Decapod Crustacean Larvae. Crustacean Issues 14.* Lisse: A.A. Balkema.

Anger, K. 2006. Contributions of larval biology to crustacean research: a review. *Invertebrate Reproduction and Development* 49:175–205.

Anger, K., G. A. Lovrich, S. Thatje, and J. A. Calcagno. 2004. Larval and early juvenile development of *Lithodes santolla* (Molina, 1782) (Decapoda: Anomura: Lithodidae) reared at different temperatures in the laboratory. *Journal of Experimental Marine Biology and Ecology* 306:217–30.

Baba, K., Y. Fujita, I. S. Wehrtmann, and G. Scholtz. 2001. Developmental biology of squat lobsters. In *The Biology of Squat Lobsters*, eds. G. C. B. Poore, S. T. Ahyong, and J. Taylor, pp. 105–48. Boca Raton, FL: CRC Press, Crustacean Issues 20.

Bahamonde, N., and M. T. López. 1961. Estudios biológicos en la población de *Aegla laevis laevis* (Latreille) de El Monte (Crustacea, Decapoda, Anomura). *Investigaciones Zoológicas Chilenas* 7:19–58.

Barria, E. M., E. E DaForno, and C. G. Jara. 2006. Larval development of the hermit crab *Pagurus edwardsii* (Decapoda: Anomura: Paguridae) under laboratory conditions. *Journal of Crustacean Biology* 26(2):154–67.

Bond-Buckup, G., and L. Buckup. 1994. A Família Aeglidae (Crustacea, Decapoda, Anomura). *Arquivos de Zoologia do Museu de Zoologia da Universidade de São Paulo* 32:1–346.

Bond-Buckup, G., A. A. P. Bueno, and K. A. Keunecke. 1996. Primeiro estágio juvenil de *Aegla prado* Schmitt (Crustacea, Decapoda, Anomura, Aeglidae). *Revista Brasileira de Zoologia* 13(4):1049–61.

Bond-Buckup, G., A. P. Bueno, and K. A. Keunecke. 1999. Morphological characteristics of juvenile specimens of *Aegla* (Decapoda, Anomura, Aeglidae). In *Crustaceans and the Biodiversity Crisis*, eds. F. R. Schram, and C. V. Klein, pp. 371–81.Leiden: Brill.

Brodie, R., and A. W. Harvey. 2001. Larval development of the land hermit crab *Coenobita compressus* H. Milne Edwards reared in the laboratory. *Journal of Crustacean Biology* 21(3):715–32.

Bueno, A. A. P., and G. Bond-Buckup 1996. Os estágios juvenis de *Aegla violacea* Bond-Buckup and Buckup (Crustacea, Anomura, Aeglidae). *Nauplius* 4:39–47.

Bueno, A. A. P., and G. Bond-Buckup. 2000. Dinâmica populacional de *Aegla platensis* Schmitt (Crustacea, Decapoda, Aeglidae). *Revista Brasileira de Zoologia* 17:43–49.

Bueno, A. A. P., G. Bond-Buckup, and L. Buckup. 2000. Crescimento de *Aegla platensis* Schmitt em ambiente natural (Crustacea, Decapoda, Aeglidae). *Revista Brasileira de Zoologia* 17:51–60.

Bueno, S. L. S., A. L. Camargo, and C. B. Moraes. 2017. A new species of stygobitic aeglid from lentic subterranean waters in southeastern Brazil, with an unusual morphological trait: short pleopods in adult males. *Nauplius* 25:e201700021.

Bueno, S. L. S., and R. M. Shimizu. 2008. Reproductive biology and functional maturity in females of *Aegla franca* (Decapoda: Anomura: Aeglidae). *Journal of Crustacean Biology* 28(4):652–62.

Bueno, S. L. S., R. M. Shimizu,and J. C. B. Moraes. 2016. A remarkable anomuran: The taxon *Aegla* Leach, 1820. Taxonomic remarks, distribution, biology, diversity and conservation. In *A Global Overview of the Conservation of Freshwater Decapod Crustaceans*, eds. T. Kawai, and N. Cumberlidge, pp. 23–64. Cham: Springer International Publishing.

Bueno, S. L. S., B. F. Takano, F. P. A. Cohen, J. C. B. Moraes, P. I. Chiquetto-Machado, L. C. M. Vieira, and R. M. Shimizu. 2014. Fluctuations in the population size of the highly endemic *Aegla perobae* (Decapoda: Anomura: Aeglidae) caused by disturbance event. *Journal of Crustacean Biology* 34:165–73.

Campodonico, G. I. 1971. Desarrollo larval de la centolla *Lithodes antarctica* Jacquinot en condiciones de laboratorio (Crustácea Decapoda, Anomura: Lithodidae). *Anales del Instituto de la Patagonia Serie Ciencias Naturales* 2:181–90.

Chiquetto-Machado, P. I., L. C. M. Vieira, R. M. Shimizu, and S. L. S. Bueno. 2016. Life cycle of the freshwater anomuran *Aegla schmitti* Hobbs, 1978 (Decapoda: Anomura: Aeglidae) from southeastern Brazil. *Journal of Crustacean Biology* 36:39–45.

Clark, P. F., and P. K. L. Ng. 2008. The lecithotrophic zoea of *Chirostylus ortmanni* Miyake and Baba, 1968 (Crustacea: Anomura: Galatheoidea: Chirostylidae) described from laboratory hatched material. *The Raffles Bulletin of Zoology* 56(1):85–94.

Cohen, F. P. A., B. F. Takano, R. M. Shimizu, and S. L. S. Bueno. 2011. Life cycle and population structure of *Aegla paulensis* (Decapoda: Anomura: Aeglidae). *Journal of Crustacean Biology* 31:389–95.

Colpo, K. D., L. O. Ribeiro, and S. Santos. 2005. Population biology of the freshwater anomuran *Aegla longirostri* (Aeglidae) from south Brazilian streams. *Journal of Crustacean Biology* 25:495–99.

Copatti, C. E., R. P. Legramanti, A. Trevisan, and S. Santos. 2016 Growth, sexual maturity and sexual dimorphism of *Aegla georginae* (Decapoda: Anomura: Aeglidae) in a tributary of the Ibicuí River in southern Brazil. *Zoologia* 33:e20160010.

Cormie, A. K. 1993. The morphology of the first zoea stage of *Lomis hirta* (Lamarck, 1818) (Decapoda, Lomisidae). *Crustaceana* 64(2):249–55.

da Silva, A. R., M. R. Wolf, and A. L. Castilho. 2016. Reproduction, growth and longevity of the endemic South American crab *Aegla marginata* (Decapoda: Anomura: Aeglidae). *Invertebrate Reproduction & Development* 60(1):59–72.

Fagetti, E., and I. Campodonico. 1971. Larval development of the red crab *Pleuroncodes monodon* (Decapoda Anomura: Galatheidae) under laboratory conditions. *Marine Biology* 8(1):70–81.

Feldmann, R. M. 1984 *Haumuriaegla glaessneri* n. gen. and sp. (Decapoda; Anomura; Aeglidae) from Haumurian (Late Cretaceous) rocks near Cheviot, New Zealand. *New Zealand Journal of Geology and Geophysics* 27:379–85.

Feldmann, R. M. 1986. Paleogeography of two decapods taxa in the Southern Hemisphere; global conclusions with sparse data. In *Crustacean Biogeography*, eds. R. H. Gore, and K. L. Keck, pp. 5–19. Rotterdam: A.A. Balkema, Crustacean Issues.

Feldmann, R. M., F. J. Vega, S. P. Applegate, and G. A. Bishop. 1998. Early Cretaceous arthropods from the Tlayúa formation at Tepexi de Rodríguez, Puebla, México. *Journal of Paleontology* 72:79–90.

Francisco, D. A., S. L. S. Bueno, and T. C. Kihara 2007. Description of the first juvenile of *Aegla franca* Schmitt, 1942 (Crustacea, Decapoda, Aeglidae). *Zootaxa* 1509:17–30.

Fransozo, A., R. C. Costa, A. L. D. Reigada, and J. M. Nakagaki. 2003. Population structure of *Aegla castro* Schmitt, 1942 (Crustacea: Anomura: Aeglidae) from Itatinga (SP), Brazil. *Acta Limnologica Brasiliensia* 15:13–20.

Fujita, Y. 2010. Larval stages of the crinoids-associated squat lobster, *Allogalathea elegans* (Adams and White, 1848) (Decapoda: Anomura: Galatheidae) described from laboratory-reared material. *Crustacean Research* 39:37–53.

Fujita, Y., and P. F. Clark. 2010. The larval development of *Chirostylus stellaris* Osawa, 2007 (Crustacea: Anomura: Chirostylidae) described from laboratory-reared material. *Crustacean Research* 39:55–66.

Fujita, Y., and M. Osawa. 2005. Complete larval development of the rare porcellanid crab, *Novorostrum decorocrus* Osawa, 1998 (Crustacea: Decapoda: Anomura: Porcellanidae), reared under laboratory conditions. *Journal of Natural History* 39(10):763–78.

Fujita, Y., and S. Shokita. 2005. The complete larval development of *Sadayoshia edwardsii* (Miers, 1884) (Decapoda: Anomura: Galatheidae) described from laboratory-reared material. *Journal of Natural History* 39(12):865–86.

Fujita, Y., S. Shokita, and M. Osawa. 2002. Complete larval development of *Petrolisthes unilobatus* reared under laboratory conditions (Decapoda: Anomura: Porcellanidae). *Journal of Crustacean Biology* 22(3):567–80.

Gherardi, F., and P. A. McLaughlin. 1995. Larval and early juvenile development of the tube-dwelling hermit crab *Discorsopagurus schmitti* (Stevens) (Decapoda: Anomura: Paguridae) reared in the laboratory. *Journal of Crustacean Biology* 15(2):258–79.

Gonçalves, R. S., D. S. Castiglioni, and G. Bond-Buckup. 2006. Ecologia populacional de *Aegla franciscana* (Crustacea, Decapoda, Anomura) em São Francisco de Paula, RS, Brasil. *Iheringia, série Zoologia* 96:109–14.

Grabowski, R. C., S. Santos, and A. L. Castilho. 2013. Reproductive ecology and size of sexual maturity in the anomuran crab *Aegla parana* (Decapoda: Aeglidae). *Journal of Crustacean Biology* 33:1–7.

Jara, C. 1977. *Aegla rostrata* n. sp., (Decapoda, Aeglidae), nuevo crustáceo dulceacuícola del sur de Chile. *Studies on Neotropical Fauna and Environment* 12:165–76.

Jara, C. 1980. Dos nuevas especies de *Aegla* Leach (Crustacea, Decapoda, Anomura) del sistema hidrográfico del Rio Valdivia. *Anales del Museo de Historia Natural* 13:255–66.

Jara, C. G., and V. L. Palacios. 2001. Occurrence of conjoined twins in *Aegla abtao* Schmitt, 1942 (Decapoda, Anomura, Aeglidae). *Crustaceana* 74(10):1059–65.

Knight, M. D. 1967. The larval development of the sand crab *Emerita rathbunae* Schmitt (Decapoda, Hippidae). *Pacific Science* 21:58–76.

Lizardo-Daudt, H. M., and G. Bond-Buckup. 2003. Morphological aspects of the embryonic development of *Aegla platensis* (Decapoda, Aeglidae). *Crustaceana* 76(1):13–25.

López, M. T. 1965. Estudios biológicos en *Aegla odebrechtti paulensis*, Schmitt. *Boletim da Faculdade de Filosofia, Ciências e Letras da Universidade de São Paulo, Série Zoologia* 25:301–14.

López-Greco, L. S., V. Viau, M. Lavolpe, G. Bond-Buckup, and E. M. Rodríguez. 2004. Juvenile hatching and maternal care in *Aegla uruguayana* (Anomura, Aeglidae). *Journal of Crustacean Biology* 24:309–13.

Marçal, I. C., L. M. Ioshimura, J. J. S. Rosa, and G. M. Teixeira. 2018. Population structure and sexual maturity of *Aegla castro* (Decapoda, Anomura), an endemic freshwater crab from Brazil. *Invertebrate Reproduction and Development* 62(1):35–42.

Martin, J. W., and L. G. Abele. 1988. External morphology of the genus *Aegla* (Crustacea: Anomura: Aeglidae). *Smithsonian Contributions to Zoology* 453:1–46.

Martin, J. W., and B. E. Felgenhauer. 1986. Grooming behavior and the morphology of grooming appendages in the endemic South American crab genus *Aegla* (Decapoda, Anomura, Aeglidae). *Journal of Zoology* 209:213–24.

McLaughlin, P. A., K. Anger, A. Kaffenberger, and G. A. Lovrich. 2001. Megalopal and early juvenile development in *Lithodes santolla* (Molina, 1782) (Decapoda: Anomura: Paguroidea: Lithodidae), with notes on zoeal variations. *Invertebrate Reproduction and Development* 40(1):53–67.

McLaughlin, P. A., K. Anger, A. Kaffenberger, and G. A. Lovrich. 2003. Larval and early development in *Paralomis granulosa* (Jachinot) (Decapoda: Anomura: Paguroidea: Lithodidae), with emphasis on abdominal changes in megalopal and crab stages. *Journal of Natural History* 37(12):1433–52.

Moraes, J. C. B., and S. L. S. Bueno. 2013. Description of the newly-hatched juvenile of *Aegla paulensis* (Decapoda, Anomura, Aeglidae). *Zootaxa* 3635(5):501–19.

Moraes, J. C. B., and S. L. S. Bueno. 2015. Description of the newly-hatched juvenile of *Aegla perobae* (Crustacea: Decapoda: Aeglidae). *Zootaxa* 3973(3):491–510.

Moraes, J. C. B., M. Terossi, R. C. Buranelli, M. Tavares, F. L. Mantelatto, and S. L. S. Bueno. 2016. Morphological and molecular data reveal the cryptic diversity among populations of *Aegla paulensis* (Decapoda, Anomura, Aeglidae), with descriptions of four new species and comments on dispersal routes and conservation status. *Zootaxa* 4193(1):1–48.

Mouchet, S. 1932. Notes sur la biologie du galathéide *Aeglea laevis* (Latr.). *Bulletin de La Societé Zoologique de France* 57:316–40.

Müller, F. 1892. O camarão preto, *Palaemon potiuna*. Primeira parte: descrição do animal adulto. Segunda parte: A metamorphose dos filhos. *Archivos do Museu Nacional* 8:179–206, + plates XI–XIII.

Noro, C. K., and L. Buckup. 2002. Biologia reprodutiva e ecologia de *Aegla leptodactyla* Buckup and Rossi (Crustacea, Anomura, Aeglidae). *Revista Brasileira de Zoologia* 19:1063–74.

Noro, C. K., and L. Buckup. 2003. O crescimento de *Aegla leptodactyla* Buckup and Rossi (Crustacea, Anomura, Aeglidae). *Revista Brasileira de Zoologia* 20:191–98.

Pike, R. B., and R. G. Wear. 1969. Newly-hatched larvae of the genera *Gastroptychus* and *Uroptychus* (Crustacea, Decapoda, Galatheidae) from New Zealand waters. *Transactions of the Royal Society of New Zealand – Biological Sciences* 11(13):189–95.

Rodrigues, W., and N. J. Hebling. 1978. Estudos biológicos em *Aegla perobae* Hebling and Rodrigues, 1877 (Decapoda, Anomura). *Revista Brasileira de Biologia* 38:383–90.

Santos, S., G. Bond-Buckup, A. S. Gonçalves, M. L. Bartholomei-Santos, L. Buckup, and C. G. Jara. 2017. Diversity and conservation status of *Aegla* spp. (Anomura, Aeglidae): an update. *Nauplius* 25:e2017011.

Schnabel, S. K., and S. T. Ahyong. 2010. A new classification of the Chirostyloidea (Crustacea, Decapoda, Anomura). *Zootaxa* 2687:56–64.

Schmitt, W. L. 1942. The species of *Aegla*, endemic South American fresh-water crustaceans. *Proceedings of the United States National Museum* 91:431–524.

Silva, L. S. A., C. M. Guerrero-Ocampo, M. L. Negreiros-Fransozo, and G. M. Teixeira. 2017. Description of the newly-hatched juvenile of *Aegla castro* Schmitt, 1942 (Crustacea, Anomura, Aeglidae). *Zootaxa* 4237(1):167–80.

Stuck, K. C., and F. M. Truesdale. 1986. Larval and early postlarval development of *Lepidopa benedicti* Schmitt, 1935 (Anomura: Albuneidae) reared in the laboratory. *Journal of Crustacean Biology* 6(1):89–110.

Swiech-Ayoub, B. P., and S. Masunari. 2001a. Biologia reprodutiva de *Aegla castro* Schmitt (Crustacea, Anomura, Aeglidae) no Buraco do Padre, Ponta Grossa, Paraná, Brasil. *Revista Brasileira de Zoologia* 18:1019–30.

Swiech-Ayoub, B. P., and S. Masunari. 2001b. Flutuações temporal e espacial de abundância e composição de tamanho de *Aegla castro* Schmitt (Crustacea, Anomura, Aeglidae) no Buraco do Padre, Ponta Grossa, Paraná, Brasil. *Revista Brasileira de Zoologia* 18:1003–17.

Teodósio, E. A. F. M. O., and S. Masunari. 2007. Description of first two juvenile stages of *Aegla schmitti* Hobbs III, 1979 (Anomura: Aeglidae). *Nauplius* 15(2):73–80.

Teodósio, E. A. F. M. O., and S. Masunari. 2009. Estrutura populacional de *Aegla schmitti* (Crustacea: Anomura: Aeglidae) nos reservatórios dos mananciais da Serra, Piraquara, Paraná, Brasil. *Zoologia* 26:19–24.

Thiel, M. 2003. Extended parental care in crustaceans – an update. *Revista Chilena de Historia Natural* 76:205–18.

Trevisan, A., and S. Santos. 2014. Population dynamics of *Aegla manuinflata* Bond-Buckup and Santos 2009 (Decapoda: Aeglidae), a threatened species. *Acta Limnologica Brasiliensia* 26:154–62.

Verdi, A. C. 1985. Estudio del desarrollo embrionario en *Aegla prado* Schmitt (Crustacea, Decapoda, Anomura). *Actas de la Jornada de Zoología del Uruguay* 1:36–37.

Viau, V. E., L. S. López-Greco, G. Bond-Buckup, and E. M. Rodríguez. 2006. Size at the onset of sexual maturity in the anomuran crab, *Aegla uruguayana* (Aeglidae). *Acta Zoologica (Stockholm)* 87:253–64.

Vogt, G. 2016. Direct development and posthatching brood care as key features of the evolution of freshwater Decapoda and challenges for conservation. In *A Global Overview of the Conservation of Freshwater Decapod Crustaceans*, eds. T. Kawai, and N. Cumberlidge, pp. 169–98. Cham: Springer International Publishing.

CHAPTER 7

Intra- and Interspecific Behavioral Interactions of Aeglidae with a Comparison to Other Decapods

Marcelo M. Dalosto and Alexandre V. Palaoro

CONTENTS

7.1 INTRODUCTION

Aeglidae is the only family within the Infraorder Anomura that successfully invaded continental waters (Bond-Buckup et al. 2008) thereby isolating it from its closest relatives, the marine Anomura. The comparison between Aeglidae and their close marine relatives can enlighten us on the behavioral adaptations needed to invade freshwater environments. And comparing with other freshwater decapods, such as crayfish, can help us to understand convergent behavioral adaptations for freshwater. If we look at both comparisons, we may get a better understanding on how the environment and phylogenetics shape behavior. In this chapter, we will use this scenario as a guideline. First, we establish what is known for aeglids regarding

mating and fighting behavior, also reviewing how different species of aeglids interact among themselves. Second, we review how aeglids interact with other freshwater decapods. Lastly, we compare the behavior of Aeglidae with that of other
Anomura and crayfish. By providing these comparisons within *Aegla* and among
decapod crustaceans, we hope to broaden our current understanding of the evolution of Aeglidae.

7.2 BEHAVIOR OF AEGLIDAE

7.2.1 Mating Behavior

Mating encompasses how individuals find each other and the behavioral acts
performed until the offspring becomes independent (Andersson 1994). For aeglids,
there are no records of how individuals find each other in nature. Evidence gathered
from movement studies suggested that male aeglids are relatively active within a
short range (~5 m) with sudden bursts of movement, moving up to 121 m in a day
(Ayres-Peres et al. 2011a; Baumart et al. 2015). Hence, it is possible that males stay
in areas favorable for mating (e.g., areas where females are drawn to, see Chapter 5),
and defend their mates or the areas once there. If males do defend females or mating
areas, then aeglids would show a resource-defense polygyny mating system (Emlen
and Oring 1977), although we do not know that for certain.

Resource-defense polygyny is divided in the process of mating, which corresponds
to defending a resource, and the pattern of mating, which in this case is polygyny.
Defending a resource implies excluding conspecifics from the said resource, usually by fighting. To be good at excluding individuals (namely, to be a good fighter),
males need to be larger than their opponents—larger males with larger weapons win
more fights than smaller males (Vieira and Peixoto 2013). Consequently, males are
selected for size, which leads to sexual dimorphism: males are typically larger than
females (Palaoro and Beermann in press). Although male aeglids do not have larger
bodies than females (Barría et al. 2014), their weapons—the claws—are larger than
in the females (Bond-Buckup et al. 2008).

Resource-defense also implies that males need to fight for females or the resources
that females use to reproduce. Since males do not fight more fiercely for females
(Palaoro et al. 2013), they should fight for other resources, such as areas where food
accumulates. Currently, there is, however, no clear understanding of what males are
fighting for (see Section 7.2.2). In a polygynous mating system, males mate with
multiple females while females mate with only one male (Andersson 1994). Since
aeglid males mate with more than one female and considering that males use their
large claws during fights (see later in the section), it is reasonable to assume that a few
males with large claws may monopolize the females in a population. Such a behavior
would increase the variance in mating success among males, and sexual selection
would act on aeglids (Palaoro and Beermann in press). Unfortunately, we do not
have data on mating success for any aeglid species; there is, however, evidence that
females sire the offspring of only one male (S. Santos and M. Bartholomei-Santos,

personal communication), which may be a first step towards understanding the elusive mating system of aeglids.

The male of *A. platensis* approaches the located female with his claws raised, at least in laboratory environments (Almerão et al. 2010). If the female is receptive, the male moves closer while lifting himself on the tip of his pereopods with claws raised and starts beating his uropods. When the female displays a submissive posture, the male climbs on her cephalothorax and vibrates his abdomen, which is thought to stimulate egg release. Afterwards, the male positions himself underneath the female. Both individuals slightly open their abdomens, and then the female releases oocytes inside the abdomen and copulation ensues (Almerão et al. 2010). However, sperm transfer has never been observed, which raises the question: how do male aeglids transfer their sperm?

Although the female abdomen may contain a white mass after the copula, no spermatozoa were detected in this mass (Almerão et al. 2010), which raises the question if this white mass is really a spermatophore. There is another possibility that does not involve the white mass: males may use a sexual tube located on the fifth pair of pereopods to transfer sperm to females. The fifth pair of pereopods remains hidden inside the branchial chamber of both sexes, and it is supposedly used as a grooming structure (Bauer 2013). If it is only used for grooming, we would not expect sexual dimorphism in the structure. The male's coxa of the fifth pair, however, differs from the female's coxa in *A. uruguayana* (Viau et al. 2006). The male's sexual tube grows into a prominent structure that could be used to transfer sperm directly into the female's gonopore (Figure 7.1G, Viau et al. 2006). The sexual tube may also be used as a taxonomic character to distinguish between species, which highlights its importance in reproduction (Moraes et al. 2016, 2017; see Chapter 2). Alternatively, this sexual tube may also be used to deposit sperm on top of the oocytes deposited inside the female's abdomen. Unfortunately, there are no studies on these possible functions in aeglids, which might allow us to distinguish between the possible scenarios. After copulation, the male releases the female and stays near her until egg attachment is completed (Almerão et al. 2010). The male then continues looking for other females, while females care for the eggs until the hatching of the juveniles (see Chapter 6).

7.2.2 Contest Behavior

Animal contests can be separated into two decisions (Hardy and Briffa 2013): should I engage in a contest? When should I give up? Deciding whether to fight or not, as predicted by evolutionary game theory, depends on the value of the resource being contested and the costs to obtain that resource: when the resource value (V) outweighs the contest costs (C), the individual may fight (Maynard Smith and Price 1973). Thus, what are the resources aeglids are fighting for? And what are the costs?

When two aeglids are released into the same aquarium, they will fight, regardless of sex (Ayres-Peres et al. 2011b). However, the reasons why males engage in fighting are currently unknown. Since V should outweigh C for a contest to begin, the first hypothesis is that resources are highly valuable. When male fitness is limited by the

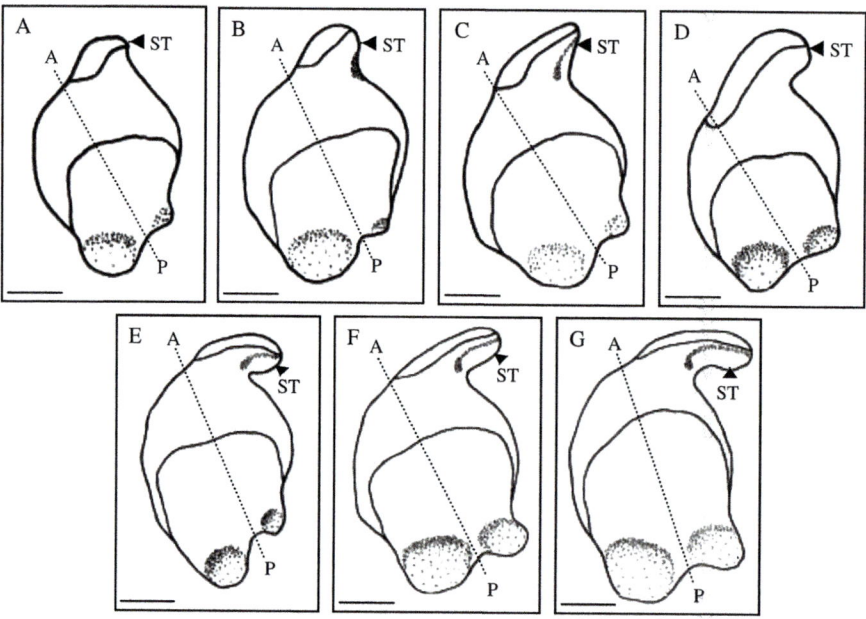

Figure 7.1   The sexual tube on the coxa of the fifth pair of pereopods of *Aegla uruguayana* increases during male growth. Schematic drawings A and B show sexually imma- ture juveniles. Individuals from C to E have mature gonads, but spermatozoa were not found. F and G are sexually mature males with mature gonads with sper- matozoids. ST = sexual tube; A and P: (A)ntero and (P)osterior axis of the coxa. (Reprinted with permission from Wiley, *Acta Zoologica*, vol. 87 (2006): 253–264.)

number of females they mate, females become a valuable resource to contest (Emlen and Oring 1977). Based on aeglid sexual dimorphism, males should fight intensely for females. If they are fighting for females, then the presence of receptive females should make males fight for longer and/or more intensely than when immature females are present. Fights in the presence of a mature female, however, are shorter and less intense than fights in the presence of an immature female (Palaoro et al. 2013); the prediction is thus not supported. Given that females are not a valuable resource for male aeglids, then they might fight for food (see Section 7.2.3), territories, or for dominance effects that may grant access to resources (Herberholz et al. 2007; Fero and Moore 2008). Future research needs to disentangle what aeglids are fighting for.

An alternative explanation usually not considered is that fighting costs may be low (Hardy and Briffa 2013). Under such a scenario, acquiring any resource would outweigh the costs. But can fighting costs be low? Fights have two sources of costs: (1) indirect costs, such as increased exposure to predators, and (2) direct costs, such as physiological costs and potential physical damage (Palaoro and Briffa 2017). Aeglids live in headwater streams that contain few large potential predators within the stream (Bond-Buckup et al. 2008). The main threat, thus, is located outside the stream: predatory mammals (Bond-Buckup et al. 2008). Considering that aeglids may be inconspicuous against the background of the stream due to their similar color pattern (S. Santos, personal communication), fighting may not necessarily increase predation risk unless aeglids fight during full moon or other high visibility situations (Palaoro et al. 2017).

Physiological costs of the contest behavior of aeglids have not been investigated so far, but there is ample evidence from other crustaceans and arthropods that fighting is physiologically costly (Briffa and Sneddon 2007). Levels of circulating glucose increase during fighting in the crab *Carcinus maenas* (see Sneddon et al. 1999), which suggests that fighting is indeed a physiologically demanding activity. Thus, it is hard to conceive that fighting may not be costly. In the velvet crab *Necora puber*, males increase their ventilation rates, which indicate higher metabolic rates (Smith and Taylor 1993). However, *N. puber* males do not increase levels of lactate during fighting (Thorpe et al. 1995) suggesting that individuals are not spending much energy although metabolic rates are higher. Another physiological variable that could relate to the costs of aggression is oxidative stress; research in this topic, however, is still in its infancy (Garratt and Brooks 2015). These contrasting pieces of evidence suggest that fighting may not necessarily be costly for all species, and that we need more information on aeglid physiology to understand the energetic demands of contest behavior.

The physical damage provided by the opponent is another source of direct costs in fights (Palaoro and Briffa 2017). However, it does not seem an important source of costs in aeglid fights: there are no reports of damaged, punctured, or broken claws in the literature (Ayres-Peres et al. 2011b; Palaoro et al. 2013; Palaoro et al. 2014; Ayres-Peres et al. 2015). Interestingly, behavioral acts correlated to physical damage are important predictors of contest outcome. In *A. longirostri*, males with taller and stronger claws that spend more time grabbing their opponent are more likely to win fights (Palaoro et al. 2014; Figure 7.2a). Hence, aeglids do not seem to be damaging

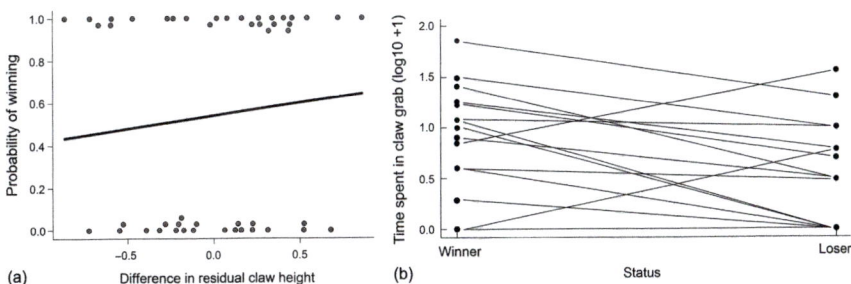

Figure 7.2 Male *Aegla longirostri* with taller claws are more likely to win fights (a) and winners spend more time grabbing the opponent's claws than losers (b). In (a), the residual claw height for winners and losers was retrieved from a linear regression between claw height and body size; then, the difference between winners and losers was calculated. The line represents the positive correlation between both variables according to a logistic regression. In (b), lines connect the fighting pair to show the dependency relation used in the paired *t*-test. (Reprinted with permission from Elsevier, *Animal Behaviour*, vol. 95 (2014): 71–79.)

opponents, but behavioral acts correlated to it are important in deciding a winner or, in other words, helping in the decision to give up.

A fight begins when two individuals face each other and start whipping their antennae (Ayres-Peres et al. 2011b). The fight then escalates to striking the opponent with the tips of the large claw until it reaches a grabbing contest. Now, individuals horizontally extend their large claw and bend their minor claw perpendicularly to the substrate. A series of attempts to grab the major claw of the opponent ensues, until one (or both) of the individuals is successful (Figure 7.3a). Contests usually terminate after a successful grab occurs, but when both individuals successfully grab their claws, the fight may escalate to unrestrained aggression (Ayres-Peres et al. 2011b). Most species apparently follow this pattern with some minor modifications (Ayres-Peres et al. 2015). For instance, *A. abtao* sometimes display their claws to the opponent after the physical contest (Ayres-Peres et al. 2015), and *A. denticulata* rarely grabs the opponent's claws. Aeglid fighting style differs from the typical fighting of decapod crabs, where brachyuran crabs typically have a "claw lock" act, in which both individuals grab each other claws at the same time and squeeze (Figure 7.3b, Dennenmoser and Christy 2013). Crayfish, on the other hand, grab each other by the base of the clawed pereopod to push the opponent backwards while squeezing its claw (Figure 7.3c; Moore 2007). Considerable damage is reported in the literature

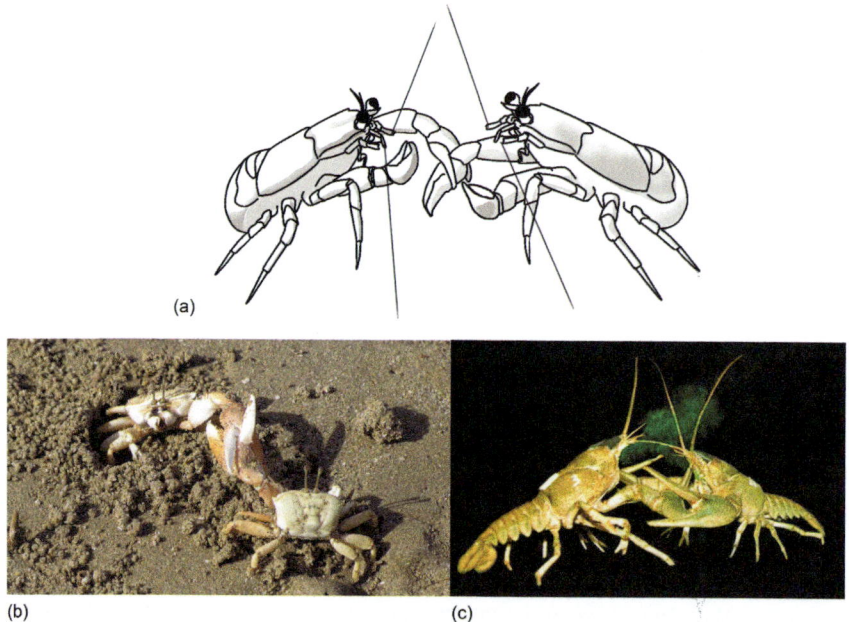

(a)

(b)   (c)

Figure 7.3 Aeglids fight differently from other decapods. (a) Drawing based on a fight between two males of *Aegla longirostri*. (b) *Uca monilifera* males fighting over a burrow. Note that both claws grab simultaneously (Photo courtesy of Denise Pope). (c) *Pontastacus leptodactylus* males trying to grab each other's claws at the base to begin a pushing contest. (Photo courtesy of Thomas Breithaupt.)

for both crabs and crayfish fighting styles (Rojas et al. 2012; Swanson et al. 2013), but not for aeglids. Therefore, even though aeglids have strong claws, they do not seem to generate significant physical damage.

If fighting costs are indeed low for aeglids, then it could influence how the animals make the decision to give up on a fight. Current theory divides the decision to give up into two main evolutionary models based on how much costs individuals are willing to bear, and how these costs accumulate during the fight (Palaoro and Briffa 2017). The "mutual assessment model" (Enquist and Leimar 1983) refers to when individuals assess their opponent's fighting ability before (and during) the fight to decide how much costs it is willing to bear; individuals are exchanging information prior (and possibly during) the fight, and hence, assessing each other. However, the decision to give up may also be based only on an individual's own fighting ability, with no information exchange between opponents (Bishop and Cannings 1978; Payne 1998). Under this model, the decision is based solely on how the actions of the focal individual accumulate costs, or how much damage he is receiving. These sets of models in which there is no information exchanged between opponents are all joined together under the same type model, the "self-assessment models" (Arnott and Elwood 2009).

Interestingly, in *A. longirostri* we found evidence for both models (Palaoro et al. 2014): males exchange information prior to the fight, which suggests mutual assessment, but winners and losers behaved according to the predictions of self-assessment models. The fact that winners grab the opponent's claws for a longer period of time than losers (Figure 7.2b) suggests that self-assessment models may represent the decision-making in aeglids better than the mutual assessment model. Although models are hard to distinguish regardless of taxa (Palaoro and Briffa 2017), fighting costs for aeglids have not been identified yet. Since fighting costs are central to contest theory and we still do not know much about them in aeglids, additional studies are required to enlighten the physiological costs and the role of claw-grabbing in aeglid fights.

7.2.3 Interactions among Aeglid Species

Contests are common in an aeglid's life: crabs are usually fighting one another in the stream. However, aeglids commonly occur in sympatry, which means that aeglids may also have to fight other aeglid species that share the same stream (Table 7.1). Additionally, most species probably have a high niche overlap: they are generalist macroconsumers with broad diets that inhabit clean and well-oxygenated headwater streams, and are nocturnal (Dalosto and Santos 2011; Cogo and Santos 2013; Baumart et al. 2015). However, in 87.5% of the sympatric instances where information on the abundance of each species is available, one species is always at least twice as abundant as the other/others (Table 7.1). Since aeglid species are ecologically similar, then how do they interact with each other?

Unfortunately, there is insufficient information for a robust answer. There are very few studies that analyzed competition and/or aggressive behavior between aeglid species (Parra et al. 2011; Ayres-Peres unpublished results), and only Parra et al. (2011) studied the interspecific aggression between three Chilean species, *A. araucaniensis*, *A. abtao*, and *A. denticulata*. These authors used *A. araucaniensis* from

Table 7.1   Records of Sympatric *Aegla* Species Showing the Most Abundant Species within the Stream

Aegla Species Occurring Sympatrically	Most Abundant Species	Reference
A. platensis, A. grisella, A. ludwigii	A. platensis	Copatti et al. (2015)
A. marginata, A. leptochela	A. marginata	Maia et al. (2013)
A. abtao, A. denticulata, A. araucaniensis	A. abtao	Parra et al. (2011)
A. manuinfalta, A. longirostri*	A. manuinflata	Santos et al. (2009)
A. platensis*, A. itacolomiensis	A. itacolomiensis	Bücker et al. (2008)
A. leptodactyla, A. camargoi	Similar abundances**	Castro-Souza and Bond-Buckup (2004)
A. jarai, A. muelleri	A. jarai	Boos Jr. (2003)
A. leptodactyla, A. camargoi	A. leptodactyla	Noro and Buckup (2002)
A. platensis, A. inermis	Data unavailable	Estevan (2015)
A. brevipalma, A. jarai	Data unavailable	Santos et al. (2012)
A. jarai, A. odebrecthii, A. oblata, A. leachi	Data unavailable	Santos et al. (2012)
A. jarai, A. oblata	Data unavailable	Santos et al. (2012)
A. platensis, A. spinipalma	Data unavailable	Machado (2012)
A. platensis, A. uruguayana	Data unavailable	Giri and Collins (2004)
A. jarai, A. spinosa	Data unavailable	Bond-Buckup and Buckup (1994)
A. spinipalma, A. longirostri	Data unavailable	Bond-Buckup and Buckup (1994)

Data for abundance varies from pure observations to quantitative methods. *=Uncertain taxonomic status; **=Abundance data of allopatric and sympatric populations were pooled, but results suggested similar abundances.

either allopatric or sympatric populations, and both dyadic (a pair of individuals, one of each species) and triadic (three individuals, one of each species) interactions. *Aegla abtao* dominated both *A. araucaniensis* and *A. denticulata* in all interactions, except in dyadic interactions against allopatric *A. araucaniensis* (Figure 7.4). Conversely, *A. denticulata* was always the subordinate species, regardless of the interaction (Figure 7.4). It is interesting that the dominant species (*A. abtao*) was also the most abundant species in these sympatric populations, whereas the subordinate was the least abundant species, indicating that interspecific aggression may play a role in regulating the co-occurrence patterns of aeglid species.

One of the unanswered questions is the role of claws during fights (see also Section 7.2.2). Claws become particularly important when species differ in their morphologies. For example, *A. abtao* has proportionally larger claws than *A. denticulata* (Ayres-Peres et al. 2015). Since large claws increase the chance of winning fights (Palaoro et al. 2014), large *A. denticulata* would need to fight small *A. abtao* to compensate for the claw asymmetry (Vieira and Peixoto 2013; Palaoro et al. 2014). However, this would add other asymmetries, such as different motivation levels due to different ages and/or body sizes (Vieira and Peixoto 2013). This problem is consistent in interspecific aggression studies: how can we compensate for asymmetries

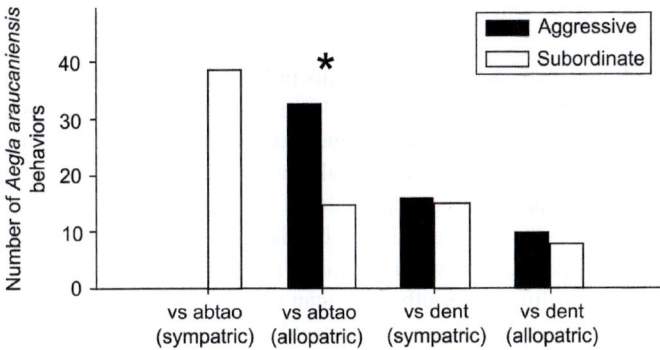

Figure 7.4 *Aegla araucaniensis* was the dominant species in most treatments, while *Aegla denticulata* was always subordinate. (*) Significant deviation in the number of both behavioral responses with respect to a hypothetical equal number of dominant and subordinate responses. (Redrawn with permission from the Taylor & Francis, New Zealand *Journal of Marine and Freshwater Research*, vol. 45 (2) (2011): 249–262.)

in overall body shape when studying interspecific contests? Alternative approaches could include, for example, a factorial experimental design where contestants would be matched by size in one treatment, strength (measured using muscle cross-sectional area and mechanical advantage as proxies) in other, and age in a separate treatment (e.g., Hardy et al. 2013; Palaoro et al. 2014).

Different fighting motivations are another frequent source of asymmetry in interspecific contests. How different species perceive the same resource may differ, resulting in asymmetrical fighting motivations (Hazlett et al. 1975; Kelly 2008). In the case of aeglids, food seems to be a better choice as a disputed resource than shelters, because aeglids do not invest in shelter construction, and food can be manipulated to standardize motivation, even among different species (e.g., Dalosto et al. 2015). Furthermore, different species seldom fight if there is no resource being disputed (Ord and Stamps 2009; Peiman and Robinson 2010). Parra et al. (2011) mentioned that one or more aeglids of different species occasionally ignored the others during the interspecific experiment, but this may have happened because they did not offer food to the fighting individuals. Conversely, when food was present, individuals of *A. manuinflata* and *A. longirostri* readily fought, and *A. longirostri* was the dominant species in most cases (Ayres-Peres unpublished data).

One final important question is: how do we define winners and losers? One option is to calculate a score based on the amount of aggressive/submissive acts, but this approach may be biased because some acts may be performed by just one individual (e.g., Berrill 1970 for *Munida sarsi*), even though they are quantified in the same manner. A more parsimonious approach would be to determine the animals' behavioral status according to the result of the fight: the individual that retreats is labeled the loser and its opponent the winner (e.g., Palaoro et al. 2014; Ayres-Peres et al. 2015). All these factors must be considered when studying interspecific aggression, and aeglids seem to be good candidates to investigate this type of interaction.

7.2.4 Interactions with Other Decapods

Aeglids do not share headwater streams only with other aeglids, they also share and interact with trichodactylid crabs, palaemonid shrimps, and parastacid crayfish (Morrone and Lopretto 1994; Dalosto and Santos 2011; Zimmermann et al. 2016). Moreover, whenever aeglids co-occur with one or more of these decapods, they are reported as the most abundant group (Dalosto and Santos 2011; Baumart et al. 2015; Zimmermann et al. 2016). It remains to be clarified why aeglids are apparently so dominant over other decapod species. One hypothesis comes from the observations of Riek (1971) regarding the South American crayfish fauna.

Riek (1971) hypothesized that it is unusual that the highly diverse crayfish (more than 500 species in North America and Australia) have 19 species described in South America (Ribeiro et al. 2017), a limited distribution (only occurring in southern South America), and low abundances (Buckup 1999; Breinholt et al. 2009). The life-style of this relatively poor diverse crayfish fauna is also unusual: crayfish are usually lotic animals, with some species inhabiting lakes and ponds; but all retain their primarily aquatic lifestyle (Vogt 2002; Moore 2007). Less than 10% of crayfish species, though, are considered semiterrestrial burrowers (Richardson 2007; Breinholt et al. 2009), but 18 of the 19 species described for South America are considered semiterrestrial burrowers, with only one (*Samastacus spinifrons*) inhabiting streams and lakes (Buckup and Rossi 1980; Rudolph and Crandall 2012; Ribeiro et al. 2017; Huber et al. 2018; Miranda et al. 2018). In comparison, South American streams harbor a diverse fauna of more than 87 species of aeglids (see Chapter 2).

The high abundance and diversity of aeglids versus the low abundance and diversity of crayfish in South America is the core of Riek's hypothesis. He argues that aeglids might have competitively excluded crayfish from streams, favoring a semiterrestrial burrowing lifestyle unavailable to aeglids due to physiological constraints (Riek 1971; Dalosto and Santos 2011). Favoring Riek's hypothesis is the large number of convergent ecological similarities between aeglids and non-burrowing crayfish: both groups are common in streams and rivers from subtropical and temperate climates (Morrone and Lopretto 1994; Crandall and Buhay 2008), are generalist omnivores (Nyström 2002; Colpo et al. 2012), and are key players for cycling of organic matter in their ecosystems (Usio 2000; Cogo and Santos 2013). Both groups also exhibit similar behavioral patterns, including primarily nocturnal activity (Gherardi 2002; Sokolowicz et al. 2007) and similar displacement patterns in streams (Bubb et al. 2006; Baumart et al. 2015).

While challenging to test, some studies provide circumstantial evidence for the competitive exclusion hypothesis: (1) burrowing crayfish (*Parastacus* spp.) have shorter and less intense fights than aeglids (Ayres-Peres et al. 2011b; Dalosto et al. 2013); (2) aeglids also have an advantage over burrowing crayfish when competing for food: *A. longirostri* was more likely to win a fight than *Parastacus brasiliensis* and held the food for longer periods when individuals were matched for claw strength (Figure 7.5; Dalosto 2016). Overall, *P. brasiliensis* won more fights than *A. longirostri* in this experiment, but crayfish were considerably larger than the aeglids used (M.M. Dalosto, personal obs.). Crayfish, however, need

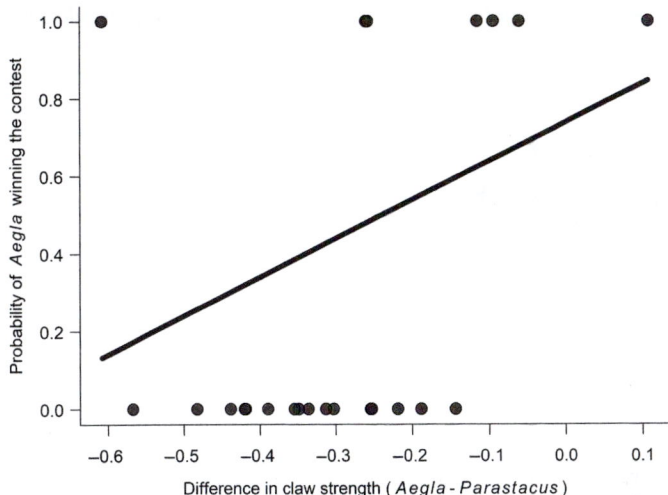

Figure 7.5 *Aegla longirostri* has a higher probability of winning a contest against *Parastacus brasiliensis* as the difference in weapon strength nears zero (Binomial GLM; $\chi^2_{1,19} = 0.5470$, $p = 0.0296$). Negative values indicate that *P. brasiliensis* is stronger than *A. longirostri*, while positive values indicate the opposite. When values reach zero, aeglids and parastacids are matched for strength.

approximately three years to attain large sizes in natural environments (Fontoura and Buckup 1989). Since *Aegla* species grow faster and occur in higher population densities than *Parastacus* (Fontoura and Buckup 1989; Silva-Gonçalves et al. 2009), crayfish would interact with many larger aeglids for about three years until they reached a size that assures a definite competitive advantage over aeglids. A similar scenario was proposed for the interactions between the American lobster *Homarus americanus* and the green crab *Carcinus maenas*: crabs dominate over small lobsters but are preyed upon by older, larger lobsters (Lynch and Rochette 2009). When individuals of *A. longirostri* and *P. brasiliensis* were matched for carapace size, aeglids won the majority of fights and held the resource for longer (Dalosto et al. 2012).

A final evidence for the competitive hypothesis is that aeglids use the stream more than crayfish when in sympatry (Baumart et al. 2015). If we consider the total area used by the individuals, aeglids and crayfish have similar home ranges; however, aeglids remain within the stream for longer times than crayfish, which spend most of their time within underground galleries outside the stream (Baumart et al. 2015). Additionally, crayfish activity peaks do not coincide with those of aeglids (Figure 7.6). Hence, these species are not only using different microhabitats, but they also move around at different times, which decreases the number of encounters in nature. According to Dalosto (pers. comm.), both species fought each other a few times, but fights were brief, since either the aeglid or the crayfish was markedly larger than its opponent, and the smaller animal fled almost immediately after the approach of the other.

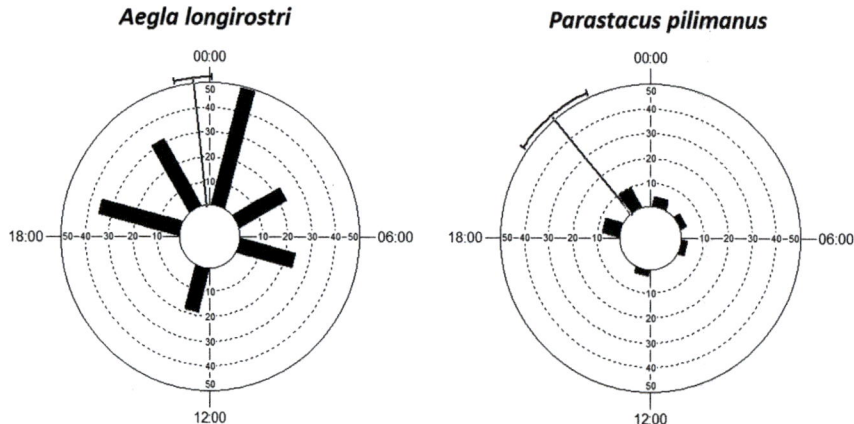

Figure 7.6 Diel activity of *Aegla longirostri* and *Parastacus pilimanus* monitored with radio-tags. Bars represent mean displacement (m), and vectors indicate the means and 95% CI. Data were pooled by species. (Reprinted with permission from Springer, *Hydrobiologia*, vol. 762 (2015): 29–39.)

While we can make educated guesses concerning the interactions between aeglids and parastacids based on a few studies, there is much less knowledge regarding how *Aegla* interacts with shrimps and crabs. There are no reports on shrimp interactions with aeglids. However, given that freshwater shrimps (genera *Macrobrachium* and *Palaemonetes*) are usually smaller than aeglids, have slender chelae, and different microhabitat preferences (they usually dwell in ponds, or in plant-covered, slow-flowing areas of streams), it is unlikely that interactions with aeglids are frequent and/or of major ecological importance (Bond-Buckup and Buckup 1989; Buckup and Bond-Buckup 1999). On the other hand, trichodactylid crabs seem to exhibit a similar niche as that of aeglids and are often encountered in sympatry with them (Dalosto and Santos 2011; Burress et al. 2013; Diawol et al. 2016). Nevertheless, direct behavioral interactions between these groups have not been studied yet. Based on the observations of Zimmermann et al. (2009) and Ayres-Peres et al. (2011b), *Trichodactylus panoplus* is less active than aeglids, and rarely displays aggressive behaviors. Additionally, the occurrence of *T. panoplus* is negatively correlated to the presence of adult individuals of *A. longirostri*, further suggesting a superior competitive ability of aeglids (Zimmermann et al. 2016). Thus, aeglids seem to interact mostly with crabs and crayfish, and the evidence so far suggests that aeglids should dominate over these other decapods in competitive interactions.

7.2.5 Aeglids as Behavioral Models

7.2.5.1 Comparisons with Other Anomura

As we have shown in this chapter, studies on aeglid behavior started quite recently. To expand our knowledge, we will draw parallels with phylogenetically close groups—after all, behavior also has phylogenetic constraints (Caetano and Machado 2013). The closest relatives are the marine anomurans, which include squat lobsters (Galatheidae), porcelain crabs (Porcellanidae), and hermit crabs (Paguridae, Diogenidae). Squat lobsters do physically resemble aeglids, but their behavior is distinct, as squat lobsters are typically gregarious, sedentary animals (Thiel and Lovrich 2011). Some species seldom display aggression towards each other, and when they do, their aggressive levels are very low, never escalating into physical contests, and they do not form dominance hierarchies (Antonsen and Paul 1997). Even though other squat lobster species engage in physical contests more often, these encounters are random sequences of behaviors (i.e., there is no escalation), and also do not result in the formation of dominance hierarchies (Berrill 1970; Claverie and Smith 2007). Conversely, aeglids are solitary animals that readily display aggression towards other conspecifics and seem to maintain dominance hierarchies (supplementary material in Ayres-Peres et al. 2015).

The behavioral pattern of the aeglids also differs from that of porcellanids, whose aggressive encounters involve short bouts of pushing followed by a rapid escalation to unrestrained physical combat with both claws (Baeza et al 2002). Fights are usually associated with the possession of a host on which porcenallids live, such as sea urchins and anemones (Baeza et al. 2002). The distribution of these hosts, both in space and time, usually correlates with the aggressiveness level of the species (Baeza and Thiel 2003). For instance, *Allopetrolisthes spinifrons* inhabits the sea anemone *Phymactis clematis*, which is rare to find and so small that only one crab can live on it (Baeza et al. 2002). Since the resource is valuable for the crab's survival, aggression is high, and individuals may rip the claws of their opponents during fights (Baeza et al. 2002). Other porcenallids that live on more abundant hosts, such as *Petrolisthes cinctipes*, usually show less aggression toward conspecifics, if any at all (Baeza and Thiel 2003; Donahue 2004; Rypien and Palmer 2007). Unfortunately, we do not know which resources modulate aeglids' fighting behavior, and future studies should tackle this issue so that more comparisons can be made.

Agonistic encounters of hermit crabs involve specialized acts that are most often related to taking over the shell of another individual (Imazu and Asakura 2006). At the onset of the fight, one individual hides within its shell (the defender). Its function is to endure the attacks of the individual that tries to pull him out of its shell (the attacker). The attacker usually grabs the defender by the shell's rim and raps its shell against the defender's, sometimes stopping and trying to pull the defender out of the shell with its claws (Hazlett 1981; Imazu and Asakura 2006). When each individual has a defined role within a fight, we call it an asymmetric contest (Arnott and Elwood 2008). Aeglids do not have asymmetric contests as far as we know—both individuals simultaneously attack and defend, and neither is defending a position or a burrow (Ayres-Peres et al. 2011a). Maybe if males defend a stream area to attract females, as hypothesized in Subsection 7.2.1, then we could have asymmetric contests in aeglids. Considering the lack of information, comparisons with hermit crabs are currently tentative and weak.

7.2.5.2 Comparisons with Crayfish

Similar environmental conditions can select for similar behavioral patterns in different groups (i.e., convergent adaptation) (Losos 2011). The group that bears the most ecological resemblance to aeglids are the North American and Australian crayfishes (with the exception of the specialized burrowers; see Richardson 2007). As we have mentioned in Section 7.2.4, there is a significant number of convergent ecological traits between aeglids and stream-dwelling crayfish that may allow to draw parallels from crayfish behavioral patterns.

A key aspect of crayfish aggression is hierarchy formation: crayfish establish long-lasting dominance hierarchies primarily based on social status recognition (Issa et al. 1999; but see Jiménez-Morales et al. 2018). These hierarchies are characterized by a decrease in aggression levels after the first encounter and can occur even between different species of crayfish (e.g., Gherardi and Daniels 2004). There is no published work investigating hierarchy formation in aeglids, but every study regarding their aggressive behavior reports that a "winner" and a "loser" were always clearly defined, with the loser consistently exhibiting subordinate behaviors (Ayres-Peres et al. 2011a, b; Palaoro et al. 2013, 2014). Furthermore, contest duration decreases after the third fight, which also agrees with the hypothesis that a hierarchy has been formed (Ayres-Peres et al. 2015). This pattern indicates that it is likely that aeglids exhibit some hierarchy formation, although confirmation of this potential hierarchy and its mechanisms remains to be investigated.

The issue of hierarchy formation also raises the question: how do aeglids communicate to maintain a hierarchy? Again, comparing aeglids with crayfish provides interesting insights. Crayfish communicate during social interactions, relying mainly on chemical cues liberated in the urine (Breithaupt and Eger 2002). Urine release increases during social interactions and blocking urine release prevents crayfish from mating and forming dominance hierarchies (Zulandt-Schneider et al. 2001; Berry and Breithaupt 2010). Thus, crayfish can communicate information related to both dominance status and sex (Breithaupt 2011). We know that allowing chemical communication between males prior to contests can alter fight dynamics in aeglids (Palaoro et al. 2014), and that fight duration decreases after the first fight (Ayres-Peres et al. 2015). These facts suggest that information is exchanged between fighting aeglid males and that there is a mechanism mitigating aggression levels in fights. The most likely hypothesis is that aeglids, like crayfish, form dominance hierarchies based on social status recognition, with information conveyed via urine. The main obstacle to answer how aeglids communicate is that aeglids are significantly smaller than crayfish, with current urine blocking and fluorescein injection protocols conceived for crayfish being unsuitable for smaller decapods (Breithaupt 2011). However, there are alternative options: experiments comparing fights between familiar and unfamiliar pairs of winners/losers could demonstrate if aeglids exhibit stable dominance hierarchies and, if they occur, whether they are based on individual or social status recognition (e.g., Zulandt-Schneider et al. 2001). Once this is established, the role of urine-borne communication could be assessed by impairing the chemical receptors in the antennules (e.g., Horner et al. 2008).

Analyzing the differences between crayfish and aeglid behavior, however, can also provide useful insights. An important difference between both decapod groups lies in the relationship between claw size and claw strength. For aeglids, claw size correlates strongly to claw strength, whereas the same correlation is weaker in crayfish, suggesting that claw size may be a dishonest signal of strength in crayfish, but not in aeglids (Wilson et al. 2007; Bywater et al. 2008; Palaoro et al. 2014). Considering that claw strength is important for aeglids and that claw grabbing is associated with winners, we may look for physiological traits that relate to both dominance status and claw use. The most logical step would be looking for possible physiological costs suffered by the eventual losers. Although there is little evidence for direct costs, one untested possibility is that the costs are instead determinant for those who inflict the force. Prolonged muscular contraction usually results in oxidative stress and potential tissue damage (Powers and Jackson 2008). Thus, a localized physiological response in the claws might mean that individuals with better anti-oxidant systems (i.e., able to mitigate the costs of prolonged contractions) might be able to grasp their opponents for longer periods, thus increasing their chances of winning. There is, however, little known about the relationship between oxidative stress and aggression. No studies have addressed this issue in decapod crustaceans, and future studies are certainly needed to clarify this relationship.

7.3 OPEN RESEARCH QUESTIONS

A. How are males and females spatially distributed? How do the sexes find each other in the natural environment?
B. What are the resources aeglids fight for? Do these resources modulate their aggression?
C. What are the costs (both direct and indirect) of aggression for aeglids? Do these costs differ between winners and losers? If so, how?
D. Do aeglids form dominance hierarchies and, if so, are they based on social status or individual recognition?
E. Aeglids exchange information during social interactions, but what sort of information is being conveyed?
F. We know that aeglids commonly occur in sympatry, both among different aeglid species and/or with other decapods, but are there any behavioral mechanisms that enable coexistence? Currently, studies only report that they coexist, none of them deal with how they are able to coexist.

REFERENCES

Almerão, M., G. Bond-Buckup, and M. D. S. Mendonça. 2010. Mating behavior of *Aegla platensis* (Crustacea, Anomura, Aeglidae) under laboratory conditions. *Journal of Ethology* 28(1):87.
Andersson, M. B. 1994. *Sexual Selection*. Princeton, NJ: University Press.

Antonsen, B. L., and D. H. Paul. 1997. Serotonin and octopamine elicit stereotypical agonistic behaviors in the squat lobster *Munida quadrispina* (Anomura, Galatheidae). *Journal of Comparative Physiology A* 181(5):501–10.

Arnott, G., and R. W. Elwood. 2008. Information gathering and decision making about resource value in animal contests. *Animal Behaviour* 76(3):529–42.

Arnott, G., and R. W. Elwood. 2009. Assessment of fighting ability in animal contests. *Animal Behaviour* 77(5):991–1004.

Ayres-Peres, L., P. B. Araujo, C. G. Jara, A. V. Palaoro, and S. Santos. 2015. How variable is agonistic behavior among crab species? A case study on freshwater anomurans (Crustacea: Decapoda: Aeglidae). *Journal of Zoology* 297(2):115–22.

Ayres-Peres, L., P. B. Araujo, and S. Santos. 2011b. Description of the agonistic behavior of *Aegla longirostri* (Decapoda: Aeglidae). *Journal of Crustacean Biology* 31(3):379–88.

Ayres-Peres, L., C. Coutinho, J. S. Baumart, A. S. Gonçalves, P. B. Araujo, and S. Santos. 2011a. Radio-telemetry techniques in the study of displacement of freshwater anomurans. *Nauplius* 19(1):41–54.

Baeza, J. A., W. Stotz, and M. Thiel. 2002. Agonistic behaviour and development of territoriality during ontogeny of the sea anemone dwelling crab *Allopetrolisthes spinifrons* (H. Milne Edwards, 1837) (Decapoda: Anomura: Porcellanidae). *Marine and Freshwater Behaviour and Physiology* 35(4):189–202.

Baeza, J., and M. Thiel. 2003. Predicting territorial behavior in symbiotic crabs using host characteristics: a comparative study and proposal of a model. *Marine Biology* 142(1):93–100.

Barría, E. M., S. Santos, C. G. Jara, and C. J. Butler. 2014. Sexual dimorphism in the cephalothorax of freshwater crabs of genus *Aegla* Leach from Chile (Decapoda, Anomura, Aeglidae): an interspecific approach based on distance variables. *Zoomorphology* 133(4):379–89.

Bauer, R. T. 2013. Adaptive modification of appendages for grooming (cleaning, antifouling) and reproduction in the Crustacea. In *Natural History of Crustacea v.1, Functional Morphology and Diversity*, eds. L. Watling, and M. Thiel, pp. 327–64. Oxford: Oxford University Press.

Baumart, J. S., M. M. Dalosto, A. S. Gonçalves, A. V. Palaoro, and S. Santos. 2015. How to deal with a bad neighbor? Strategies of sympatric freshwater decapods (Crustacea) for coexistence. *Hydrobiologia* 762(1):29–39.

Berrill, M. 1970. The aggressive behavior of *Munida sarsi* (Crustacea: Galatheidae). *Sarsia* 43(1):1–12.

Berry, F. C., and T. Breithaupt. 2010. To signal or not to signal? Chemical communication by urine-borne signals mirrors sexual conflict in crayfish. *BMC Biology* 8(1):25.

Bishop, D. T., and C. Cannings. 1978. A generalized war of attrition. *Journal of Theoretical Biology* 70(1):85–124.

Bond-Buckup, G., and L. Buckup. 1989. Os Palaemonidae das águas continentais do Brasil meridional (Crustacea, Decapoda). *Revista Brasileira de Biologia* 49(4):883–96.

Bond-Buckup, G., and L. Buckup. 1994. A família Aeglidae (Crustacea, Decapoda, Anomura). *Arquivos de Zoologia* 32:159–347.

Bond-Buckup, G., C. G. Jara, M. Pérez-Losada, L. Buckup, and K. A. Crandall. 2008. Global diversity of crabs (Aeglidae: Anomura: Decapoda) in freshwater. *Hydrobiologia* 595(1):267–73.

Boos Júnior, H. 2003. Crustáceos límnicos e aspectos da biologia de Aegla jarai Bond-Buckup & Buckup e Aegla sp. (Decapoda, Aeglidae) no Parque Natural Municipal das

Nascentes do Ribeirão Garcia, Blumenau, SC. MSc. diss., Universidade Federal do Rio Grande do Sul, Brazil.

Breinholt, J. W., M. Pérez-Losada, and K. A. Crandall. 2009. The timing of the diversification of the freshwater crayfishes. In *Decapod Crustacean Phylogenetics*, eds. J. W. Martin, K. A. Crandall, and D. L. Felder, pp. 343–56. New York: CRC Press/Taylor & Francis.

Breithaupt, T. 2011. Chemical communication in crayfish. In *Chemical Communication in Crustaceans*, eds. T. Breithaupt, and M. Thiel, pp. 257–76. New York: Springer.

Breithaupt, T., and P. Eger. 2002. Urine makes the difference: chemical communication in fighting crayfish made visible. *Journal of Experimental Biology* 205(9):1221–31.

Briffa, M., and L. U. Sneddon. 2007. Physiological constraints on contest behaviour. *Functional Ecology* 21(4):627–37.

Bubb, D. H., T. J. Thom, and M. C. Lucas. 2006. Movement, dispersal and refuge use of co-occurring introduced and native crayfish. *Freshwater Biology* 51(7):1359–68.

Bücker, F., R. Gonçalves, G. Bond-Buckup, and A. S. Melo. 2008. Effect of environmental variables on the distribution of two freshwater crabs (Anomura: Aeglidae). *Journal of Crustacean Biology* 28(2):248–51.

Buckup, L. 1999. Família Parastacidae. In *Os crustáceos do Rio Grande do Sul*, eds. L. Buckup, and G. Bond-Buckup, pp. 319–27. Porto Alegre: Editora da Universidade federal do Rio Grande do Sul.

Buckup, L., and G. Bond-Buckup. 1999. *Os crustáceos do Rio Grande do Sul*. Porto Alegre: Editora da Universidade/UFRGS.

Buckup, L., and A. Rossi. 1980. O Gênero *Parastacus* no Brasil (Crustacea, Decapoda, Parastacidade). *Revista Brasileira de Biologia* 40:663–81.

Burress, E. D., M. M. Gangloff, and L. Siefferman. 2013. Trophic analysis of two subtropical South American freshwater crabs using stable isotope ratios. *Hydrobiologia* 702(1):5–13.

Bywater, C. L., M. J. Angilletta, and R. S. Wilson. 2008. Weapon size is a reliable indicator of strength and social dominance in female slender crayfish (*Cherax dispar*). *Functional Ecology* 22(2):311–16.

Caetano, D. S., and G. Machado. 2013. The ecological tale of Gonyleptidae (Arachnida, Opiliones) evolution: phylogeny of a Neotropical lineage of armoured harvestmen using ecological, behavioural and chemical characters. *Cladistics* 29(6):589–609.

Castro-Souza, T., and G. Bond-Buckup. 2004. The trophic niche of two sympatric *Aegla* Leach species (Crustacea, Aeglidae) in a tributary of hydrographic basin of Pelotas River, Rio Grande do Sul Brazil. *Revista Brasileira de Zoologia* 21(4):805–13.

Claverie, T., and I. P. Smith. 2007. Functional significance of an unusual chela dimorphism in a marine decapod: specialization as a weapon? *Proceedings of the Royal Society of London B: Biological Sciences* 274(1628):3033–38.

Cogo, G. B., and S. Santos. 2013. The role of aeglids in shredding organic matter in neotropical streams. *Journal of Crustacean Biology* 33(4):519–26.

Colpo, K. D., L. C. Ribeiro, B. Wesz, and L. O. Ribeiro. 2012. Feeding preference of the South American endemic anomuran *Aegla platensis* (Decapoda, Anomura, Aeglidae). *Naturwissenschaften* 99(4):333–36.

Copatti, C. E., J. V. D. V. Machado, and A. Trevisan. 2015. Morphological variation in the sexual maturity of three sympatric aeglids in a river in southern Brazil. *Journal of Crustacean Biology* 35(1):59–67.

Crandall, K. A., and J. E. Buhay. 2008. Global diversity of crayfish (Astacidae, Cambaridae, and Parastacidae – Decapoda) in freshwater. *Hydrobiologia* 595(1):295–301.

Dalosto, M., and S. Santos. 2011. Differences in oxygen consumption and diel activity as adaptations related to microhabitat in Neotropical freshwater decapods (Crustacea). *Comparative Biochemistry and Physiology Part A: Molecular & Integrative Physiology* 160(4):461–66.

Dalosto, M. M. 2016. Agressão interespecífica em lagostins: Invasões biológicas, dominância e exclusão competitiva (Crustacea: Astacidea). PhD thesis, Universidade Federal de Santa Maria, Brazil.

Dalosto, M. M., A. V. Palaoro, J. R. Costa, and S. Santos. 2013. Aggression and life underground: the case of burrowing crayfish. *Behaviour* 150(1):3–22.

Dalosto, M. M., A. V. Palaoro, and S. Santos. 2012. Agressão interespecífica entre Aeglidae e Parastacidae. Presented in the VII Congresso Brasileiro sobre Crustáceos, 2012, Belém – PA (Brazil). Programação e resumos do VII Congresso Brasileiro sobre Crustáceos – VII CBG 11 a 14 de Novembro de 2012. Porto Alegre – RS: Sociedade Brasileira de Carcinologia, 2012, vol. 1. p. 83–83.

Dalosto, M. M., A. V. Palaoro, C. Souty-Grosset, S. L. S. Bueno, T. G. Loureiro, M. P. Almerão, P. B. Araujo, and S. Santos. 2015. One step ahead of the enemy: investigating aggressive interactions between invasive and native crayfish before the contact in nature. *Biological Invasions* 17(12):3503–15.

Dennenmoser, S., and J. H. Christy. 2013. The design of a beautiful weapon: compensation for opposing sexual selection on a trait with two functions. *Evolution* 67(4):1181–88.

Diawol, V. P., M. V. Torres, and P. A. Collins. 2016. Field evaluation of oxygen consumption by two freshwater decapod morphotypes (Trichodactylidae and Aeglidae): the effect of different times of the day, body weight and sex. *Marine & Freshwater Behaviour & Physiology* 49(4):251–63.

Donahue, M. J. 2004. Size-dependent competition in a gregarious porcelain crab *Petrolisthes cinctipes* (Anomura: Porcellanidae). *Marine Ecology Progress Series* 267:219–31.

Emlen, S. T., and L. W. Oring. 1977. Ecology, sexual selection, and the evolution of mating systems. *Science* 197(4300):215–23.

Enquist, M., and O. Leimar. 1983. Evolution of fighting behaviour: decision rules and assessment of relative strength. *Journal of Theoretical Biology* 102(3):387–410.

Estevan, C. 2015. Ecomoformologia de espécies de Aegla Leach, 1820: efeitos do isolamento pela distância e variáveis ambientais sobre a forma da carapaça. MSc diss., Universidade Regional Integrada do Alto Uruguai e das Missões – Campus Erechim, Brazil.

Fero, K., and P. A. Moore. 2008. Social spacing of crayfish in natural habitats: what role does dominance play? *Behavioral Ecology and Sociobiology* 62(7):1119–25.

Fontoura, N. F., and L. Buckup. 1989. Dinâmica populacional e reprodução em *Parastacus brasiliensis* (von Martens, 1869) (Crustacea, Decapoda, Parstacidae). *Revista Brasileira de Biologia* 49(4):911–21.

Garratt, M., and R. C. Brooks. 2015. A genetic reduction in antioxidant function causes elevated aggression in mice. *Journal of Experimental Biology* 218(2):223–27.

Gherardi, F. 2002. Behaviour. In *Biology of Freshwater Crayfish*, ed. D. M. Holdich, pp. 258–90. Oxford: Blackwell Scientific Press.

Gherardi, F., and W. H. Daniels. 2004. Dominance hierarchies and status recognition in the crayfish *Procambarus acutus acutus*. *Canadian Journal of Zoology* 81(7):1269–81.

Giri, F., and P. A. Collins. 2004. A geometric morphometric analysis of two sympatric species of the family Aeglidae (Crustacea, Decapoda, Anomura) from the La Plata basin. *Italian Journal of Zoology* 71(1):85–88.

Hardy, I. C., M. Goubault, and T. P. Batchelor. 2013. Hymenopteran contests and agonistic behavior. In *Animal Contests*, eds. I. C. W. Hardy, and M. Briffa, pp. 147–77. Cambridge: Cambridge University Press.

Hardy, I. C. W., and M. Briffa. 2013. *Animal Contests*. Cambridge: Cambridge University Press.

Hazlett, B., D. Rubenstein, and D. Rittschof. 1975. Starvation, energy reserves, and aggression in the crayfish *Orconectes virilis* (Hagen, 1870) (Decapoda, Cambaridae). *Crustaceana* 28(1):11–16.

Hazlett, B. A. 1981. The behavioral ecology of hermit crabs. *Annual Review of Ecology and Systematics* 12(1):1–22.

Herberholz, J., C. McCurdy, and D. H. Edwards. 2007. Direct benefits of social dominance in juvenile crayfish. *Biological Bulletin* 213(1):21–27.

Horner, A. J., M. Schmidt, D. H. Edwards, and C. D. Derby. 2008. Role of the olfactory pathway in agonistic behavior of crayfish, *Procambarus clarkii*. *Invertebrate Neuroscience* 8(1):11–18.

Huber, A. F., F. B. Ribeiro, and P. B. Araujo. 2018. New endemic species of freshwater crayfish *Parastacus* Huxley, 1879 (Crustacea: Decapoda: Parastacidae) from the Atlantic forest in southern Brazil. *Nauplius* 26:e2018015.

Imazu, M., and A. Asakura. 2006. Agonistic, aggressive and sexual behavior of five species of hermit crabs from Japan (Decapoda: Anomura: Paguridae, Diogenidae). *Crustacean Research* 6:95–107.

Issa, F. A., D. J. Adamson, and D. H. Edwards. 1999. Dominance hierarchy formation in juvenile crayfish *Procambarus clarkii*. *Journal of Experimental Biology* 202(24):3497–506.

Jiménez-Morales, N., K. Mendoza-Ángeles, M. Porras-Villalobos, E. Ibarra-Coronado, G. Roldán-Roldán, and J. Hernández-Falcón. 2018. Who is the boss? Individual recognition memory and social hierarchy formation in crayfish. *Neurobiology of Learning and Memory* 147:79–89.

Kelly, C. D. 2008. The interrelationships between resource-holding potential, resource-value and reproductive success in territorial males: how much variation can we explain? *Behavioral Ecology and Sociobiology* 62(6):855–71.

Losos, J. B. 2011. Convergence, adaptation, and constraint. *Evolution* 65(7):1827–40.

Lynch, B. R., and R. Rochette. 2009. Spatial overlap and interaction between sub-adult American lobsters, *Homarus americanus*, and the invasive European green crab *Carcinus maenas*. *Journal of Experimental Marine Biology & Ecology* 369:127–35.

Machado, J. V. D. V. 2012. Efeito da simpatria sobre a diversidade genética de Aegla platensis (Crustacea, Decapoda). MSc diss., Universidade Federal de Santa Maria, Brazil.

Maia, K. P., S. L. Bueno, and E. Trajano. 2013. Ecologia populacional e conservação de eglídeos (Crustacea: Decapoda: Aeglidae) em cavernas da área cárstica do Alto Ribeira, em São Paulo. *Revista da Biologia* 10(2):40–45.

Miranda, I., K. M. Gomes, F. B. Ribeiro, P. B. Araujo, C. Souty-Grosset, and C. D. Schubart. 2018. Molecular systematics reveals multiple lineages and cryptic speciation in the freshwater crayfish *Parastacus brasiliensis* (von Martens, 1869) (Crustacea: Decapoda: Parastacidae). *Invertebrate Systematics* 32:1265–81.

Moore, P. A. 2007. Agonistic behavior in freshwater crayfish: the influence of intrinsic and extrinsic factors on aggressive behavior and dominance. In *Evolutionary Ecology of Social and Sexual Systems: Crustacea as Model Organisms*, eds. J. E. Duffy, and M. Thiel, pp. 90–114. Oxford: Oxford Univerity Press.

Moraes, J. C. B., M. Tavares, and S. L. S. Bueno. 2017. Taxonomic review of *Aegla marginata* Bond-Buckup & Buckup, 1994 (Decapoda, Anomura, Aeglidae) with description of a new species. *Zootaxa* 4323(4):519–33.

Moraes, J. C. B., M. Terossi, R. C. Buranelli, M. Tavares, F. L. Mantelatto, and S. L. S. Bueno. 2016. Morphological and molecular data reveal the cryptic diversity among populations of *Aegla paulensis* (Decapoda, Anomura, Aeglidae), with descriptions of four new species and comments on dispersal routes and conservation status. *Zootaxa* 4193(1):1–48.

Morrone, J. J., and E. C. Lopretto. 1994. Distributional pattern of freshwater Decapoda (Crustacea: Malacostraca) in Southern South America: a panbiogeographic approach. *Journal of Biogeography* 21:97–109.

Noro, C. K., and L. Buckup. 2002. Biology and ecology of *Aegla leptodactyla* Buckup & Rossi (Crustacea, Anomura, Aeglidae). *Revista Brasileira de Zoologia* 19(4):1063–79.

Nyström, P. 2002. Ecology. In *Biology of Freshwater Crayfish*, ed. D. M. Holdich, pp. 192–224. Oxford: Blackwell Scientific Press.

Ord, T. J., and J. A. Stamps. 2009. Species identity cues in animal communication. *The American Naturalist* 174:585–93.

Palaoro, A. V., L. Ayres-Peres, and S. Santos. 2013. Modulation of male aggressiveness through different communication pathways. *Behavioral Ecology and Sociobiology* 67(2):283–92.

Palaoro, A. V., and J. Beermann. (in press). An overview of crustacean mating systems. In *The Natural History of Crustacea – Reproductive Biology*, eds. R. Cothran, and M. Thiel, vol. 6. Oxford: Oxford University Press.

Palaoro, A. V., and M. Briffa. 2017. Weaponry and defenses in fighting animals: how allometry can alter predictions from contest theory. *Behavioral Ecology* 28(1):328–36.

Palaoro, A. V., M. M. Dalosto, J. R. Costa, and S. Santos. 2014. Freshwater decapod (*Aegla longirostri*) uses a mixed assessment strategy to resolve contests. *Animal Behaviour* 95:71–79.

Palaoro, A. V., M. Velasque, S. Santos, and M. Briffa. 2017. How does environment influence fighting? The effects of tidal flow on resource value and fighting costs in sea anemones. *Biology Letters* 13(5).

Parra, C. A., E. Barría, and C. G. Jara. 2011. Behavioural variation and competitive status in three taxa of *Aegla* (Decapoda: Anomura: Aeglidae) from two-community settings in Southern Chile. *New Zealand Journal of Marine and Freshwater Research* 45(2):249–62.

Payne, R. J. 1998. Gradually escalating fights and displays: the cumulative assessment model. *Animal Behaviour* 56(3):651–62.

Peiman, K. S., and B. W. Robinson. 2010. Ecology and evolution of resource-related heterospecific aggression. *Quarterly Review of Biology* 85:133–58.

Powers, S. K., and M. J. Jackson. 2008. Exercise-induced oxidative stress: cellular mechanisms and impact on muscle force production. *Physiological Reviews* 88(4):1243–76.

Ribeiro, F. B., A. F. Huber, C. D. Schubart, and P. B. Araujo. 2017. A new species of *Parastacus* Huxley, 1879 (Crustacea, Decapoda, Parastacidae) from a swamp forest in southern Brazil. *Nauplius* 25:e2017008.

Richardson, A. M. M. 2007. Behavioral ecology of semiterrestrial crayfish. In *Evolutionary Ecology of Social and Sexual Systems: Crustaceans as Model Organisms*, eds. J. E. Duffy, and M. Thiel, pp. 319–38. Oxford: Oxford University Press.

Riek, E. F. 1971. The freshwater crayfishes of South America. *Proceedings of the Biological Society of Washington* 84:129–36.

Rojas, R., M. C. Morales, M. M. Rivadeneira, and M. Thiel. 2012. Male morphotypes in the Andean river shrimp *Cryphiops caementarius* (Decapoda: Caridea): morphology, coloration and injuries. *Journal of Zoology* 288(1):21–32.

Rudolph, E. H., and K. Crandall. 2012. A new species of burrowing crayfish, *Virilastacus jarai* (Crustacea, Decapoda, Parastacidae) from central-southern Chile. *Proceedings of the Biological Society of Washington* 125:258–75.

Rypien, K. L., and A. R. Palmer. 2007. The effect of sex, size and habitat on the incidence of puncture wounds in the claws of the porcelain crab *Petrolisthes cinctipes* (Anomura: Porcellanidae). *Journal of Crustacean Biology* 27(1):59–64.

Santos, S., G. Bond-Buckup, L. Buckup, M. Pérez-Losada, M. Finley, and K. A. Crandall. 2012. Three new species of *Aegla* (Anomura) freshwater crabs from the Upper Uruguay River hydrographic basin in Brazil. *Journal of Crustacean Biology* 32(4):529–40.

Santos, S., G. Bond-Buckup, M. Perez-Losada, M. L. Bartholomei-Santos, and L. Buckup. 2009. *Aegla manuinflata*, a new species of freshwater anomuran (Decapoda: Anomura: Aeglidae) from Brazil, determined by morphological and molecular characters. *Zootaxa* 2088(1):31–40.

Silva-Gonçalves, R., G. Bond-Buckup, and L. Buckup. 2009. Crescimento de *Aegla itacolomiensis* (Crustacea, Decapoda) em um arroio da Mata Atlântica no sul do Brasil. *Iheringia, Série Zoologia* 99(4):397–402.

Smith, I. P., and A. C. Taylor. 1993. The energetic cost of agonistic behaviour in the velvet swimming crab, *Necora* (= *Liocarcinus*) *puber* (L.). *Animal Behaviour* 45(2):375–91.

Smith, J. M., and G. R. Price. 1973. The logic of animal conflict. *Nature* 246(5427):15.

Sneddon, L. U., A. C. Taylor, and F. A. Huntingford. 1999. Metabolic consequences of agonistic behaviour: crab fights in declining oxygen tensions. *Animal Behaviour* 57(2):353–63.

Sokolowicz, C. C., L. Ayres-Peres, and S. Santos. 2007. Diel activity and digestion time of *Aegla longirostri* (Crustacea, Decapoda, Anomura). *Iheringia. Série Zoologia* 97(3):235–38.

Swanson, B. O., M. N. George, S. P. Anderson, and J. H. Christy. 2013. Evolutionary variation in the mechanics of fiddler crab claws. *BMC Evolutionary Biology* 13(1):137.

Thiel, M., and G. A. Lovrich. 2011. Agonistic behaviour and reproductive biology of squat lobsters. In *The Biology of Squat Lobsters*, eds. G. C. B. Poore, S. T. Ahyong, and J. Taylor, pp. 223–47. Boca Raton, FL: CRC Press.

Thorpe, K. E., A. C. Taylor, and F. A. Huntingford. 1995. How costly is fighting? Physiological effects of sustained exercise and fighting in swimming crabs, *Necora puber* (L.) (Brachyura, Portunidae). *Animal Behaviour* 50(6):1657–66.

Usio, N. 2000. Effects of crayfish on leaf processing and invertebrate colonisation of leaves in a headwater stream: decoupling of a trophic cascade. *Oecologia* 124(4):608–14.

Viau, V. E., L. S. López Greco, G. Bond-Buckup, and E. M. Rodríguez. 2006. Size at the onset of sexual maturity in the anomuran crab, *Aegla uruguayana* (Aeglidae). *Acta Zoologica* 87(4):253–64.

Vieira, M. C., and P. E. C. Peixoto. 2013. Winners and losers: a meta-analysis of functional determinants of fighting ability in arthropod contests. *Functional Ecology* 27(2):305–13.

Vogt, G. 2002. Functional anatomy. In *Biology of Freshwater Crayfish*, ed. D. M. Holdich, pp. 53–151. Oxford: Blackwell Scientific Press.

Wilson, R. S., M. J. Angilletta Jr, R. S. James, C. Navas, and F. Seebacher. 2007. Dishonest signals of strength in male slender crayfish (*Cherax dispar*) during agonistic encounters. *The American Naturalist* 170(2):284–91.

Zimmermann, B. L., A. W. Aued, S. Machado, D. Manfio, L. P. Scarton, and S. Santos. 2009. Behavioral repertory of *Trichodactylus panoplus* (Crustacea: Trichodactylidae) under laboratory conditions. *Revista Brasileira de Zoologia* 26:5–11.

Zimmermann, B. L., C. S. Dambros, and S. Santos. 2016. Association of microhabitat variables with the abundance and distribution of two Neotropical freshwater decapods (Anomura: Brachyura). *Journal of Crustacean Biology* 36(2):198–204.

Zulandt-Schneider, R. A., R. Huber, and P. A. Moore. 2001. Individual and status recognition in the crayfish, *Orconectes rusticus*: the effects of urine release on fight dynamics. *Behaviour* 138(2):137–53.

CHAPTER **8**

Physiological Ecology: Osmoregulation and Metabolism of the Aeglid Anomurans

John Campbell McNamara and Samuel Coelho Faria

CONTENTS

8.1 INTRODUCTION

The freshwater squat lobsters of the anomuran family Aeglidae have been little studied physiologically, perhaps owing to their limited geographical distribution throughout the southern regions of South America, and to their discrete presence, restricted to well-preserved lotic habitats and often species-specific drainage or river basins (Santos et al. 2017). Many investigations have focused on their palaeo-bio-geographical origins and subsequent radiation, current geographical distributions, systematics and taxonomic status, the search for new and cryptic species, geometric morphometry and morphology, ecology and population biology, trophic niche, feeding and predation, behavior, life cycle and history, and more recently, threatened conservation status (Bueno et al. 2016).

In contrast to this plethora of biological and ecological interests, the bulk of physiological studies on the aeglids are much more circumscribed and have concentrated on elucidating their intermediate metabolism with regard to the accumulation and use of energetic substrates and high energy macromolecules in various tissues (Oliveira et al. 2003, 2007). Other investigations have focused on oxygen consumption as related to seasonal and diel behavioral activity patterns, and on the influence of body mass and sex (Dalosto and Santos 2011; Diawol et al. 2016). Very few studies have examined osmotic and ionic regulation (Freire et al. 2008a; Faria et al. 2011; Freire et al. 2013), albeit in some depth. These physiological studies are limited mainly to those species that occur within the geographical regions of the investigators, which has further restricted a broader comprehension of the physiology of the group. Clearly, the extant Aeglidae constitute a taxon in which physiological investigation is still incipient, but one in which many avenues for novel research using current methodologies are apparent, including comprehensive phylogenetic analyses of physiological traits, as illustrated in this chapter.

Here, we review and collate information from the main areas on which physiological studies have focused in detail, i.e., intermediate energy and oxidative metabolism, and osmotic and ionic regulation, with a view to providing an appreciation of current knowledge, and to offer a guide for future studies on this fascinating yet neglected group of poorly known anomuran decapods. Given the marine origins of the Aeglidae, and their current geographical distribution limited to cold fresh waters, we particularly address the physiological adaptations that have enabled the occupation of this new niche from a phylophysiological perspective.

8.2 OSMOTIC AND IONIC REGULATION

8.2.1 The Phylophysiological Origins of the Aeglidae and the Palaeo-Environment

Although the 87 currently known species of extant aeglids, or freshwater squat lobsters, are now distributed exclusively within a diversity of freshwater habitats ranging throughout southern South America (Santos et al. 2017), there is little doubt that their ancestral lineages originated within a marine setting (Schmitt 1942; Feldmann 1984; Feldmann et al. 1998; see also Chapter 1 of this book). This was likely in the Indo-Pacific region from which the aeglids radiated eastwards into southern neotropical lotic habitats some 74 million years ago (Pérez-Losada et al. 2004), effectively dispersing throughout southern South America in the late Oligocene around 25 Mya (Collins et al. 2011; Bueno et al. 2016). The westerly distributed taxa radiated some 40–45 Mya while speciation in the central and eastern taxa occurred later around 23–25 Mya (Bond-Buckup et al. 2008).

Palaeontological and sedimentological studies on the two known fossil species of Aeglidae, *Haumuriaegla glaessneri*, from a putatively Upper Cretaceous (Haumurian = Maastrichtian) horizon in eastern southern New Zealand (Feldmann 1984), and *Protaegla miniscula*, from the Tlayúa Formation (middle to late Albian, late Lower Cretaceous) (Feldmann et al. 1998) in southern central Mexico, have revealed the marine nature of the ancestral palaeo-environments. The fauna associated with *H. glaessneri* were predominantly marine in nature and included Cnidaria, Bivalvia, Gastropoda, Ammonoidea, and Chordata (Feldmann 1984), very suggestive of a marine facies. In the case of *Protaegla miniscula*, palaeo-environmental analysis suggests that the Tlayúa Formation likely represents a shallow marine lagoon (Applegate 1992) with restricted water circulation resulting in an anaerobic and/or hypersaline environment (Feldmann et al. 1998). Besides strong marine, lagoonal, and reefal influences, the Tlayúa lagoon also appears to have undergone bouts of periodic freshwater inflow (Feldmann et al. 1998) based on the presence of diagnostic terrestrial and freshwater organisms.

Aeglids have been considered the most basal lineage of the anomuran superfamily Galatheoidea (Martin and Abele 1986; Martin and Davis 2001; Pérez-Losada et al. 2002). Some (McLaughlin et al. 2007; Ahyong et al. 2011) maintain that they constitute a separate superfamily, the Aegloidea, closest to the crab-like Lomisoidea, and to the Chirostyloidea and Paguroidea. Their phylogenetic relationships constitute an issue of on-going discussion. Given their ancestral marine palaeo-environment and probable phylogenetic relationships, it seems reasonable to suppose that the physiology, and particularly the osmoregulatory physiology of extant, shallow water, marine galatheoids, and paguroids would reflect that of the primitive marine aeglids. To illustrate as regards salinity tolerance and osmoregulatory ability, the marine galatheoid squat lobster, *Galathaea squamifera*, survives in dilute seawater of 26 ‰ salinity (S, g/L) but not in 17 ‰ S. The species is an osmoconformer, highly permeable to water, K^+, and cesium, and unable to regulate hemolymph osmolality and Na^+ and Cl^- concentrations (Bryan 1965). Similarly, the paguroid hermit crab *Clibanarius erythropus* also is an osmoconformer, maintaining its hemolymph isosmotic between 14 and 88 ‰ S, becoming slightly hyperosmotic at 5 ‰ S (Castillo et al. 1988). Although strongly euryhaline, *C. erythropus* adjusts cell volume through mechanisms of isosmotic intracellular regulation, using free amino acids, Na^+, and K^+ as osmotic effectors. The paguroid hermits *Pagurus longicarpus* and *P. pollicaris* are strict osmoconformers, maintaining their hemolymph isosmotic over salinities from 5 to 40 ‰ S (Young 1979). Although likewise essentially an osmoconformer, *Pagurus maclaughlinae* hyper-osmoregulates very slightly from 10 to 45 ‰ S (Rhodes-Ondi and Turner 2010).

These findings derived from related extant anomuran species suggest that the ancestral marine aeglids were likely osmoconformers, maintaining hemolymph osmolality similar to that of their surrounding *milieu*. Given the absence of an osmotic gradient against the external medium, when challenged by dilute media they would be able to adjust cell volume to some extent by means of isosmotic intracellular regulation in lieu of mechanisms of anisosmotic extracellular regulation (see Section 8.2.2). Thus, they would not be able to penetrate directly into dilute media.

The question now posited is: how might an isosmotic marine osmoconformer evolve physiologically into a freshwater osmoregulator, capable of maintaining strong osmotic and ionic gradients against this osmotically challenging medium? While this issue cannot be addressed directly, we can infer likely changes and possible pathways from the findings disclosed by the few osmoregulatory studies undertaken on extant aeglids (Freire et al. 2008a; Faria et al. 2011; Freire et al. 2013). These investigations have examined both isosmotic intracellular, and anisosmotic extracellular regulation in three aeglid species, *Aegla franca*, *A. schmitti*, and *A. parana*, from the southeastern region of Brazil.

8.2.2 Physiological Challenges of the Extant Environment: Osmotic and Ionic Regulation in Fresh Water

Living in fresh water is a physiologically challenging task for crustaceans (Mantel and Farmer 1983; Péqueux 1995; Freire et al. 2008b). Water is necessarily gained passively by osmosis while major ions such as Na^+ and Cl^- are lost across the permeable body surfaces like the gills and arthrodial membranes, and via the urine—often isosmotic to the hemolymph. Such osmotically gained water must be excreted to avoid cell and tissue swelling and expansion of the extracellular space. Ions must be actively transported against strong gradients from the surrounding fresh water across epithelial interfaces, particularly in the branchiostegites and gills, but also by the antennal gland tubules and intestine, to replace those lost by passive diffusional efflux (Freire et al. 2008b). These mechanisms are metabolically costly and are driven by ATP consuming processes that can become limiting especially since the summed effects of adverse environmental parameters likely make synergistic demands on aerobic metabolism.

The physiological processes that maintain this dynamic equilibrium of osmotically active particles and water in the extracellular space are collectively known as "anisosmotic extracellular regulation" (Mantel and Farmer 1983; Péqueux 1995). The interspecific variability seen among the constituent mechanisms that preserve hemolymph osmolality, ionic concentration and composition, and extracellular volume depends on cell and tissue type and function, which in turn are a consequence of phylogenetic history and environment (McNamara and Faria 2012; McNamara et al. 2015). Thus, the decapod Crustacea exhibit a plethora of anisosmotic regulatory abilities and patterns, ranging from hyper/hypo-osmoregulating, euryhaline, intertidal and estuarine species, through migratory, diadromous freshwater species, to strongly hyper-osmoregulating, stenohaline, hololimnetic species whose life cycles are entirely independent of salt water. Regulation of the intracellular fluid, a process evolutionarily likely more ancient than extracellular regulation given the division of cellular function inherent to multicellularity, encompasses those mechanisms that maintain cell water and ionic composition (Na^+, K^+, Cl^-) and the content of organic osmolytes such as non-essential free amino acids, compatible with regulatory cell volume changes. This allows some degree of osmotic flexibility at the cellular level, particularly when mechanisms of anisosmotic extracellular regulation are absent or exceeded by external hyper- or hypo-osmotic challenge. Such cellular mechanisms are collectively known as "isosmotic intracellular regulation" (Mantel and Farmer 1983; Péqueux 1995) and maintain the osmotic equilibrium, i.e., obligatory isosmoticity, between the intra- and extracellular fluids consequent to transport driven, regulatory cell volume decrease or increase.

The extant Aeglidae fall into the category of hololimnetic hyper-osmoregulators that maintain strong osmotic and ionic gradients against their surrounding medium, although their hemolymph osmolalities are reduced compared to marine taxa. Their typical freshwater environments include small watercourses, streams, rivers, lakes, and cave waters, characterized mainly by clear, cold, fast-flowing, well-oxygenated water (Bond-Buckup and Buckup 1994). Water temperatures range from 8–11°C in the southernmost habitats (Miserendino 2001; Oliveira et al. 2007) to 16–23°C in the northernmost habitats (Swiech-Ayoub and Masunari 2001). There are no records of extant estuarine species. A second major issue to address then concerns why these strictly freshwater inhabitants have retained the ability to tolerate saline media after such a lengthy evolutionary period in fresh water.

As a preliminary appraisal to investigations of osmoregulatory physiology, mortality analyses of salinity challenge in three species of aeglids (Figure 8.1) reveal ≈85% survival in media of 14 and 21 ‰ S in *Aegla franca* after 10 days exposure, dropping quickly to 25% in 28 ‰ S, and to just 7% after 24 h in full strength seawater (Faria et al. 2011). The upper lethal salinity limit (=50% mortality) is 26.4 ‰ S after 6.5 days. Coherently, *A. schmitti* shows no mortality in 15 ‰ S after 7 h exposure (Freire et al. 2008a), while after 4 days exposure, A. *longirostri* exhibits no mortality below 15 ‰ S, 20% mortality in 15 ‰ S, 60% in 20 ‰ S, and survives

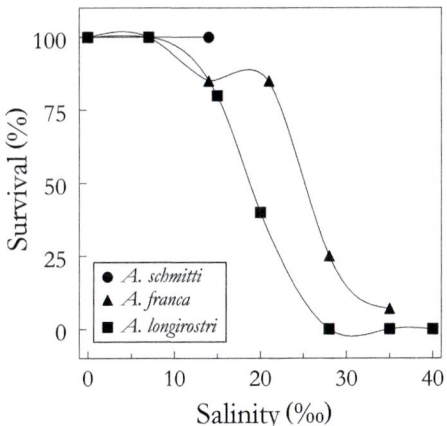

Figure 8.1 Comparative survival rates (%) as a function of direct salinity exposure in three species of hololimnetic, freshwater aeglids. Data for *Aegla schmitti* are from Freire et al. (2008a) and refer to seven hours salinity exposure; data for *A. franca* are from Faria et al. (2011), ten days acclimation; and data for *A. longirostri* are from Cogo and Santos (2007), four days exposure. Survival becomes markedly reduced around 20 S (≈600 mOsm/kg H_2O), an osmolality approaching isosmoticity. The lethal salinity limits (LS_{50}) for *A. franca* and *A. longirostri* are 26.4 ‰S and 19 ‰S, respectively.

in 25 ‰ S for two days. Mortality is 100% above 25 ‰ S up to 40 ‰ S in which this species survives just 12 h (Cogo and Santos 2007). Evidently these aeglids can survive moderate osmotic challenge well, but clearly reach an abrupt transition zone around 19–21 ‰ S ($\approx$600 mOsm/kg H_2O) at which the limits of anisosmotic extracellular regulatory ability are approached and isosmoticity sets in. Clearly, isosmotic intracellular regulation is incapable of sustaining cellular physiological processes when isosmoticity with the external medium is reached.

The extensive data available for anisosmotic extracellular regulation particularly in *Aegla franca* (see Faria et al. 2011), *A. schmitti* (see Freire et al. 2008a), and in *A. parana* (see Freire et al. 2013) reveal that extant aeglids maintain strong osmotic ($\approx$30: 1) and ionic (Cl^-, 25: 1) gradients against fresh water. Hemolymph osmolalities in fresh water are 409 ± 5.9, 453 ± 6.3, and 454 ± 10 mOsm/kg H_2O, for *A. franca*, *A. schmitti*, and *A. parana*, respectively, while hemolymph chloride concentration is 203.5 ± 8.6 mM Cl^- in *A. franca*. While *A. franca* is not a functionally euryhaline species, and never encounters saline media in its natural osmotic setting, its ability for anisosmotic extracellular regulation is remarkable and reveals ample, conserved potential euryhalinity. *Aegla franca* hyper-regulates hemolymph osmolality up to 30 ‰ S (=900 mOsm/kg H_2O), becoming isosmotic at 30.5 ‰ S (=916 mOsm/kg H_2O; Faria et al. 2011). Its overall hyper-osmoregulatory capability (Δ hemolymph osmolality/Δ external osmolality) is weak (0.57) since hemolymph osmolality tends to increase quickly in salinities above 20 ‰ S, reflecting the transition zone to isosmoticity. In distinct contrast, hemolymph chloride regulatory ability in *A. franca* is arresting. In fresh water, [Cl^-] is held at 203.5 ± 8.6 mM Cl^-, maintained up to 15 ‰ S (=240 mM Cl^-), the isochloride point being 249 mM Cl^-. Chloride concentration then increases with salinity increase but is distinctly hypo-regulated, even at 30 ‰ S (Faria et al. 2011). Overall chloride regulatory ability is strong (0.19) compared to osmoregulatory ability. In *A. schmitti*, hemolymph osmolality increases to 512 ± 20.7 mOsm/kg H_2O after 7 h in 15 S (Freire et al. 2008a), a regulatory ability of 0.13 (Δ hemolymph osmolality/Δ external osmolality). The corresponding value for *A. franca* for 6 h in 25 ‰ S is 0.5, revealing a species-specific difference in short-term osmoregulatory ability.

Acclimation time course studies (Faria et al. 2011) show that on exposure to 25 ‰ S, hemolymph osmolality in *A. franca* rapidly reaches a maximum overshoot of 784.8 ± 5.2 mOsm/kg H_2O after 6 h, decreasing to 741.0 ± 12.9 mOsm/kg H_2O after 24 h exposure, remaining stable and isosmotic thereafter up to 240 h. Hemolymph [Cl^-] increases quickly to 272.6 ± 8.4 mM Cl^- by 12 h, reaching a maximum at 339.8 ± 10.0 mM Cl^- after 24 h, remaining stable and strongly hypo-chloremic thereafter up to 240 h. Evidently, while aeglids quickly become isosmotic in 75% seawater, hemolymph [Cl^-] is hypo-regulated despite considerable initial Cl^- influx from the medium.

Capability for isosmotic intracellular regulation is not well developed in the extant aeglids compared to various marine and other freshwater decapod taxa (Faria et al. 2011), and anisosmotic extracellular regulation predominates on osmotic challenge. *Aegla franca* shows 82% tissue hydration in fresh water *in vivo* (Faria et al. 2011). When challenged at 25 ‰ S [=750 mOsm/kg H_2O, 400 mM Cl^-], tissue water is held at ≈80% for five days, decreasing to 75.2% after 10 days exposure (Δ 8%) (Faria et al. 2011). *Aegla schmitti* also shows ≈80% tissue hydration in fresh water, maintained after seven hours exposure *in vivo* to 15 ‰ S; hemolymph osmolality is essentially unchanged (Freire et al. 2008a). Expressing change in tissue water content (Δ % hydration) as a function of change in hemolymph osmolality (Δ mOsm/kg H_2O) on hyperosmotic challenge (Foster et al. 2010) when corrected for the osmotic gradient applied suggests that *A. schmitti* regulates tissue water better than *A. franca*, the respective indices being 5 and 20 (% hydration/mOsm/kg H_2O [×10^3]). These findings attest to effective extracellular regulation at the whole animal and systemic levels.

However, muscle tissue slices of *A. schmitti* (Freire et al. 2008a) and *A. parana* (Freire et al. 2013) when challenged with hyper-osmotic saline solutions *in vitro* (607–613 mOsm/kg H_2O = 20 ‰ S) lose water (11%) within 15–30 min, declining to 17–19% after 1.5 to 2 h, with no sign of volume recovery. In contrast, on hypo-osmotic challenge (209 mOsm/kg H_2O), muscle slices from *A. parana* gain water (17%) within 15 min, fully recovering their volume within 45 min (Freire et al. 2013). A sodium-potassium-two chloride symporter may play a role in muscle fiber volume regulation in this species (Freire et al. 2013). These data disclose an asymmetrical volume regulatory response to osmotic challenge at the tissue level in extant aeglids. They appear to have lost the ability to conserve tissue water when hemolymph osmolality increases, but have retained the capacity to recover cell volume on hemolymph dilution such as occurs during molting. Compared to extant marine osmoconformers and the putative ancestral aeglids, well-developed, anisosmotic extracellular capability appears to constitute a coherent adaptation within the context of their incursion into fresh water.

With regard to the organic effectors of isosmotic intracellular regulation, the concentrations of total free amino acids (FAA) in the abdominal muscle of *A. franca* increase by 3.5-fold, from 18.2 ± 1.7 mmol/kg wet mass in fresh water to 63.3 ± 6.9 mmol/kg wet mass, after 48 h osmotic challenge in 25 ‰ S, and are maintained stable for 10 days (Faria et al. 2011). In fresh water, the main muscle FAA are glycine, taurine, arginine, alanine, and proline, which make up ≈85% of the FAA pool; taurine, arginine, alanine each contribute ≈22% (Faria et al. 2011). Alanine predominates the cellular FAA response, increasing progressively by 5.4-fold from 3.5 ± 0.5 mmol/kg wet mass to 19.4 ± 0.8 mmol/kg wet mass after 10 days at 25 ‰ S. Proline is the most labile FAA, and increases 12-fold. The remaining FAA also increase steadily by ≈3-fold. However, in *A. franca*, total muscle FAA represent just 4.5% of cellular osmolality in fresh water, increasing 1.6-fold to 7.0% in 25 ‰ S (Faria et al. 2011). Similarly, in the hololimnetic trichodactylid crab *Dilocarcinus pagei*, total muscle FAA contribute even less, just 2.6% in fresh water, unchanged on acclimation to 25 ‰ S (Augusto et al. 2007). Corresponding values for hololimnetic freshwater Astacidea and diadromous Palaemonidae range from 25–30%, slightly less than marine Caridea and Dendrobranchiata (30–31%) (Faria et al. 2011). Marine Brachyura show the highest contribution of total FAA to cellular osmolality (37%) (Faria et al. 2011). Thus, the aeglids, like some hololimnetic decapods, show little ability for isosmotic intracellular regulation based on the mobilization of an organic osmolyte pool. However, volume regulation in response to hyper-osmotic challenge *in vitro* via the sodium-potassium-two chloride symporter seems likely (Freire et al. 2013).

Non-phylogenetic correlations of muscle total FAA titers with hemolymph osmolalities across the principal decapod sub/infraorders (Faria et al. 2011) reveal that marine taxa show high hemolymph osmolalities and elevated intracellular FAA concentrations while freshwater representatives maintain low hemolymph osmolalities and exhibit reduced intracellular FAA titers (Figure 8.2, *left panel*; R= +0.820, P<0.0001). Apparently, during their evolutionary sojourn into fresh water, the putative dependence of the ancestral aeglids on FAA as intracellular osmotic effectors has diminished greatly as witnessed in extant freshwater aeglids like *A. franca* and as also seen in hololimnetic palaemonids (Faria et al. 2011) and brachyurans (Augusto et al. 2007). This inability to effect significant isosmotic intracellular regulation on hyperosmotic challenge *in vivo* using FAA has been supplanted by mechanisms of anisosmotic regulation of the extracellular fluid that limit alterations in cell volume and ionic composition, thus buffering to a certain degree changes in the intracellular fluid.

This trade-off between anisosmotic extracellular and isosmotic intracellular regulation (Figure 8.2, *left panel*) typifies the homoplastic nature of the osmoregulatory evolution that has taken place during the conquest of fresh water by decapod groups (Figure 8.2, *right panel*). Increases in extracellular osmoregulatory capability together with reduction in intracellular regulation likely occurred independently several times among the "shrimp-like" decapods, particularly within the Dendrobranchiata (Sergestidae) and Caridea (Alpheidae, Atyidae, Desmocarididae,

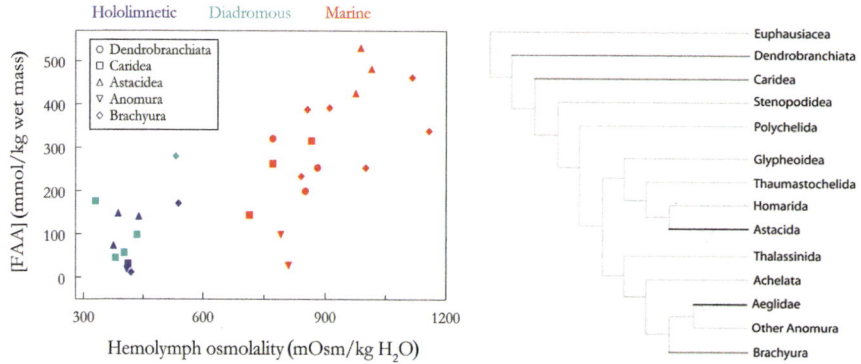

Figure 8.2 *Left panel.* Positive correlation (R= +0.82, P<0.0001) between muscle free amino acids (FAA, mmol/kg wet mass) and hemolymph osmolalities (mOsm/kg H_2O) in marine, diadromous, and hololimnetic decapod species selected among the Dendrobranchiata and Pleocyemata [Data from Faria et al. (2011)]. Both osmoregulatory traits decrease with increasing degree of colonization of freshwater. *Right panel.* A hypothesis for decapod phylogeny [Dixon et al. (2003), modified by Faria et al. (2011)]. Black lines indicate exclusively freshwater taxa; gray lines denote taxa habiting marine, brackish, or fresh waters, while light gray lines signify taxa restricted to seawater.

Kakudicarididae, Palaemonidae, and Typhlocarididae) as well as among brachyuran crabs (Deckeniidae, Gecarcinucidae, Paratelphusidae, Platytelphusidae, Potamidae, Pseudotelphusidae, Potamonautidae, Trichodactylidae, and Xyphocarididae) (Von Sternberg 1997; Von Sternberg et al. 1999; De Grave et al. 2008; Tavares et al. 2009). However, the monophyly of the astacid crayfish and aeglids (Dixon et al. 2003) suggests that various elements of intra- and extracellular osmoregulation are homologous among the species within each group (Faria et al. 2011) (Figure 8.2, *right panel*). Apparently, the diminished osmotic gradient seen in fresh water results from natural selection, an adaptation found at the outset of Astacida and *Aegla*, followed by a reduction in intracellular osmotic effectors like FAA concentrations that have coevolved, constrained by reduced hemolymph osmolalities. Increased anisosmotic regulatory ability such as seen in *Aegla* species is phylogenetically inherited, while reduced dependence on isosmotic intracellular regulation is an inevitable corollary of isosmoticity at lower hemolymph osmolalities.

8.2.3 The Principal Effectors of Osmotic and Ionic Regulation: The Trichobranchiate Gills

A central tenet to comprehending crustacean physiology concerns the intimate relationship maintained between structure and function, at all levels of biological organization, from the systemic through organ, tissue, and cellular levels to the molecular level. Mechanisms of anisosmotic extracellular regulation lend themselves admirably to such analyses, which are necessarily based on an adequate knowledge of the microanatomy and ultrastructure of the main effector organs of osmotic and ionic regulation, i.e., the gills and antennal glands. Based on histological, ultrastructural, ultracytochemical, immunocytochemical, electrophysiological, biochemical, and gene expression investigations, various models for ion transport across the gill epithelia of freshwater decapods like palaemonid shrimps and brachyuran crabs have been proposed and refined (Freire and McNamara 1995; Péqueux 1995; McNamara and Lima 1997; McNamara and Torres 1999; Freire et al. 2008b; Henry et al. 2012; McNamara and Faria 2012). Such models allow hypothesis testing with regard to the transport of specific ions, and importantly, the precise localization of transport-related molecules such as the sodium-potassium ATPase, the V(H$^+$)-ATPase, or proton pump, and the sodium-potassium-two chloride symporter, in the membranes of specialized, characteristic epithelial cells known as ionocytes. Together with antiporters like the sodium-proton and the chloride-bicarbonate exchangers, and ion and water channels, these few molecules underlie much of the transport taking place across the crustacean gill epithelium. Nevertheless, much refinement is necessary before a more complete picture of epithelial ion transport in the decapod Crustacea can be developed.

Very little is known of the microanatomy of the aeglid gill (Martin and Abele 1986, 1988), and ultrastructural studies of their epithelia are lacking entirely. For this reason, a brief analysis of the trichobranchiate gill microanatomy and overall chloride transport ability, and of the epithelial architecture and ultrastructure of the pleurobranch gill filaments in *A. franca* are provided here (Antunes and McNamara 2010).

These findings are then examined with regard to a putative mechanism of ion transport across the aeglid gill epithelium.

Aeglids possess 13 gill pairs consisting of nine pairs of arthrobranchiae and four pairs of pleurobranchiae (Martin and Abele 1986; Figure 8.3). Each trichobranchiate gill consists of a small, flat, plate-like base inserted into the body wall and from which the long, tubular gill filaments extend forwards in a posterior-anterior orientation into the branchial chamber in a fan-like fashion (Figure 8.4). Filament length varies with animal size, and gill size and type, and filament insertion. The ≈30 shorter external filaments measure about 3.8 mm in length while the 40–60 internal filaments measure around 5.5 mm in length (Figure 8.4). Silver nitrate staining to reveal regions of chloride transport shows that the gills are not functionally homogeneous—there are marked differences in chloride deposition between each gill type and among the filaments within a single gill. The posterior and outer-most gills stain more intensely than do the antero-dorsal gills (Figure 8.5A) as do the posterior-most filaments of the arthrobranchiae (Figure 8.5B), reflecting their function in ion transport. All the filaments in the pleurobranchiae stain equally intensely (Figure 8.5C).

Each gill filament extends individually from the plate-like base or gill axis, and is covered by a fine cuticle overlaying the filament epithelium. In histological sections, filament transects measure ≈100 μm in diameter, are oval to circular in profile, and reveal a single epithelial layer consisting of pillar cells that exhibit thick apical flanges arrayed beneath the cuticle (Figure 8.6A). The columnar pillar cell perikarya contain the cell nucleus, and abut onto extensive lateral expansions of the filament septum that extend partially around the filament. This fine annular septum is not fenestrated and lies transversely across the filament along its length (Figure 8.6A,B),

Anterior
Dorsal

Figure 8.3 Scanning electron micrograph taken after removal of the left branchial chamber from *Aegla franca* revealing the arrangement of the trichobranchiate gill tufts protruding from the pleural wall above the insertions of the pereopods. The upper larger tufts are pleurobranchiae while the lower, smaller tufts are arthrobranchiae. The tips of individual gill filaments within each tuft are oriented in an anterior direction; water flows through the chamber over the gill filaments from posterior to anterior. Scale bar = 1.0 mm. (Modified from Antunes [2011].)

Figure 8.4 Scanning electron micrograph showing an individual arthrobranch from the fourth pereopod of *Aegla franca*. The gill filaments radiate fan-like from the central point of insertion and are disposed in two layers: the shorter external trichae and the longer internal filaments (see also Figure 8.3). Scale bar=2.5 mm. (Modified from Antunes [2011].)

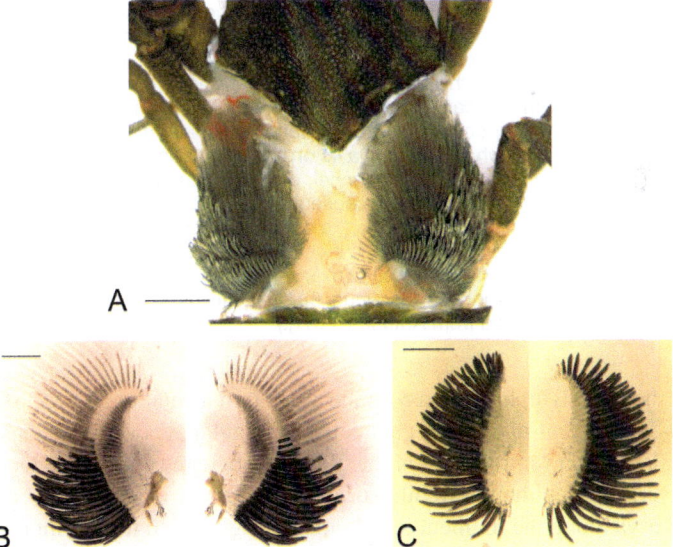

Figure 8.5 (A) Photomicrograph of the cephalothorax of *Aegla franca* from which the carapace and gill chambers have been dissected after silver nitrate staining to localize free chloride, deposited as silver chloride, indicating ion transport. The posterior and outer-most gills stain more intensely than do the anterior-dorsal gills, revealing a differential ion transport function between gills. Scale bar=4 mm. (B) The filaments of individual gills are also differentiated, the posterior-most dorsal filaments of the arthrobranchiae staining more intensely than the ventral filaments. (C) All filaments of the pleurobranchiae stain equally intensely. Scale bars= 1.0 mm. (Modified from Antunes [2011].)

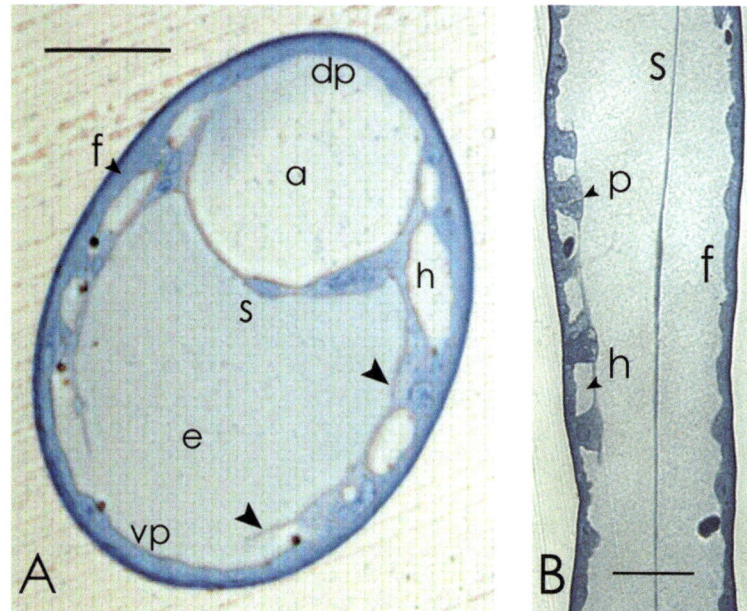

Figure 8.6 Photomicrographs of epoxy resin thick sections through gill filaments from the fourth arthrobranch of *Aegla franca* revealing the epithelial architecture. (A) In transverse sections, the filaments are oval to circular in profile, exhibiting a fine epithelium below the cuticle. A transverse septum (s) divides the hemolymph space asymmetrically into dorsal afferent (a) and ventral efferent (e) channels. The epithelium consists of the juxtaposed lateral expansions or flanges (f) of the pillar cells whose perikarya lie on annular extensions of the septum (arrows), particularly in the efferent channel, creating lacunae through which hemolymph (h) also flows. The septum is absent from the dorsal (dp) and ventral (vp) poles of the filament; at these poles, the epithelium consists of flattened epithelial cells alone. Scale bar=25 μm. (B) Longitudinal section of a filament showing the continuous septum (s), pillar cell perikarya (p) lying on the annular septal extensions, interposing hemolymph lacunae (h), and the pillar cell flanges (f). Scale bar= 40 μm. (Modified from Antunes [2011].)

dividing the hemolymph space asymmetrically into a smaller, dorsal afferent channel and a larger, ventral efferent channel. The annular septal expansions are absent from the dorsal and ventral poles of the filament where the epithelium forms a simple single layer. This epithelial architecture creates lateral lacunae between the pillar cell perikarya along the filament length and through which hemolymph flows.

Transmission electron microscopy reveals that in the lateral regions of the filament, the apical membrane of the pillar cell flanges is augmented by tufts of numerous short evaginations that abut onto the overlying cuticle, forming a discrete subapical space (Figure 8.7). The underlying apical cytoplasm is replete with abundant mitochondria and cisternae of the rough endoplasmic reticulum. The lower flange membrane exhibits a system of infoldings associated with mitochondria (Figure 8.7). Adjacent flanges interdigitate extensively forming complex junctional regions, characterized by lengthy septate junctions.

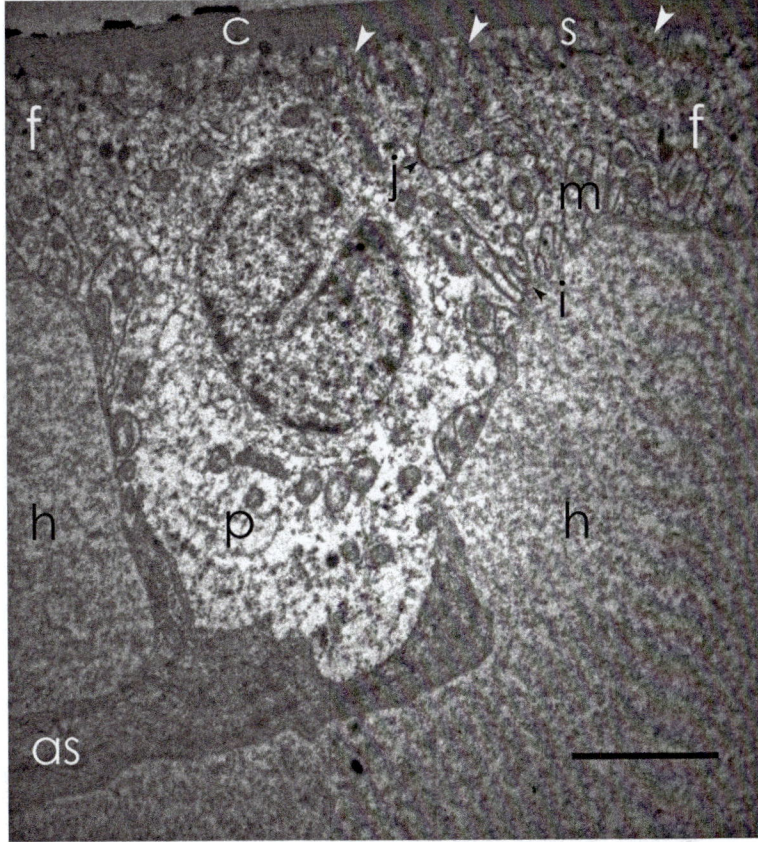

Figure 8.7 Transmission electron micrograph of an epithelial pillar cell from a pleurobranch gill filament of *Aegla franca* revealing the apical flanges (f) extending laterally below the cuticle (c) and above the perikaryon (p), the basal region of which rests on the annular septum (as). The apical flange membrane is augmented by short evaginations (arrows) forming a discrete subapical space (s). The lower flange membrane exhibits deep invaginations (i) associated with mitochondria (m) also present throughout the cytoplasm. Junctional regions (j) are evident between adjacent flanges. The perikaryon and lower flanges are bathed by the hemolymph (h). Scale bar=2 μm. (Modified from Antunes [2011].)

In the dorsal and ventral poles of the filament, the flat epithelial cells are replete with organelles, particularly mitochondria and cisternae of endoplasmic reticulum (Figure 8.8). The membrane surface area is augmented by apical evaginations and basal invaginations whose lumens are continuous with the hemolymph space. The basal pillar cell membrane can be thrown into numerous deep invaginations each of which lies adjacent to a mitochondrion in a highly structured association (Figure 8.9), forming a basal cytoplasmic region dominated by membrane infoldings and mitochondria, typical of an active ion transport zone.

The overall ultrastructure of the epithelial pillar cells is emblematic of an ion transporting epithelium with a distinct apical-basal polarity, and is notably similar to

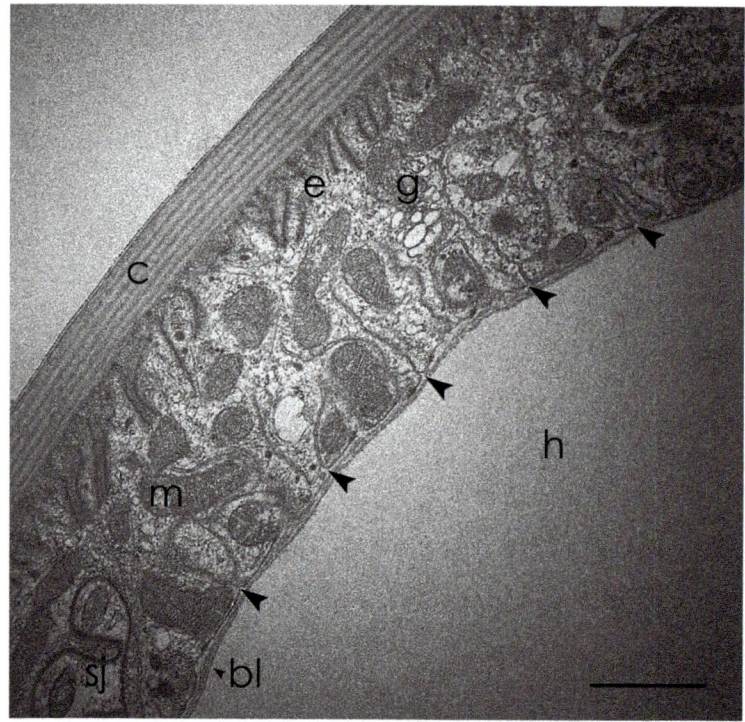

Figure 8.8 Transmission electron micrograph from a polar region of a pleurobranch gill filament from *Aegla franca* showing an epithelial cell arrayed beneath the cuticle (c). The flattened cell lies on the basal lamina (bl), and reveals a cytoplasm replete with mitochondria (m), rough endoplasmic reticulum, and other organelles like Golgi bodies (g). The apical membrane exhibits numerous evaginations (e) while the basal membrane shows invaginations (arrows) in continuity with the hemolymph space (h). This ultrastructure is typical of an ion transporting cell. Scale bar = 1 μm. (Modified from Antunes [2011].)

that of the trichobranchial filaments of the freshwater astacoid crayfish *Procambarus clarkii* (Dickson et al. 1991). These ultrastructural features are also reminiscent of the ionocytes in ion transporting epithelia well known from the gills of palaemonid shrimps and brachyuran crabs (Taylor and Taylor 1992; Freire et al. 2008b).

The physiological and ultrastructural characteristics disclosed here for the few aeglid species investigated, i.e., elevated osmotic and ionic gradients, poorly developed capability for isosmotic intracellular regulation, and highly differentiated epithelial cells that possess apical and basal membrane infoldings associated with abundant mitochondria, suggest that salt uptake by the aeglid gill likely follows the transport pathways proposed for other strictly freshwater decapods such as palaemonid shrimps and brachyuran crabs (Freire et al. 2008b; Henry et al. 2012; McNamara and Faria 2012; McNamara et al. 2015). The flattened polar cells and apical pillar cell flanges, tightly linked by lengthy septate junctions, constitute high resistance epithelia with low permeability to ions and water, each possibly housing

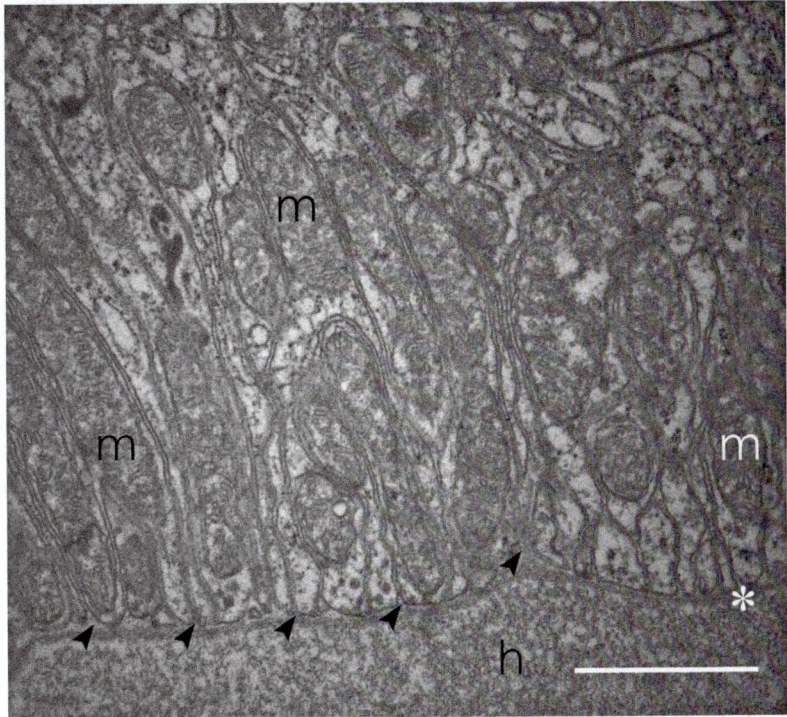

Figure 8.9 Transmission electron micrograph revealing detail of the basal region of a flattened epithelial cell from the polar region of a pleurobranch gill filament from *Aegla franca*, above the hemolymph space (h). The basal cell membrane is extensively invaginated by infoldings (arrows), each of which is intimately associated with a mitochondrion (m) in a highly structured manner and continuous with the hemolymph across the basal lamina (*). Scale bar = 1 µm. (Modified from Antunes [2011].)

distinct transport mechanisms. Salt uptake across the polar epithelium would proceed via a suite of transporters located in the apical and basal cell membranes. Initial sodium influx would likely take place through apical Na^+ channels, driven by the electrical gradient created by proton extrusion into the subapical space via an apical $V(H^+)$-ATPase. $Na^+/H^+(NH_4^+)$ antiporters also may be present. Chloride would be exchanged for bicarbonate via an apical Cl^-/HCO_3^- antiporter. Both H^+ and HCO_3^- would derive from the hydration of metabolic CO_2 by cytosolic carbonic anhydrase. Sodium would be actively transported from the cytosol into the hemolymph by the Na^+/K^+-ATPase located in the abundant basal membrane infoldings, while chloride would flow through basal Cl^- channels or the sodium-potassium two-chloride symporter. In the lateral, flanged pillar cell epithelium, ion transport appears to proceed directly across the invaginated, basal flange membrane into the hemolymph lacunae.

In this regard, several ion transporters and reference proteins for quantitative gene expression have been cloned from aeglid tissues, and their partial sequences deposited with GenBank, i.e., gill (KP125888) and antennal gland (KP125889) sodium-potassium ATPase alpha subunit mRNA from *A. franca*; gill (JN190492) and antennal gland (KM364041) V(H⁺)-ATPase B subunit mRNA from *A. franca*; genomic DNA sequences for sodium-potassium ATPase alpha-subunits from *A. alacalufi* (GU383035) and *A. neuquensis* (GU383036); ribosomal protein L10 mRNA (KU991254) from *A. franca* gill; genomic DNA sequences for arginine kinase from *A. alacalufi* (GU382856) and *A. neuquensis* (GU382857); and genomic DNA sequences for glyceraldehyde 3-phosphate dehydrogenase from *A. alacalufi* (GU382953) and *A. neuquensis* (GU382954).

Future studies of osmotic and ionic regulation in aeglids should focus on locating the various transporters hypothesized in this section in the epithelial cells of the gill filaments, on quantifying their gene and protein expression under different osmotic challenges, particularly salt secretion in hyperosmotic media, using both individual ion transporter genes and a broader transcriptomics approach. Other fundamental aspects to elucidate include the microanatomy, ultrastructure, and physiology of the antennal glands, which are absolutely unknown, including urine production and osmolality with regard to the hemolymph.

8.3 PATTERNS OF OXYGEN CONSUMPTION AND INTERMEDIATE METABOLISM

Aeglids generally inhabit cold, swift, well-oxygenated waters to which they appear to be restricted in their distribution. This suggests an elevated demand for respiratory oxygen uptake and an intolerance of hypoxic environments. The very few experimental studies of oxygen consumption performed in aeglids appear to confirm these predictions.

Rate specific oxygen consumption (MO_2) and diel activity as a function of oxygen availability have been examined in *A. longirostri* (see Dalosto and Santos 2011). MO_2 is higher under normoxic (3.72 mgO_2 L^{-1}) than hypoxic (2.01 mgO_2 L^{-1}, $\Delta= -46\%$) conditions, decreasing from about 40 μgO_2 g h^{-1} to 25 μgO_2 g h^{-1} after 48 h exposure to hypoxia. However, MO_2 does not decrease between 24 and 48 h exposure ($\approx$20 μgO_2 g h^{-1}), although activity increases considerably, possibly related to augmented gill ventilation or to greater muscle motion as part of an escape reflex from hypoxic media. MO_2 shows no clear diel activity pattern. These findings suggest that MO_2 is partially oxygen-independent in *A. longirostri*, i.e., the species is neither an oxyconformer nor an oxyregulator, exhibiting a mixed pattern of oxygen consumption depending on the degree and duration of hypoxia. In both *A. platensis* and *A. singularis*, MO_2 as measured in the field declines with fresh body mass over the range of approximately 0.25 to 2.0 g (Diawol et al. 2016). There is no evidence of a diel pattern in MO_2 in these species, or of differences between males and females.

Compared to other sympatric freshwater decapods, *A. longirostri* exhibits a higher metabolic rate than the crayfish *Parastacus brasiliensis* and the freshwater crab *Trichodactylus panoplus*; this demand may underlie the restriction of *A. longirostri* and aeglids in general to niches in which oxygen availability is elevated or saturated (Dalosto and Santos 2011). It is interesting that oxygen availability in the marine palaeo-environment of the ancestral aeglids is thought to be restricted and hypoxic, if not anaerobic (Feldmann et al. 1998). Clearly, profound alterations in metabolic profile and in tissue oxygen demands have accompanied the Aeglidae during their radiation into and occupation of oxygen-rich fresh waters from the ancestral habitat.

In contrast to the dearth of studies on oxygen consumption, investigations of the mobilization and use of energetic substrates and intermediate metabolism in aeglids are more readily available. These have focused particularly on the profiles of glucose, glycogen, protein, lipid, and triglycerides in different storage and transport tissues as a function of diel and seasonal variations, sex, and body mass.

Glucose is the predominant carbohydrate in crustacean hemolymph and plays a role in the synthesis of glycogen, mucopolysaccharides, chitin, ribose, and reduced nicotinamide adenine dinucleotide phosphate, and pyruvate (Waterman 1960; Morris and Airriess 1998). In contrast, its storage macromolecule, glycogen, predominates in muscle, hepatopancreas, gills, and hemocytes (Morris and Airriess 1998). Glycogen contributes to the metabolic events underpinning the molting cycle and fuels osmoregulation, growth, and adjustment to hypoxia and starvation (Jimenez and Kinsey 2015). Muscle is the principal tissue of protein storage in crustaceans (Claybrook 1983), and one in which intracellular free amino acids also predominate. These contribute to cell volume regulation as labile organic osmolytes, participating in isosmotic intracellular regulation (Mantel and Farmer 1983; Augusto et al. 2007, 2009). Protein metabolism varies particularly during ovarian development and gametogenesis, owing to increased synthesis of enzymes, hormones, and lipoproteins (Saborowski 2015). Lipid concentrations are generally elevated in crustaceans, particularly in the hepatopancreas and in the muscle tissue (O'Connor and Gilbert 1968). Lipid degradation in the hepatopancreas is elevated during molting and gametogenesis, periods of increased energy demand. Thus, the intermediate metabolism of crustaceans reflects a diversity of anabolic and catabolic processes, particularly related to the molting cycle, reproduction and growth, diel and seasonal variations, starvation and oxygen availability (Chang and O'Connor 1983; Claybrook 1983; Jimenez and Kinsey 2015; Saborowski 2015). Some of these processes and their metabolic substrates have been investigated in aeglids.

Findings for *A. ligulata* reveal that glycogen levels in the hepatopancreas, gills, and muscle, and hemolymph glucose titers, are much like those of other crustaceans (Oliveira et al. 2003), including *A. platensis* (see Oliveira et al. 2007). These energetic substrates show diel and seasonal variations. While hemolymph glucose does not vary diurnally, highest glucose titers are found in spring, likely owing to the reproductive period and food availability (Oliveira et al. 2003). Tissue glycogen concentrations do not vary, and seasonal variations are absent in females (Oliveira et al. 2003). However, winter males show hepatopancreas glycogen levels ≈3-fold greater than those seen in summer and autumn, possibly reflecting decreased foraging activity and glycogen storage (Oliveira et al. 2003).

In *A. platensis*, there is no diel variation in the titers of glucose, glycogen, total protein, lipids, or triglycerides in the hemolymph, hepatopancreas, abdominal muscle, or gills (Oliveira et al. 2007). However, storage and mobilization of these substrates fluctuate seasonally. Hemolymph glucose is lowest during autumn and increases ≈5-fold in winter females, decreasing gradually during spring and summer. In males, glucose also increases during winter, reaching levels ≈3-fold greater than those of autumn in spring and summer. Overall hemolymph glucose titers range from 0.36 ± 0.03 mM to 1.15 ± 0.29 in males, and from 0.40 ± 0.04 to 1.98 ± 0.63 mM in females (Oliveira et al. 2003). Hepatopancreas glycogen varies seasonally in females, reaching a maximum during autumn, decreasing gradually to very low levels during summer, while muscle and gill glycogen are fairly constant (Oliveira et al. 2003). This suggests glycogen storage in the hepatopancreas during the autumn reproductive period when glycogen-derived carbon is transferred to the gonads. Hemolymph total protein levels show clear seasonal fluctuations in *A. platensis*. Titers are ≈2-fold higher in spring and summer than in winter, which may reflect a decrease in tissue protein. The autumn decrease likely correlates with vitellogenic protein synthesis in the ovary, and with gametogenesis and reproductive behavior in males.

These findings for freshly caught *A. platensis* have been corroborated in laboratory-held animals (Ferreira et al. 2005) where, independently of consuming a protein- or carbohydrate-rich diet, hemolymph total protein is higher in summer than winter. Total lipids in *A. platensis* do not vary seasonally in females, but in males are higher in spring and summer, particularly in muscle and to a lesser degree in the hepatopancreas and gill. Triglycerides also increase during the summer in all tissues, reaching ≈3-fold the level seen in autumn. Such findings suggest increased energy demand in summer, likely for gamete production, and incubation and egg laying in autumn and winter (Oliveira et al. 2007). Starvation studies in *A. platensis* (Silva-Castiglioni et al. 2016) reveal that food deprivation for up to 30 days induces a decrease in hemolymph glucose titers, a transitory decrease in cholesterol, and an increase in total protein, normal values recovering on re-feeding. Together with starvation-induced modulation of gastric emptying and digestive processes, these metabolic responses may facilitate survival during periods of acute fasting *in natura*.

Seasonal and size-related variations in hemolymph glucose, triglycerides, cholesterol, and total protein, and hepatopancreas and muscle glycogen, lipid, and protein have been examined in *A. uruguayana* (see Musin et al. 2017), revealing that all these energy macromolecules vary seasonally. Hemolymph glucose and cholesterol titers are maximal in spring and summer, respectively, as are triglycerides and total protein. Hepatopancreas glycogen, lipid, and protein levels reach maxima in summer, while muscle glycogen and lipid titers attain maxima in spring, and protein in summer. These findings suggest the storage and mobilization of energy macromolecules mainly for reproductive processes such as gonadal maturation, oogenesis, and spermatogenesis, and vitellogenesis, including hormone and enzyme synthesis, and parental care (Musin et al. 2017).

Increases in and mobilization of energy macromolecules such as lipids and cholesterol are intimately related to reproductive events like ovarian development and vitellogenesis. Such mobilization also can be induced *in vivo* in the ovaries and hepatopancreas of aeglids such as *A. uruguayana* (see Silva-Castiglioni et al.

2009) and *A. platensis* (see Cahansky et al. 2008), employing neuro-regulators such as naloxone, a dopaminergic receptor inhibitor, and spiperone, an endogenous opioid inhibitor, when experimentally delivered via the diet. These antagonists apparently inhibit the effects of endogenous neurotransmitters that act on eyestalk and other neurosecretory centers, leading to the secretion of gonad-stimulating hormone and the inhibition of secretion of gonad-inhibiting hormone, resulting in ovarian development.

Based on these data from *A. ligulata*, *A. platensis*, and *A. uruguayana*, the catabolism and anabolism of energy substrates in aeglids clearly follow seasonal patterns (Oliveira et al. 2007; Silva-Castiglioni et al. 2016; Musin et al. 2017). During summer, lipids and triglycerides are stored and are mobilized, particularly during autumn when glycogen is stocked for use during winter, spring, and summer. Proteins available in the hemolymph during spring and summer are probably used during autumn and winter. The seasonally regulated intermediate metabolism of the aeglids appears to underlie their ability to exploit nutritional resources and to survive when exposed to less favorable conditions or fasting, and particularly for reproductive events (Oliveira et al. 2007; Silva-Castiglioni et al. 2016; Musin et al. 2017).

8.4 ENVIRONMENTAL CHARACTERISTICS AND THE EXTANT AEGLID HABITAT: A BRIEF HISTORY

The distribution of families of freshwater decapods throughout South America has been strongly influenced by regional environmental factors such as air and water temperatures, dissolved oxygen and ion contents, and pH, as revealed by quantitative analyses of biogeographical data (Tumini et al. 2016). Comparative analyses among *Aegla* species have revealed the impact of environmental parameters on physiological and reproductive traits, in addition to demonstrating that some environmental characters are phylogenetically structured (Bueno and Shimizu 2008; Faria et al. 2018).

The more basal aeglid species are distributed throughout colder climes, mainly in Chile (clades A and B, Figure 8.10, *left panel*) while the most derived species occur in warmer regions, notably in Brazil (clades C, D, and E). This phylogenetic history is consonant with the South American palaeodrainage hypothesis, since the Andean orogeny shifted the major drainage systems towards the Atlantic Ocean (Coney and Evenchick 1994). Thus, during the radiation of the aeglids into the northern, warmer regions, mean annual air temperature has increased (Figure 8.10, *right panel*) while thermal amplitude has decreased. There is no pattern of reduction in dissolved O_2 content, however (Faria et al. 2018).

The mean annual air temperature for the habitat of the northernmost aeglid species, *A. franca* (20° 60' S) is 20.2±2.0°C (EMBRAPA/ESALQ-USP 2003) while that for one of the southernmost species, *A. denticulata* (40° 49' S) is 10.5±3.1°C (National Oceanic and Atmospheric Administration and World Climate Database 2008) (Figure 8.10, *right panel*, Bueno and Shimizu 2008). Interestingly, there is no

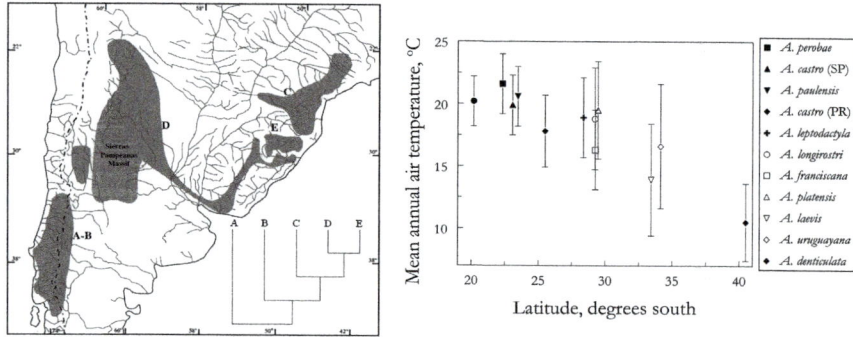

Figure 8.10 *Left panel.* Distribution of *Aegla* lineages A, B, C, D, and E throughout southern South America (Modified from Pérez-Losada et al. 2004). Note that clades A and B are more basal and are distributed closer to the original marine setting (Pacific Ocean); clades C, D, and E are more derived and have radiated eastwards. *Right panel.* Mean annual air temperature has increased while thermal amplitude (i.e., standard deviation) has decreased with the colonization of the northeastern regions. (Data from Bueno and Shimizu [2008].)

correlation between water temperatures and phylogenetic relationships among the species from the warmer clades C, D, and E (Faria et al. 2018). Thus, closely related aeglid species do not share similar thermal niches, which in summer vary from 12.0°C in *A. franciscana* (see Gonçalves et al. 2006) to 24.0°C in *A. castro* (see Faria et al. 2018). However, aeglid species in clades C and D do tend to occupy similar water temperatures (21.2±0.8 and 22.3±0.9°C), both being higher than the mean water temperature (17.8±1.9°C) for the southernmost species of clade E (Faria et al. 2018).

In contrast to the plastic interspecific variability seen in habitat water temperature, water pH and conductivity are phylogenetically structured (Faria et al. 2018). Owing to their effects on ion transport, both of these environmental traits affect physiological mechanisms like osmoregulation and acid-base balance equilibrium (Anger et al. 1994; Miserendino 2001). Such effects appear to be similar between closely related species, especially among the clade C aeglids (Faria et al. 2018). Assuming the precepts of phylophysiology (McNamara and Faria 2012), i.e., that physiological evolution accompanies phylogenetic history, and given the marine origin of the Aeglidae, an important question begging investigation is: are aeglids from the colder clades (A and B) weaker osmoregulators and more salt tolerant than those from the warmer clades (C, D, and E)? This is an open question that warrants exploration.

8.5 ENVIRONMENTAL AND ANTHROPOGENIC IMPACTS ON AEGLID PHYSIOLOGY

The monophyly of the aeglids reflects their unique evolutionary history, which has engendered an assemblage of biological features exclusive to the genus. Sadly however, the taxon is highly vulnerable to local and regional extinction since almost

70% of *Aegla* species are threatened (Bueno et al. 2016; Santos et al. 2017). Aeglid populations suffer the effects of chemical pollutants and anthropogenic activities owing to their high endemism and fragmented geographical distributions, often restricted to specific river basins (Bond-Buckup et al. 2008). The home ranges of certain species like *A. cavernicola*, *A. macrophthalma*, *A. perobae*, and *A. strinatii* measure just 0.02 km^2 in area, distributions so limited that aeglids are now considered the most threatened decapod taxon in South America (Bueno et al. 2016).

Since biological distribution and tolerance of environmental change and human impact are intrinsically linked to oxidative status (Monosson 2006), quantification of antioxidant defenses and tissue metal accumulation when seen from a phylogenetic perspective can be used as an indicator of aeglid physiological sensitivity and of metal tolerance (Faria et al. 2018). Metallothionein-like proteins (MTLP), antioxidant capacity against peroxyl radicals (ACAP), and the reduced/oxidized glutathione system (GSH-GSSG) are physiological descriptors of antioxidant defense capacity. These systems constitute essential non-specific scavengers of reactive oxygen species whose activities and titers correlate with variation in environmental parameters such as water metal and ion concentrations, temperature, alkalinity, pH, and pO$_2$ (Gamble et al. 1995; Abele and Puntarulo 2004; Monserrat et al. 2007). The inclusion of these antioxidant defense descriptors, together with various physico-chemical water traits in a phylogenetic multivariate analysis for various *Aegla* species reveals that ≈63% of total variance can be attributed to the first two eigenvectors (Figure 8.11, *left panel*). Antioxidant capacity against peroxyl radicals is driven by water conductivity and natural concentrations of Cr, Mn, Fe, Cd, and Pb in the aeglid habitat, while metallothionein-like protein titers do not correlate with metal concentration (Faria et al. 2018). Further examination of closely related species like *A. longirostri* and *A. inermis* reveals a lack of phylogenetic correlation between certain environmental parameters and antioxidant defenses, particularly antioxidant capacity against peroxyl radicals (Faria et al. 2018) (Figure 8.11, *right panel*). In contrast, tissue metal accumulation shows phylogenetic signal, signifying that this biomarker can be employed to estimate metal tolerance within a phylogenetic framework (Faria et al. 2018). Indeed, antioxidant capacity against peroxyl radicals together with tissue metal accumulation should be considered important biomarkers useful to evaluate putative environmental and anthropogenic impacts on aeglid physiology.

8.6 CONCLUSIONS AND PERSPECTIVES
FOR FUTURE INVESTIGATION

This review chapter clearly reveals that essential research in many areas of physiological investigation is sorely lacking in aeglids in general. The main areas studied to date are patterns of intermediate metabolism in different tissues in a few species, mainly as related to diel and seasonal variation, together with phylogenetic comparative analyses of antioxidant defenses; and osmoregulatory ability, also limited to a few species, focused on isosmotic intracellular and anisosmotic extracellular

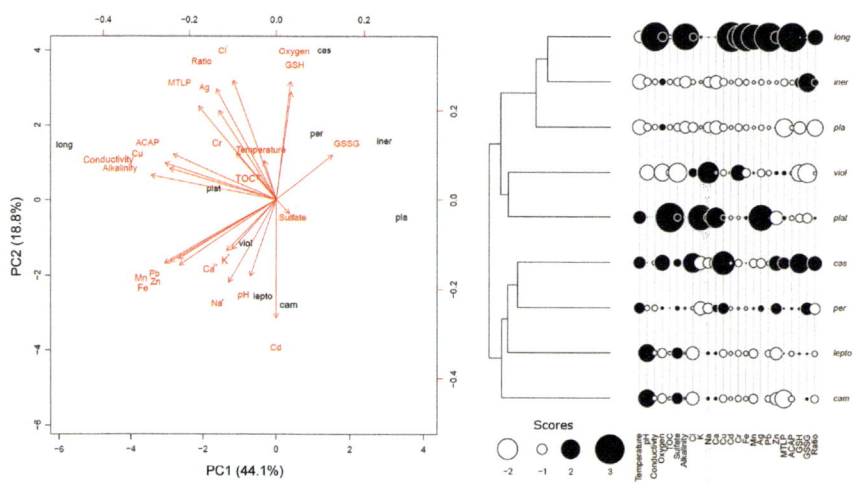

Figure 8.11 *Left panel.* Phylogenetic multivariate analysis employing antioxidant defenses [Metallothionein-like proteins (MTLP), antioxidant capacity against peroxyl radicals (ACAP), and the reduced/oxidized glutathione system (GSH-GSSG)] together with various environmental parameters for nine *Aegla* species. Considering only traits whose contribution to variance is greater than 60%, eigenvector 1 includes physicochemical water traits such as conductivity, alkalinity, Cl⁻, Cd, Cr, Fe, Mn, Pb, and Zn concentrations, and biomarkers like MTLP and ACAP. Eigenvector 2 correlates with dissolved O_2 and Na^+, K^+, Ca^{2+}, Ag, and reduced glutathione (GSH) concentrations. ACAP is associated with water conductivity and Cr, Mn, Fe, Cd, Pb, and Zn concentrations. MTLP variability does not correlate with any metal concentration. *Right panel.* Negative and positive scores generated by the multivariate analysis are indicated by white and black circles, scaled in proportion to their absolute values. Key: cam *A. camargoi*, cas *A. castro*, iner *A. inermis*, lepto *A. leptodactyla*, long *A. longirostri*, pla *A. plana*, plat *A. platensis*, per *A. perobae*, viol *A. violacea*. (From Faria et al. [2018].)

regulation. There is a smattering of information available on other unrelated pharmacological and physiological aspects such as sinus gland chromatophorotropins and pigment translocation (Castrucci and Mendes 1979), and the neuroregulatory mobilization of energy macromolecules related to ovarian development (Cahansky et al. 2008; Silva-Castiglioni et al. 2009), for example. Further, there have been very few micro-anatomical or ultrastructural studies performed on any tissue in any species of *Aegla*. Given the structure-function paradigm that underlies physiological investigation, this lack of a structural foundation for physiological processes hampers better comprehension of integrated systemic, tissue, and cellular functions in this remarkable taxon.

The biogeographical distribution of the extant aeglids renders the group ideal for comparative physiological investigations, particularly of processes that likely vary with latitude. For example, the known species of *Aegla* habit latitudes from 20°S to 50°S, with water temperatures ranging from around 8 to 25°C. There have been no studies of oxygen consumption as a function of temperature in any species, and the scope for aerobic and anaerobic metabolism as extreme temperature effects to become rate limiting is unknown. Such studies should be performed in species widely separated by latitude to comprehend how temperature affects respiratory metabolism in general and its effect on key enzyme activities such as citrate synthase and lactate dehydrogenase in particular, respective proxies for aerobic and anaerobic metabolism. These investigations should consider metabolic profiles as a function of long-term thermal acclimation to temperatures distinct from those of the habitat to examine metabolic thermal sensitivity. Control studies could be performed using natural thermal clines both within the distribution range of single species and among species habiting different thermal niches. Broader studies could consider these scenarios within a phylogenetic context, probing for the constraint between phylogenetic relationships and metabolic and other responses related to thermal niche and metabolic adaptation. Comparisons could be made with the responses of other anomuran taxa that inhibit intertidal or marine settings that vary widely and amply in thermal range.

Ontogenetic studies concerning intermediate metabolism have been restricted to a narrow range of body mass, mainly to young adults and adults. There have been no studies of embryonic metabolism or that of newly hatched juveniles. The respiratory metabolism and thermal sensitivity of the complete ontogenetic sequence (different embryonic stages of development, newly hatched juveniles, late juveniles and young, and adult specimens) is an open question. Quantitative gene and protein expression studies of key metabolic enzymes may provide clues to the metabolic and biochemical adjustments of the aeglids to their thermal habitats, as should comparative base sequence analyses with other decapods. Given the diversity of thermal niches occupied by the extant species of *Aegla*, the opportunities for investigating temperature-metabolism relationships seem boundless.

Despite some detail having been attained in osmoregulatory physiology, there have been no studies of the effects of salinity on respiratory metabolism in aeglids. This is an elemental aspect to consider, given the metabolic demands of their typically strong hyperosmoregulatory ability, their inability to hypo-osmoregulate, and their limited capacity to hypo-regulate hemolymph chloride, all mechanisms driven by ATP consuming processes at some point. Focusing specifically on osmoregulatory capability, species-wide studies may shed light on the cellular processes and mechanisms that have accompanied the conquest of fresh water by the aeglids, particularly when findings are analyzed using phylogenetic comparative methodologies. It is not known whether aeglids produce urine isosmotic or hypo-osmotic to their hemolymph, for example.

Comparisons of the base sequences coding for ion transporting proteins present in the gills and antennal glands such as the Na^+/K^+-ATPase and $V(H^+)$-ATPase, carbonic anhydrase, and the $Na^+/K^+/2Cl^-$ symporter among others would be helpful in establishing the degree of variability in these vital transporters that enable the aeglids to maintain their strong osmotic and ionic gradients in fresh water. Further, investigations of quantitative gene and protein expression together with transcriptomic analyses focusing on such transporters as a function of exposure to salinity may illuminate the mechanisms of cellular regulation of osmoregulatory processes. The localization of these transport proteins within the gill and antennal gland epithelia using immunocytochemical methodologies to establish testable models and hypotheses would be convenient. Finally, an essential question to consider spotlights the intersection of thermal biology, osmoregulatory ability, and the phylophysiological perspective: are the more basal aeglids from the colder clades weaker osmoregulators yet more salt tolerant than those from the warmer derived clades?

Such studies should help to answer the question posited at the outset of this chapter: given the overall challenges to be confronted, how might an isosmotic marine osmoconformer have evolved physiologically into a strong freshwater osmoregulator?

ACKNOWLEDGMENTS

We are indebted to Claudia Doi Antunes for generously allowing the use of unpublished photomicrographs of the gills of *Aegla franca* from her M. Sc. thesis (Comparative Biology Program, FFCLRP/USP). We are most grateful to Professor Carolina Arruda de Oliveira Freire (UFPR) for her helpful comments and suggestions on a late draft of parts of this chapter. This review was supported financially by an Excellence in Research Scholarship from the Conselho Nacional de Desenvolvimento Científico e Tecnológico to JCM (CNPq 303613/2017-3) and a post-doctoral scholarship from the Fundação de Amparo à Pesquisa do Estado de São Paulo to SCF (FAPESP 2017/05310-9).

REFERENCES

Abele, D., and S. Puntarulo. 2004. Formation of reactive species and induction of antioxidant defence systems in polar and temperate marine invertebrates and fish. *Comparative Biochemistry and Physiology* 138A:405–15.

Ahyong, S. T., J. K. Lowry, M. Alonso, et al. 2011. Subphylum Crustacea Brünnich, 1772. In *Animal Biodiversity: An Outline of Higher-Level Classification and Survey of Taxonomic Richness*, ed. Zhang, Z.-Q., vol. 3148, pp. 165–91. Zootaxa.

Anger, K., E. Spivak, C. Bas, D. Ismael, and T. Luppi. 1994. Hatching rhythms and dispersion of decapod crustacean larvae in a brackish coastal lagoon in Argentina. *Helgoländer Meeresuntersuchungen* 48:445–66.

Antunes, C. D. 2011. Relação da arquitetura branquial e ultraestrutura de células ion-transportadoras nos mecanismos fisiológicos que permitiram a invasão do meio diluido pelos Anomura (Crustacea, Decpoda): *Aegla franca* (Schmitt) e *Clibanarius vittatus* (Bosc). M Sc thesis, Comparative Biology Program, FFCLRP, Universidade de São Paulo.

Antunes, C. D., and J. C. McNamara. 2010. Structural and functional characterization of ion-transporting cells in the gill epithelium of the freshwater anomuran, *Aegla franca* (Crustacea, Anomura). In *17 IFSM International Microscopy Congress*, Rio de Janeiro. Resumos, #L16.35.

Applegate, S. P. 1992. A new genus and species of pycnodont from the Cretaceous (Albian) of central Mexico, Tepexi de Rodriguez, Puebla. Mexico. *Universidad Nacional Autónoma, Instituto de Geología, Revista* 10(2):164–78.

Augusto, A., L. J. Greene, H. J. Laure, and J. C. McNamara. 2007. Adaptive shifts in osmoregulatory strategy and the invasion of freshwater by brachyuran crabs: evidence from *Dilocarcinus pagei* (Trichodactylidae). *Journal of Experimental Zoology* 307A:688–98.

Augusto, A., A. S. Pinheiro, L. J. Greene, H. J. Laure, and J. C. McNamara. 2009. Evolutionary transition to freshwater by ancestral marine palaemonids: evidence from osmoregulation in a tide pool shrimp. *Aquatic Biology* 7:113–22.

Bond-Buckup, G., and L. Buckup. 1994. A família Aeglidae (Crustacea, Decapoda, Anomura). *Arquivos de Zoologia* 32:159–347.

Bond-Buckup, G., C. G. Jara, M. Pérez-Losada, L. Buckup, and K. A. Crandall. 2008. Global diversity of crabs (Aeglidae: Anomura: Decapoda) in freshwater. *Hydrobiologia* 595:267–73.

Bryan, G. W. 1965. Ionic regulation in the squat lobster *Galathea squamifera*, with special reference to the relationship between potassium metabolism and the accumulation of radioactive caesium. *Journal of the Marine Biological Association of the United Kingdom* 45:97–113.

Bueno, S. L. S., and R. M. Shimizu. 2008. Reproductive biology and functional maturity in females of *Aegla franca* (Decapoda: Anomura: Aeglidae). *Journal of Crustacean Biology* 28:652–62.

Bueno, S. L. S., R. M. Shimizu, and J. C. B. Moraes. 2016. A Remarkable Anomuran: the Taxon *Aegla* Leach, 1820. Taxonomic remarks, distribution, biology, diversity and conservation. In *A Global Overview of the Conservation of Freshwater Decapod Crustaceans*, eds. T. Kawai, and N. Cumberlidge, pp. 23–64. Cham: Springer International Publishing.

Cahansky, A. V., B. K. Dutra, D. Silva-Castiglioni, et al. 2008. Induction of ovarian growth in *Aegla platensis* (Crustacea, Aeglidae) by means of neuroregulators incorporated to food. *Revista de Biologia Tropical* 56:1201–07.

Castillo, R., G. Charmantier, and P. Thuet. 1988. Osmotic regulation in the hermit crab *Clibanarius erythropus*. *Biochemical Systematics and Ecology* 16:325–28.

Castrucci, A. M., and E. G. Mendes. 1979. Erythrophore studies in two anomuran crustaceans, a freshwater *Aegla castro* and an intertidal *Petrolisthes armatus*. *Boletim de Fisiologia Animal* 3:49–60.

Chang, E., and J. D. O'Connor. 1983. Metabolism and transport of carbohydrates and lipids. In *The Biology of Crustacea. Vol. 5. Internal Anatomy and Physiological Regulation*, ed. L. H. Mantel, pp. 263–87. New York: Academic Press.

Claybrook, D. L. 1983. Nitrogen metabolism. In *The Biology of Crustacea. Vol. 5. Internal Anatomy and Physiological Regulation*, ed. L. H. Mantel, pp. 163–212. New York: Academic Press.

Cogo, G. B., and S. Santos. 2007. Grau de adaptação de *Aegla longirostri* (Crustacea, Anomura) ao ambiente dulcícola. In *Anais do VIII Congresso de Ecologia do Brasil*, Caxambu, MG, pp. 1–2.

Collins, P. A., F. Giri, and V. Williner. 2011. Biogeography of the freshwater decapods in the La Plata Basin, South America. *Journal of Crustacean Biology* 31:179–91.

Coney, E. J., and C. A. Evenchick. 1994. Consolidation of the American cordilleras. *Journal of South American Earth Sciences* 7:241–62.

Dalosto, M., and S. Santos. 2011. Differences in oxygen consumption and diel activity as adaptations related to microhabitat in Neotropical freshwater decapods (Crustacea). *Comparative Biochemistry and Physiology* 160A:461–66.

De Grave, S., Y. Cai, and A. Anker. 2008. Global diversity of shrimps (Crustacea: Decapoda: Caridea) in freshwater. *Hydrobiologia* 595:287–93.

Diawol, V. P., M. V. Torres, and P. A. Collins. 2016. Field evaluation of oxygen consumption by two freshwater decapod morphotypes (Trichodactylidae and Aeglidae); the effect of different times of the day, body weight and sex. *Marine and Freshwater Behavior and Physiology* 49:251–63.

Dickson, J. S., R. M. Dillaman, R. D. Roer, and D. B. Roye. 1991. Distribution and characterization of ion transporting and respiratory filaments in the gills of *Procambarus clarkii*. *Biological Bulletin* 180:154–66.

Dixon, C. J., S. T. Ahyong, and F. R. Schram. 2003. A new hypothesis of Decapoda phylogeny. *Crustaceana* 76:935–75.

EMBRAPA/ESALQ-USP. 2003. Banco de Dados Climáticos do Brasil. Empresa Brasileira de Pesquisa Agropecuária (EMBRAPA)/Escola Superior de Agricultura "Luiz de Queiroz", Universidade de São Paulo (ESALQ-USP).

Faria, S. C., A. S. Augusto, and J. C. McNamara. 2011. Intra- and extracellular osmotic regulation in the hololimnetic Caridea and Anomura: a phylogenetic perspective on the conquest of fresh water by the decapod Crustacea. *Journal of Comparative Physiology* 181B:175–86.

Faria, S. C., R. D. Klein, Costa, P. G., et al. 2018. Phylogenetic and environmental components of inter-specific variability in the antioxidant defense system in freshwater anomurans *Aegla* (Crustacea, Decapoda). *Scientific Reports* 8:2850.

Feldmann, R. M. 1984. *Haumuriaegla glaessneri* n. gen. and sp. (Decapoda; Anomura; Aeglidae) from Haumurian (Late Cretaceous) rocks near Cheviot, New Zealand. *New Zealand Journal of Geology and Geophysics* 27:379–85.

Feldmann, R. M., F. J. Vega, S. B. Applegate, and G. A. Bishop. 1998. Early Cretaceous arthropods from the Tlayúa formation at Tepexi de Rodríguez, Puebla, México. *Journal of Paleontology* 72:79–90.

Ferreira, B. D. P., C. S. Hack, G. T. Oliveira, and G. Bond-Buckup. 2005. Perfil metabólico de *Aegla platensis* Schmitt, 1942 (Crustacea, Anomura) submetida a dietas ricas em carboidratos ou proteínas. *Revista Brasileira de Zoologia* 22:161–68.

Foster, C., E. M. Amado, M. M. Souza, and C. A. Freire. 2010. Do osmoregulators have lower capacity of muscle water regulation than osmoconformers? A study on decapod crustaceans. *Journal of Experimental Zoology* 313:80–94.

Freire, C. A., E. M. Amado, L. R. Souza, et al. 2008a. Muscle water control in crustaceans and fishes as a function of habitat, osmoregulatory capacity, and degree of euryhalinity. *Comparative Biochemistry and Physiology* 149A:435–46.

Freire, C. A., and J. C. McNamara. 1995. Fine structure of the gills of the freshwater shrimp *Macrobrachium olfersii* (Decapoda): effect of acclimation to high salinity medium and evidence for involvement of the lamellar septum in ion uptake. *Journal of Crustacean Biology* 15:103–16.

Freire, C. A., H. Onken, and J. C. McNamara. 2008b. A structure-function analysis of ion transport in crustacean gills and excretory organs. *Comparative Biochemistry and Physiology* 151A:272–304.

Freire, C. A., L. R. Souza-Bastos, E. M. Amado, V. Prodocimo, and M. M. Souza. 2013. Regulation of muscle hydration upon hypo- or hyper-osmotic shocks: differences related to invasion of the freshwater habitat by decapod crustaceans. *Journal of Experimental Zoology, A. Ecological Genetics and Physiology* 319:297–309.

Gamble, S. C., P. S. Goldfarb, C. Porte, and D. R. Livingstone. 1995. Glutathione peroxidase and other antioxidant enzyme function in marine invertebrates (*Mytilus edulis, Pecten maximus, Carcinus maenas* and *Asterias rubens*). *Marine and Environmental Research* 39:191–95.

Gonçalves, R. S., Silva-Castiglioni, D., and G. Bond-Buckup. 2006. Ecologia populacional de *Aegla franciscana* (Crustacea, Decapoda, Anomura) em São Francisco de Paula, RS, Brasil. *Iheringia, Série Zoologia* 96:109–14.

Henry, R. P., C. Lucu, H. Onken, and D. Weihrauch. 2012. Multiple functions of the crustacean gill: osmotic/ionic regulation, acid-base balance, ammonia excretion, and bioaccumulation of toxic metals. *Frontiers in Physiology* 3:1–33.

Jimenez, A. G., and S. T. Kinsey. 2015. Energetics and metabolic regulation. In *The Natural History of Crustacea: Physiology*, eds. E. Chang, and M. Thiel, pp. 389–417. New York: Oxford University Press.

Mantel, L. H., and L. L. Farmer. 1983. Osmotic and ionic regulation. In *The Biology of Crustacea, Vol. 5, Internal Anatomy and Physiological Regulation*, ed. D. E. Bliss, pp. 53–159. New York: Academic Press.

Martin, J. W., and L. G. Abele. 1986. Phylogenetic relationships of the genus *Aegla* (Decapoda, Anomura, Aeglidae), with comments on anomuran phylogeny. *Journal of Crustacean Biology* 6:576–616.

Martin, J. W., and L. G. Abele. 1988. External morphology of the genus *Aegla* (Crustacea: Anomura: Aeglidae). *Smithsonian Contributions to Zoology* 453:1–46.

Martin, J. W., and G. E. Davis. 2001. An updated classification of the recent Crustacea. *Natural Histroy Museum of Los Angeles County Science Series* 39:1–124.

McLaughlin, P. A., R. Lemaitre, and U. Sorhannus. 2007. Hermit crab phylogeny: a reappraisal and its "fall-out". *Journal of Crustacean Biology* 27:97–115.

McNamara, J. C., and S. C. Faria. 2012. Evolution of osmoregulatory patterns and gill ion transport mechanisms in the decapod Crustacea: a review. *Journal of Comparative Physiology* 182B:997–1014.

McNamara, J. C., C. A. Freire, A. H. Torres, and S. C. Faria. 2015. The conquest of fresh water by the palaemonid shrimps: an evolutionary history scripted in the osmoregulatory epithelia of the gills and antennal glands. *Biological Journal of the Linnean Society* 114:673–88.

McNamara, J. C., and A. G. Lima. 1997. The route of ion water movements across the gill epithelium of the freshwater shrimp *Macrobrachium olfersii* (Decapoda, Palaemonidae): evidence from ultrastructural changes induced by acclimation to saline media. *Biological Bulletin* 192:321–31.

McNamara, J. C., and A. H. Torres. 1999. Ultracytochemical location of Na^+/K^+-ATPase activity and effect of high salinity acclimation in gill and renal epithelia of the freshwater shrimp *Macrobrachium olfersii* (Crustacea, Decapoda). *Journal of Experimental Zoology* 284:617–28.

Miserendino, M. L. 2001. Macroinvertebrate assemblages in Andean Patagonian Rivers and streams: environmental relationships. *Hydrobiologia* 444:147–58.

Monosson, E. 2006. *Evolution in a toxic world*. Washington, DC: Island Press.

Monserrat, J. M., P. E. Martínez, L. A. Geracitano, et al. 2007. Pollution biomarkers in estuarine animals: Critical review and new perspectives. *Comparative Biochemistry and Physiology Part* 46C:221–34.

Morris, S., and C. N. Airriess. 1998. Integration of physiological responses of crustaceans to environmental challenge. *South African Journal of Zoology* 33:87–106.

Musin, G. E., A. Rossi, V. P. Diawol, P. A. Collins, and V. Williner. 2017. Dynamic metabolic pattern of *Aegla uruguayana* (Schmitt, 1942) (Decapoda: Anomura: Aeglidae): responses to seasonality and ontogeny in a temperate freshwater environment. *Journal of Crustacean Biology* 37:436–44.

National Oceanic and Atmospheric Administration, National Centers for Environmental Information. 2008. State of the climate: global climate report for year 2007. Available at: https://www.ncdc.noaa.gov/sotc/global/200713 (accessed January 31, 2019).

O'Connor, J. D., and L. I. Gilbert. 1968. Aspects of lipid metabolism in crustaceans. *American Zoologist* 8:529–39.

Oliveira, G. T., F. A. Fernandes, G. Bond-Buckup, A. A. P. Bueno, and R. S. M. da Silva. 2003. Circadian and seasonal variations in the metabolism of carbohydrates in *Aegla ligulata* (Crustacea: Anomura: Aeglidae). *Memoirs of Museum Victoria* 60:59–62.

Oliveira, G. T., F. A. Fernandes, A. A. P. Bueno, and G. Bond-Buckup. 2007. Seasonal variations in the intermediate metabolism of *Aegla platensis* (Crustacea, Decapoda, Aeglidae). *Comparative Biochemistry and Physiology* 147A:600–06.

Péqueux, A. 1995. Osmotic regulation in crustaceans. *Journal of Crustacean Biology* 15:1–60.

Pérez-Losada, M., G. Bond-Buckup, C. G. Jara, and K. A. Crandall. 2004. Molecular systematics and biogeography of the southern South American freshwater "crabs" *Aegla* (Decapoda, Anomura, Aeglidae) using multiple heuristic tree search approaches. *Systematic Biology* 53:767–80.

Pérez-Losada, M., C. G. Jara, G. Bond-Buckup, M. L. Porter, and K. A. Crandall. 2002. Anomuran phylogenetic relationships: on the taxonomic positioning of Aeglidae freshwater crabs. *Journal of Crustacean Biology* 22:670–76.

Rhodes-Ondi, S. E., and R. L. Turner. 2010. Salinity tolerance and osmotic response of the estuarine hermit crab *Pagurus maclaughlinae* in the Indian River Lagoon, Florida. *Estuarine, Coastal and Shelf Science* 86:189–96.

Saborowski, R. 2015. Nutrition and digestion. In *The Natural History of the Crustacea. Vol. 4: Physiology*, eds. E. S. Chang, and M. Thiel, pp. 285–319. Oxford: Oxford University Press.

Santos, S., G. Bond-Buckup, A. S Gonçalves, et al. 2017. Diversity and conservation status of *Aegla* spp. (Anomura, Aeglidae): an update. *Nauplius* 25:e2017011.

Schmitt, W. L. 1942. The species of *Aegla*, endemic South American freshwater crustaceans. *Proceedings of the United States National Museum* 91:431–524.

Silva-Castiglioni, D., A. V. Cahansky, E. M. Rodríguez, B. K. Dutra, G. T. Oliveira, and G. Bond-Buckup. 2009. Indução do crescimento ovariano em *Aegla uruguayana* (Crustacea, Anomura, Aeglidae) mediante a incorporação de neuroreguladores ao alimento. *Iheringia, Série Zoologia* 99:286–90.

Silva-Castiglioni, D., A. A. N. Valgas, I. D. Machado, B. S. Freitas, and G. T. Oliveira. 2016. Effect of different starvation and refeeding periods on macromolecules in the haemolymph, digestive parameters, and reproductive state in *Aegla platensis* (Crustacea, Decapoda, Aeglidae). *Marine and Freshwater Behaviour and Physiology* 49:27–45.

Swiech-Ayoub, B. P., and S. Masunari. 2001. Flutuações temporal e espacial de abundância e composição de tamanho de *Aegla castro* Schmitt (Crustacea, Anomura, Aeglidae) no Buraco do Padre, Ponta Grossa, Paraná, Brasil. *Revista Brasileira de Zoologia* 18:1003–17.

Tavares, C., C. Serejo, J. W. Martin. 2009. A preliminary phylogenetic analysis of the Dendrobranchiata based on morphological characters. In *Decapod Crustacean Phylogenetics*, eds. J. W. Martin, D. L. Felder, and K. A. Crandall, pp. 262–74. Boca Raton, FL: CRC Press, Crustacean issues 18.

Taylor, H. H., and E. W. Taylor. 1992. Microscopic anatomy of invertebrates. In *Decapod Crustacea'*, eds. F. W. Harrison, and A. G. Humas, vol. 10, pp. 203–93. New York: Wiley-Liss.

Tumini, G., F. Giri, V. Williner, and P. A. Collins. 2016. The importance of biogeographical history and extant environmental conditions as drivers of freshwater decapod distribution in southern South America. *Freshwater Biology* 61:715–28.

Von Sternberg, R. 1997. Cladistics of the freshwater crab family Trichodactylidae (Crustacea: Decapoda): appraising the reappraisal. *Journal of Comparative Biology* 2:49–62.

Von Sternberg, R., N. Cumberlidge, and G. Rodríguez. 1999. On the marine sister groups of the freshwater crabs (Crustacea: Decapoda). *Journal of Zoological Systematics and Evolutionary Research* 37:19–38.

Waterman, T. H. 1960. *The Physiology of Crustacea. Vol. 1, Metabolism and Growth.* New York: Academic Press.

Young, A. M. 1979. Osmoregulation in three hermit crab species, *Clibanarius vittatus* (Bosc), *Pagurus longicarpus* Say and *P. pollicaris* Say (Crustacaea: Decapoda: Anomura). *Comparative Biochemistry and Physiology* 63A:377–82.

CHAPTER 9

Conservation Status and Threats of Aeglidae: Beyond the Assessment

Harry Boos, Paula Guimarães Salge, and Marcelo A. A. Pinheiro

CONTENTS

9.1 WHY CONSERVE?

The need to prevent the extinction of species dates back to the environmental movements of the 1960s, when the first reports of widespread pollution, ecosystem disruption, and species loss began to appear. Today, as these trends have worsened on a global scale, the arguments to justify conservation actions have expanded to include the economic usefulness of species and ecosystems to society (Mace 2014). Nonetheless, assessing the risk of species extinction remains a crucial step in guiding efforts towards biodiversity conservation.

9.2 WHAT TO CONSERVE?

The protocols developed by the International Union for Conservation of Nature (IUCN) to produce the global Red Lists of endangered species in "Red Books," date back to the 1960s. Since then the categories and criteria that make up the IUCN

Red List protocols have been continuously refined (Mace et al. 2008). The result of these efforts is the most recent IUCN Red List protocols published in March 2017 (IUCN 2017a).

The IUCN Red List protocols have been adopted by governments of several countries seeking to protect species from extinction, including Brazil. Brazil stands out because its large geographical area includes species-rich tropical and subtropical ecosystems and biodiversity hotspots such as the Atlantic Forests and the Cerrado (Myers et al. 2000; Mittermeyer et al. 2004; Oliveira et al. 2017). Conservation attention has been primarily focused on endemic species that have a restricted area of occurrence, because these are typically the ones most threatened by expanding human populations (Malcolm et al. 2006; Devictor et al. 2008; Pandit et al. 2009).

Brazil hosts a significant number of species of aeglids, and most of them already receive protection under the law, since it is illegal to intentionally capture an animal threatened with extinction, and violators can face 9 to 18 months in prison and a fine of US$ 1,400 (Brazil 2008). Brazil has recently updated its list of endangered species, which now includes species of invertebrates, as well as vertebrates. The delay in including invertebrates has been attributed to a perceived lack of charisma, a point of view that included the freshwater aeglids (Bueno et al. 2016). The inclusion of aquatic invertebrates in the list of endangered species, which are targets of commercial and artisanal fisheries, has generated a series of challenges in the fishing community (Di Dario et al. 2015; Pinheiro et al. 2015). The legal disputes arising from this show that legal protection is still not extended to some threatened species of invertebrates, including the aeglids, even though they are not commonly used as a food resource.

In Brazil, the first official list of endangered animals (vertebrates) was published in 1968 (Brazil 1968). It took another 36 years for the list of endangered Brazilian fauna to include aquatic invertebrates (Brazil 2004; Pinheiro et al. 2015) such as three troglobitic species of aeglids: *Aegla cavernicola*, *A. leptochela*, and *A. microphthalma*. Pérez-Losada et al. (2002, 2009) took an alternative approach to prioritize conservation efforts and included phylogenetic methods that considered the evolutionary component of biodiversity to conserve genetically distant species.

9.3 HOW IS THE EXTINCTION RISK ASSESSED?

The IUCN Red List protocols aim to include all species found in a region, or globally, and assign each assessed species to one of eight categories, three of which are threatened categories indicating an increasingly high risk of extinction: Vulnerable (VU), Endangered (EN), and Critically Endangered (CR). The assignment to a category involves the compilation of data on population levels and trends, geographical distribution, habitat requirements, and threats, derived from the literature and first-hand field studies. These data are used to evaluate species against five criteria (A–E; see the following paragraph for explanation), depending on the quality of the available data; as a result of this evaluation a Red List category will be assigned for the species. This stage typically involves Red List Workshops that bring together group specialists pooling together their expertise to apply the criteria and assign a category, taking into account current knowledge about the biology, distribution, population trends, and current or projected threats (Mace et al. 2008).

The criteria used by IUCN (2001, 2017a) are as follows: **(A) population reduction** (past, present, and/or projected for the future); **(B) geographical distribution** (restricted and showing fragmentation, decline, or population fluctuations); **(C) small population** (and with fragmentation, decline, or fluctuations); **(D) very small population** (or very restricted distribution); and **(E) quantitative analysis of extinction risk.** Species could be assigned one of eight categories, as follows:

- **Data Deficient (DD):** When there is insufficient or inadequate information on populations, habitat, distribution, and threats to assess the risk of extinction; it indicates that more information is needed.
- **Least Concern (LC):** The probability of extinction is lowest when the data indicate that the species has a wide distribution, abundant and stable population levels, and without significant threats. A species with restricted distribution may be assessed as LC, as long as there are no significant direct threats.
- **Near Threatened (NT):** When the geographical range and/or population levels of a species are declining, but these levels are still above the threshold for any of the threatened categories. A species may be assessed as NT if it is likely to fall into a threat category in the near future.
- **Vulnerable (VU):** When all available information indicates that a species meets any of the thresholds for VU under criteria A to E listed above, the species is assessed as being at risk of extinction in nature.
- **Endangered (EN):** When all available information indicates that a species meets any of the thresholds for EN under criteria A to E listed above, a species is assessed as being at serious risk of extinction in nature.
- **Critically Endangered (CR):** When all available information indicates that a species meets any of the thresholds for CR under criteria A to E listed above, a species is assessed as being at extremely high risk of extinction in nature.
- **Extinct in the Wild (EW):** A species is considered to be EW when it is extirpated from all parts of its natural habitat, and the only living individuals are either kept in captivity, or exist as a naturalized population outside its historic range.
- **Extinct (EX):** A species is considered to be EX when the last existing member dies and it is no longer found in any part of its known range, despite searches of suitable habitat at appropriate times of the year.

The correct application of the method should be a constant concern during the extinction risk assessment, indicating the category (ies) and justification (s) for each of the criteria used. For this, it is not enough to know the taxonomy, distribution, and biology of the species. It is necessary to know and accurately quantify the threats that in fact impact the evaluated species.

The evaluation of the risk of extinction of newly described species can be problematic due to a lack of data, and many of these taxa have been listed in threatened categories (Santos et al. 2012; Moraes et al. 2016; Pinheiro and Santana 2016; Ribeiro et al. 2016, 2017; Bueno et al. 2017). New species are often described based on a few specimens from either few localities, or only a single location, and have a small extent of occurrence (EOO) and area of occupancy (AOO), that might reflect the initial sampling effort rather than the true state of affairs (IUCN 2017a).

Misunderstandings also occur in the application of "B" criteria (restricted geographical distribution, with fragmentation, decline, or population fluctuations) and "D" (very small population or very restricted distribution). In criterion "B," the categorization of a species at risk of extinction, either by subcriterion "B1" (extension of occurrence—EOO) or "B2" (area of occupation of species—AOO), must follow at least two of the following conditions (IUCN 2017a): **(a) Population severely fragmented** (or in few locations); **(b) Continued decline in at least one of the items**: (i) Extent of occurrence; (ii) area of occupation; (iii) area, extent, and/or habitat quality; (iv) number of locations or subpopulations; (v) number of mature individuals; and **(c) Extreme fluctuations**, in at least one of the items: (i) Extent of occurrence; (ii) area of occupancy; (iii) number of locations or subpopulations; (iv) number of mature individuals.

It is essential to apply the Red List protocols correctly to avoid inaccurate extinction risk assessments. For example, Moraes et al. (2016) and Santos et al. (2017) evaluated four species of *Aegla* (*A. japi*, *A. jundiai*, *A. paulensis*, and *A. vanini*) as Vulnerable (VU), according to the criteria VU B2aD2, but used only one subcriterion (a), whereas criterion B2 (EOO) requires at least two subcriteria.

The EOO calculation must be based on the presence of individuals in basins and/or microbasins. It is important to note that the EOO is not meant to represent the actual range of the species; instead the EOO represents how easily threats to a taxon may spread. A species with a small EOO that is low enough to meet the thresholds for one of the three threat categories is at a higher risk of extinction because a threat at one location may easily spread and reach individuals in other locations in the range. Contrarily, a species with a large EOO is at lower risk of extinction because a threat on one side of the range has to spread a lot further to reach individuals on the other side of the range.

The area of occupancy (AOO) of a species is the area within its EOO, which is actually occupied by the species and reflects the fact that a taxon will not usually occur throughout its EOO area because there may be numerous unsuitable or unoccupied habitats, especially in the case of freshwater organisms. The AOO can be estimated by summing the area around the point localities (e.g., 4 km^2 per locality), or it can be estimated using distribution or habitat maps (perhaps derived from remote imagery and/or analyses of spatial environmental data). The calculation of AOO is especially useful for species (such as aeglids) that may have a wide geographic distribution, but which occupy very specific habitats (IUCN 2017b).

The adoption of the hydrographic basin as a spatial unit, either to map the occurrence of species or to analyze the threats that impact them, is useful because basins are a management unit appropriate to species found in inland waters. In the case of freshwater species, geographical point localities based on field collections, scientific publications, and museum records are used to indicate in which sub-basins a species occurs (IUCN 2017b).

It is important to note that the restricted distribution of a species, as happens with most aeglids, is not enough to fit them into a category of extinction risk (criterion B).

This condition must be added to other aspects such as the number of locations, continued decline in EOO, AOO, habitat quality, and number of locations or number of mature individuals (IUCN 2017a). Likewise, populations with restricted distribution or found only in few locations (criterion D2), are at risk only when they are under plausible present or future threats, in which case they would qualify for Critically Endangered (CR), or CR "Possibly Extinct" (IUCN 2017a).

The number of locations where a species is found reflects the distribution of the species, and if these are declining over time, may also reflect the threats that are impacting it. Under the IUCN Red List terminology, a location corresponds to an area (that may include several nearby point localities) where a single event would rapidly impact all individuals of the species in the location (IUCN 2001, 2017a). The impact must be direct and must occur in the area occupied by the species, and it is essential that the threat is adequately documented. When the most severe threat to the species is habitat loss, for example, a location corresponds to the region where a single agricultural or urban occupation project can eliminate or drastically reduce the population (IUCN 2017a). Thus, the threat must be present in the geographic space under consideration, and when using AOO, the loss of habitat quality [subcriterion b(iii)] must be present in the location, rather than in a diffuse form throughout the EOO.

The reason why the online IUCN Red List is accepted globally as the industry "gold standard" of reliability is because the extinction risk of species listed there have been subjected to rigorous scrutiny at several levels by conservation specialists before publication. The correct assessment of the conservation status of species is, therefore, a fundamental prerequisite so that mitigation measures in form of conservation actions can be taken, aimed at saving threatened species from extinction.

9.4 ASSESSING THE RISK OF EXTINCTION OF AEGLIDS

The assessment of the extinction risk of 82 out of 87 species of aeglids assessed (Bueno, Camargo, and Moraes 2017; Moraes et al. 2017; Santos 2017) revealed that 57 species (67%) were threatened (CR 17, EN 24, VU 16) and at risk of extinction (Table 9.1). *Aegla intermedia* Girard 1855, was not assessed because it has never been found again and its type-series has disappeared (Bond-Buckup and Buckup 1994; Santos et al. 2017). Also, *A. quilombola* Moraes et al. (2017), is a newly described species and has not been assessed.

Bueno et al. (2016) evaluated the extinction risk of 42 species of *Aegla* from Brazil and found that 26 were at risk of extinction (8 CR, 12 EN, 6 VU) and were therefore legally protected in Brazil (Brazil 2014; ICMBio 2018) (Figure 9.1). All these threatened species are endemic and part of the Brazilian carcinofauna and are also the most at-risk group of crustaceans in Brazil (Magris et al. 2010; Boos et al. 2016) (Figure 9.1). Similarly, 11 out of 18 species of *Aegla* (61.1%) in Chile are threatened with extinction (2 CR, 6 EN, 3 VU) (Chile 2014).

There are no official government records of the extinction risk of the aeglid fauna in the other countries where this genus occurs (Argentina, Bolivia, Paraguay, and

Table 9.1   The Extinction Risk of 57 Threatened Species of *Aegla* Derived Using the
IUCN (2001, 2017a) Red List Criteria

Nº	Species	Red List Threatened Category/Criteria	Reference
1	*Aegla affinis* Schmitt (1942)	CR B1ab(iii)+2ab(iii)	Chile (2014)
2	*Aegla bahamondei* Jara (1982)	EN B1ab(iii)+2ab(iii)	Chile (2014)
3	*Aegla brevipalma* Bond-Buckup and Santos (2012)	CR B2ab(iii)	Brazil (2014)
4	*Aegla camargoi* Buckup and Rossi (1977)	EN B2ab(iii)	Brazil (2014)
5	*Aegla carinata* Bond-Buckup and Gonçalves (2014)	CRB2ab(iii,iv)	Santos et al. (2017)
6	*Aegla cavernicola* Türkay (1972)	CR B2ab(iii,v)	Brazil (2014)
7	*Aegla charon* Bueno et al. (2017)	CR B2ab(iii)	Bueno, Camargo, and Moraes (2017)
8	*Aegla cholchol* Jara and Palacios (1999)	VU B1ab(iii)+2ab(iii)	Chile (2014)
9	*Aegla concepcionensis* Schmitt (1942)	EN B1ab(iii)+2ab(iii)	Chile (2014)
10	*Aegla denticulata lacustris* Jara (1989)	CR B1ab(iii)	Chile (2014)
11	*Aegla expansa* Jara (1992)	EN B1ab(iii)+2ab(iii)	Chile (2014)
12	*Aegla franca* Schmitt (1942)	CR B2ab(iii)	Brazil (2014)
13	*Aegla georginae* Santos and Jara (2013)	CR B2ab(iii)	Santos et al. (2017)
14	*Aegla grisella* Bond-Buckup and Buckup (1994)	VU B1ab(iii)	Brazil (2014)
15	*Aegla humahuaca* Schmitt (1942)	VU B1ab(iii, iv)	Santos et al. (2017)
16	*Aegla inconspicua* Bond-Buckup and Buckup (1994)	VU B1 ab(iii)	Brazil (2014)
17	*Aegla inermis* Bond-Buckup and Buckup (1994)	EN B1ab(iii)	Brazil (2014)
18	*Aegla intercalata* Bond-Buckup and Buckup (1994)	VU B1ab(iii, iv)	Santos et al. (2017)
19	*Aegla itacolomiensis* Bond-Buckup and Buckup (1994)	EN B1 ab(iii)	Brazil (2014)
20	*Aegla japi* Moraes et al. (2016)	VU B2aD2*	Moraes et al. (2016) Santos et al (2017)
21	*Aegla jaragua* Moraes et al. (2016)	CR A4e**	Moraes et al. (2016)
22	*Aegla jundiai* Moraes et al. (2016)	VU B2aD2*	Moraes et al. (2016) Santos et al. (2017)
23	*Aegla laevis* (Latreille 1818)	EN B1ab(iii)+2ab(iii)	Chile (2014)
24	*Aegla lancinhas* Bond-Buckup and Buckup (2015)	EN B2ab(iii)	Santos et al. (2017)
25	*Aegla lata* Bond-Buckup and Buckup (1994)	CR B1ab(i,iii,iv)	Brazil (2014)
26	*Aegla leachi* Boos, Bond-Buckup and Buckup (2012)	EN B1ab(iii)+2ab(iii)	Brasil (2014)
27	*Aegla leptochela* Bond-Buckup and Buckup (1994)	CR B2ab(iii,v)	Brazil (2014)

(Continued)

Table 9.1 (Continued) The Extinction Risk of 57 Threatened Species of *Aegla* Derived Using the IUCN (2001, 2017a) Red List Criteria

N°	Species	Red List Threatened Category/Criteria	Reference
28	*Aegla leptodactyla* Buckup and Rossi (1977)	VU B1 ab(iii)	Brazil (2014)
29	*Aegla ligulata* Bond-Buckup and Buckup (1994)	VU B1 ab(iii)	Brazil (2014)
30	*Aegla loyolai* Bond-Buckup and Santos (2015)	EN B2ab(iii)	Santos et al. (2017)
31	*Aegla ludwigi* Santos and Jara (2013)	EN B2ab(iii)	Santos et al. (2017)
32	*Aegla manni* Jara (1980)	VU B1ab(iii)+2ab(iii)	Chile (2014)
33	*Aegla manuinflata* Bond-Buckup and Santos (2009)	EN B1ab(iii)+2ab(iii)	Brazil (2014)
34	*Aegla meloi* Bond-Buckup and Santos (2015)	CR B2ab(iii)	Santos et al. (2017)
35	*Aegla microphthalma* Bond-Buckup and Buckup (1994)	CR B2ab(iii,v)	Brazil (2014)
36	*Aegla oblata* Bond-Buckup and Santos (2012)	EN B1 ab(iii)	Brazil (2014)
37	*Aegla obstipa* Bond-Buckup and Buckup (1994)	EN B1ab(iii)	Brazil (2014)
38	*Aegla occidentalis* Jara et al. (2003)	EN B1ab(iii)+2ab(iii)	Chile (2014)
39	*Aegla papudo* Schmitt (1942)	EN A2ce	Chile (2014)
40	*Aegla paulensis* Schmitt (1942)	VU B2aD2*	Moraes et al. (2016) Santos et al. (2017)
41	*Aegla perobae* Hebling and Rodrigues (1977)	CR B2ab(iii)	Brazil (2014)
42	*Aegla plana* Buckup and Rossi (1977)	EN B1 ab(iii)	Brazil (2014)
43	*Aegla pomerana* Bond-Buckup and Buckup (2010)	EN B1 ab(iii)	Brazil (2014)
44	*Aegla renana* Bond-Buckup and Santos (2010)	CR B2ab(iii)	Brazil (2014)
45	*Aegla ringueleti* Bond-Buckup and Buckup (1994)	CR B2ab(iii)	Santos et al. (2017)
46	*Aegla rosanae* Campos Jr. (1998)	CR B2ab(iii)	Moraes et al. (2016)
47	*Aegla rossiana* Bond-Buckup and Buckup (1994)	EN B1 ab(iii)	Brazil (2014)
48	*Aegla saltensis* Bond-Buckup and Jara (2010)	VU B2ab(iii, iv)	Santos et al. (2017)
49	*Aegla sanlorenzo* Schmitt (1942)	EN B2ab(iii)	Santos et al. (2017)
50	*Aegla septentrionalis* Bond-Buckup and Buckup (1994)	EN B1ab(iii, iv)	Santos et al. (2017)
51	*Aegla spectabilis* Jara (1986)	VU B1ab(iii)+2ab(iii)	Chile (2014)
52	*Aegla spinipalma* Bond-Buckup and Buckup (1994)	VU B1 ab(iii)	Brazil (2014)
53	*Aegla spinosa* Bond-Buckup and Buckup (1994)	VU B1 ab(iii)	Brazil (2014)
54	*Aegla strinatii* Türkay (1972)	EN B2ab(iii)	Brazil (2014)

(Continued)

Table 9.1 (Continued) **The Extinction Risk of 57 Threatened Species of *Aegla* Derived Using the IUCN (2001, 2017a) Red List Criteria**

N°	Species	Red List Threatened Category/Criteria	Reference
55	*Aegla talcahuano* Schmitt (1942)	EN B1ab(iii)+2ab(iii)	Santos et al. (2017)
56	*Aegla vanini* Moraes et al. (2016)	VU B2aD2*	Moraes et al. (2016) Santos et al (2017)
57	*Aegla violacea* Bond-Buckup and Buckup (1994)	EN B1ab(iii,iv)	Brazil (2014)

*Aegla japi, A. jundiai, A. paulensis, and *A. vanini* were not evaluated correctly, as explained in the text. ****Aegla jaragua* (CR A4e) was evaluated initially as CR A4eB2a, but the subcriterion B2a has now been omitted because it is unnecessary for the categorization of the species as CR.

Uruguay). However, in a recent study of the aeglid fauna in three of these countries, Santos et al. (2017) found seven threatened species of *Aegla* in Argentina (2 CR, 2 EN, 3 VU), one in Bolivia (1 EN), and one in Uruguay (1 CR).

The extinction risk of practically the entire aeglid fauna has been assessed using the IUCN Red List protocols by a team of specialists who have contributed to the taxonomy, distribution, and biology of the aeglids (Bond-Buckup et al. 2009). The results presented indicate alarmingly high numbers of threatened species of aeglids (some of the highest on record). Characteristics such as restricted distribution and unique biology make them particularly vulnerable to habitat destruction and pollution. Bond-Buckup and Buckup (1994) pointed out the need to intensify studies on the aquatic fauna, especially aeglids, considering the advanced process of deterioration of the limnic environments in South America. This aspect is particularly evident in the evaluations of most of the threatened aeglids presented here that use criterion B (restricted distribution), due to rapid environmental degradation in many parts of their geographical distribution range.

9.5 THREATS TO AEGLIDS

The main risks to the 52 threatened species of aeglids (Table 9.1) are associated with the suppression of riparian forests, silting up and pollution of water bodies from agriculture, livestock, and aquaculture (Figure 9.2).

The activities listed in Figure 9.2 are responsible for the continuous decline in the quality of habitats occupied by aeglid crabs [subcriterion b(iii)] and are frequently cited as part of the justification for the extinction risk assessment of these animals. In general, the majority of these threats that impact aeglids are related to economic activities in the regions where each species occurs. Despite differences in landscape and vegetation, anthropogenic threats are well documented for some species of *Aegla* in Brazil and Chile. For example, in the state of Santa Catarina in Brazil, the breeding of cattle and pigs has caused a decline in the water quality in the single locality where *A. brevipalma* has been recorded (Bond-Buckup et al. 2008; Santos et al. 2012) (Figure 9.3). Another example is the increasing habitat degradation of

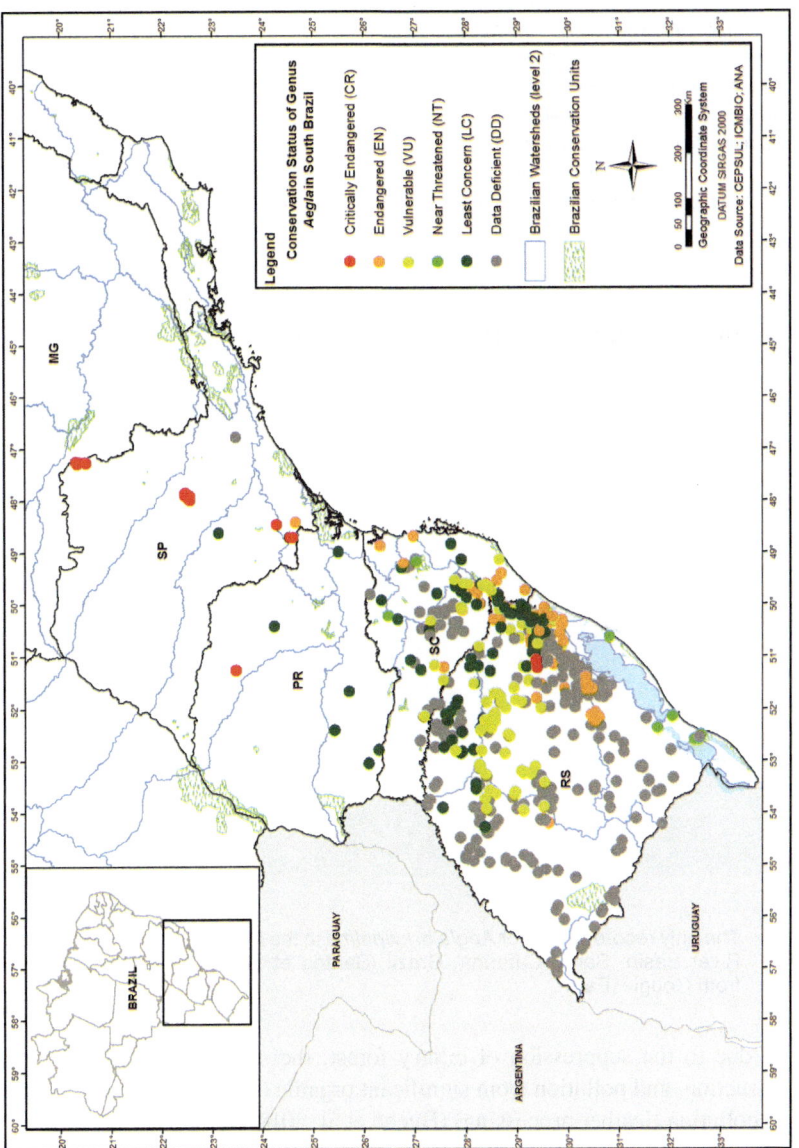

Figure 9.1 Distribution and extinction risk categories of the species assessed in Brazil. In the map the following states are indicated: MG (Minas Gerais), SP (São Paulo), PR (Paraná), SC (Santa Catarina), and RS (RS). Source: Brazil (2014), Bueno et al. (2016), and ICMBio (2018).

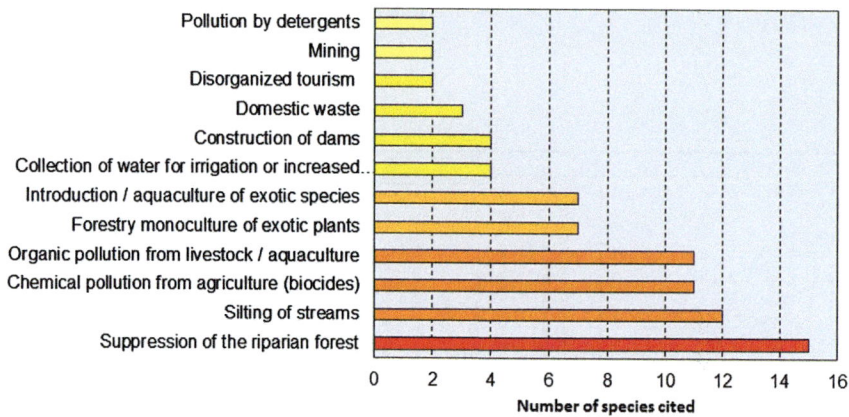

Figure 9.2   Threats indicated for species at risk of extinction.

Figure 9.3   The only recorded site for *Aegla brevipalma* in the Matator River in the Uruguay River Basin, Santa Catarina, Brazil (Santos et al. 2012). Source: Modified from Google®Earth.

A. franca due to the suppression of ciliary forest, the sedimentation of streams, road construction, and pollution from significant organic (animal husbandry) and/or industrial pollution (leather processing) (Bueno et al. 2018) (Figure 9.4).

Troglobitic species (such as *A. cavernicola*) are adapted to the relatively stable environmental conditions that characterize the interior of caves and are directly affected by changes of epigean habitats (Trajano 2000; Bueno et al. 2010). The destruction of the riparian forest, the silting of streams, and the pollution of streams by detergents are the main threats to this species (Pacca et al. 2007) (Figure 9.5).

Figure 9.4   One of the localities of *Aegla franca* in the Rio das Canoas in the Paraná River Basin, São Paulo, Brazil (Bueno et al. 2018). Source: modified from Google®Earth.

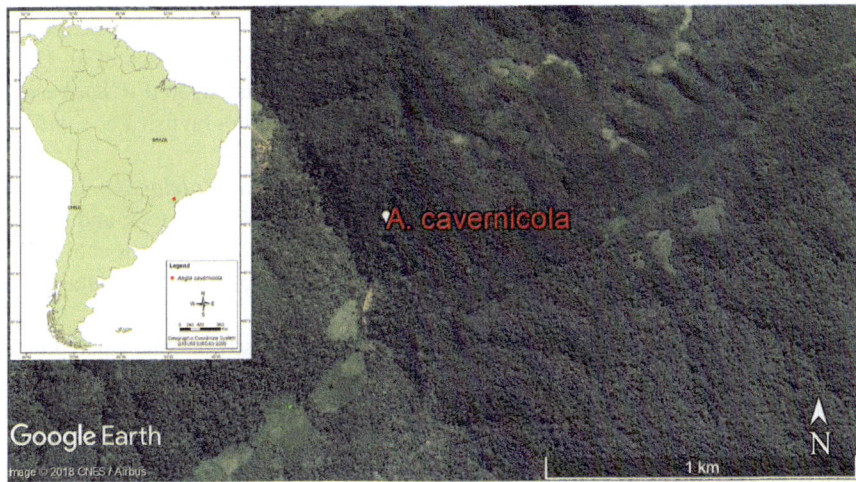

Figure 9.5   The only recorded site for the troglobitic species *Aegla cavernicola*, in the State of São Paulo, Brazil (Bueno et al. 2010). Source: Modified from Google®Earth.

In Chile, environmental degradation occurs mainly through the collection of water for irrigation, silting of streams, and organic and chemical pollution from biocides, and are threats to aeglids such as *A. laevis* (see Chile 2014) found on grape plantations for the wine industry (Bond-Buckup et al. 2008) (Figure 9.6).

Other threatened species of aeglids, such as *A. affinisi* and *A. denticulate*, are impacted by decreased habitat quality and predation by introduced species such as

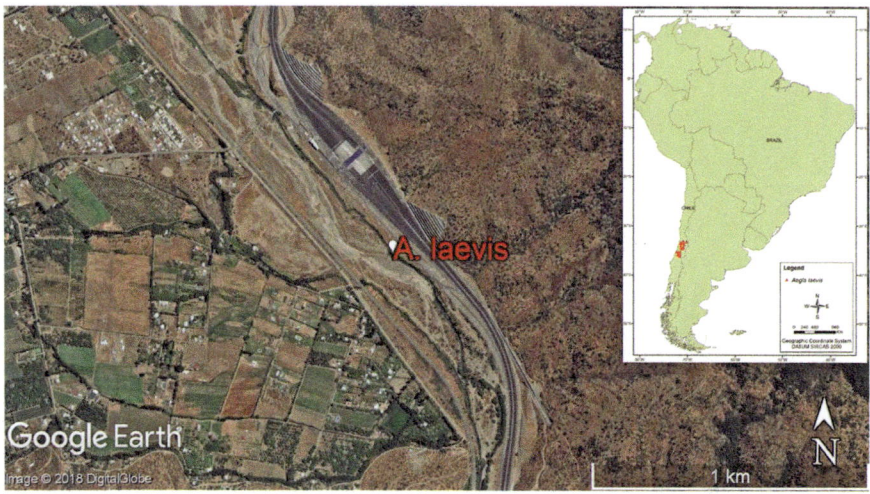

Figure 9.6   Locality of *Aegla laevis* in Angostura, Province of Huasco, Chile (Bahamonde and López 1963). Source: Modified from Google®Earth.

salmonid fish (Bond-Buckup et al. 2008; Chile 2014). The impact of exotic animals has similarly affected aeglids in other regions. For example, Moraes et al. (2016) assessed the extinction risk of *A. jaragua* as Critically Endangered (CR), because of threats from a well-established population of the exotic crayfish *Procambarus clarkii* (Girard 1852).

In Bolivia, the habitat of the endemic species *A. septentrionalis* has been degraded by pollution from organic sewage (domestic and/or livestock and agriculture) and from solid and liquid residues and debris (Flores 2010). Santos et al. (2017) evaluated this species as EN based mainly on similar threats to subpopulations of this species found in Argentina.

Anthropogenic threats to species of *Aegla* originate mostly on land where human populations generate the pollutants that impact freshwater environments (Duarte et al. 2016, 2017). These impacts are intensified along the central river basins whose waters are used for commercial transport, irrigation, animal husbandry, and subsistence farming (Liyanage and Yamada 2017). Humans have used rivers to dispose untreated urban and agricultural wastes (e.g., sewage, herbicides, insecticides) as well as industrial effluents (e.g., heavy metals, HPAs, etc.).

A number of species of aeglids (including threatened species) found in the northern and northeastern State of Rio Grande do Sul (RS) in Brazil (Guaíba watersheds) are likely to be impacted by the expansion of agricultural land used for viticulture (which has doubled in the last 20 years) (EMBRAPA 2017) (Figure 9.7). However, viticulture represents only 1% of the agriculture in this state (Pignati et al. 2017). According to these authors, in the same area, soybean cultivation predominates, accounting for 59% of the crop in RS, and coinciding with a higher human density and intense cattle activity (beef cattle and leather tanning industry).

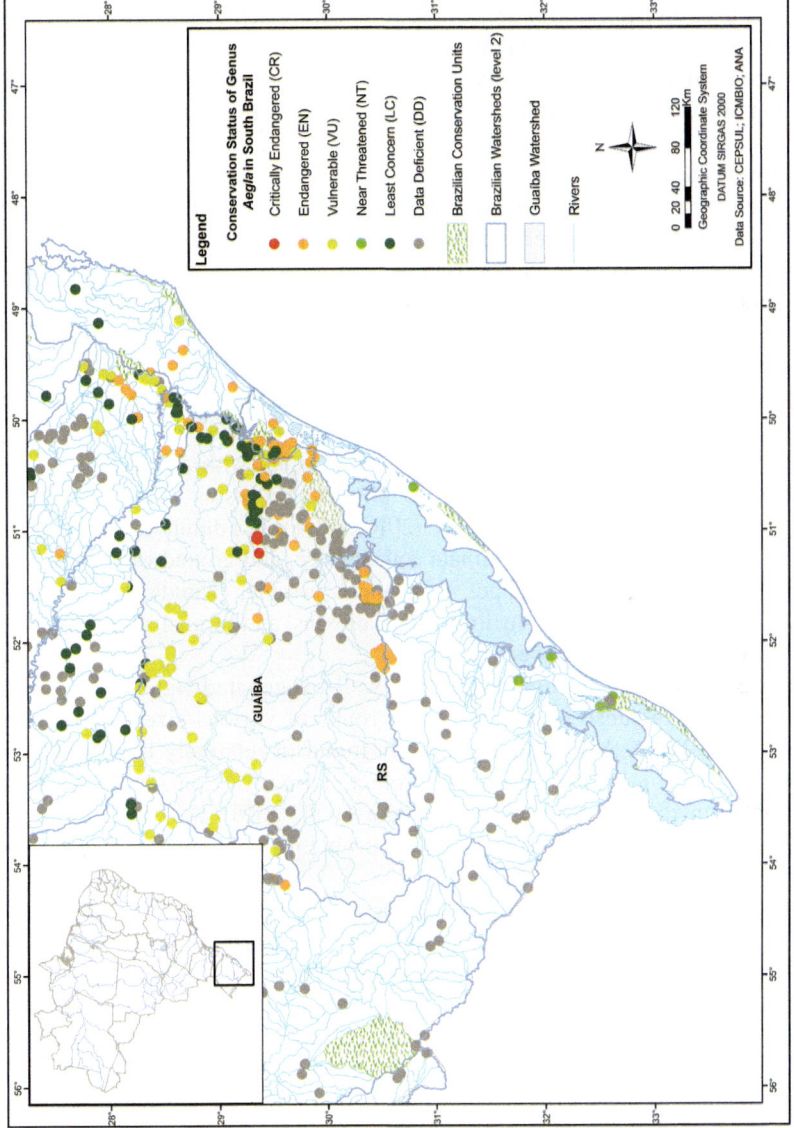

Figure 9.7 Regions with the highest number of aeglid records in Brazil (Guaíba watersheds), indicating the respective categories of extinction risk. Source: Bond-Buckup and Buckup (1994), Boos et al. (2012), Bueno et al. (2016), Santos et al. (2017).

Herbicides such as Paraquat (1,1′-dimethyl-4,4′-bipyridine-dichloride), a toxic product based on ammonium and water-miscible, are used in soybean cultivation, although Paraquat has been under restrictive use by ANVISA (GLOBO 2018). This herbicide promotes toxic and clastogenic effects in aquatic animals (Sartori and Vidrio 2018) and humans (Veríssimo et al. 2017). There are some herbicides and insecticides used in agricultural areas, which quickly reach groundwater, streams, and rivers, and pose a potential threat.

The accumulation of manure from beef cattle breeding causes organic contamination of freshwater habitats, which is worsened by the addition of untreated urban wastewater (Wen et al. 2017). According to the latter authors, the discharge of sewage from livestock and urban areas increase the level of contaminants, which are intensified by the reduction of the volume of rivers by damming and redirecting them, thus decreasing water sustainability. In this sense, precipitation gains importance as one of the leading environmental factors that affect contamination levels, causing the dispersion of pollutants in water bodies (Galloway and Cowling 1978) together with a decrease in the concentration of these pollutants.

Another source of common pollutants originates from the improper disposal and management of solid waste still primarily found in dumps. Over time, this material generates run-off (slurry), which is a dark-colored liquid with a nauseating odor, originating from biological, chemical, and physical processes of organic waste decomposition. According to Schepis et al. (2016), solid waste dumps are responsible for a range of pollutants that contaminate soil, surface, and groundwater with high levels of metal (e.g., mercury, copper, manganese, arsenic, cyanide) and phenolic compounds (polycyclic aromatic hydrocarbons, dioxins, and furans), all detrimental to animals, especially aquatic species.

Most aeglids inhabit streams with clear, well-oxygenated water, with a high flow velocity and a substrate composed by rocks of various sizes (Chiquetto-Machado et al. 2016). Aeglids are generally found in relatively large numbers associated with the most preserved and protected areas of riparian forests with minimal human disturbances (Zimmermann et al. 2016). Although environmental care has been intensifying, sanitary problems from sewage still persist (IBGE 2010).

In this sense, it is fundamental to link records of presence/absence (or abundance) of freshwater species with the quality of the environment they occupy, especially the amount and conservation status of the riparian forests (Arantes et al. 2018). Other studies have correlated physiological and reproductive responses as stress indicators in contaminated environments (Bertrand et al. 2018). Therefore, it is essential to use different approaches and techniques (e.g., genetic, physiological, and ecotoxicological ones) to develop an integrative and reliable evaluation about environmental quality, comprising alternative forms of environmental monitoring (Faria et al. 2018).

The use of alternative techniques to study well-known species was previously mentioned by Larramendi (2017), who cited several species of invertebrates in various environments. Inexpensive, rapid, and simple-to-use techniques, such as genotoxicity (e.g., micronucleus test, MN) and cytotoxicity tests (neutral red retention time, NRRT), have been useful to analyze the response of organisms to environmental quality. The application of these tests does not require the death of the specimens, since it only depends on a hemolymph sample from a few specimens, followed by their subsequent return to the natural environment (Pinheiro et al. 2017). These methods would be especially relevant for studies on aeglids because their populations are small. The simple recording or estimation of the abundance of these specimens in streams and rivers can already indicate the presence of water of excellent quality. Once again, it is crucial to conserve well-structured areas of ciliary forest, protecting streams and rivers, which are the habitat of the aeglids. In addition, the correct shadowing of streams (Gregory et al. 1991), together with biogeochemical and hydrological processes observed in riverine areas regulate the concentration of pollutants and mitigate the impact of sources of contamination from the uplands (McClain et al. 2003; de Sosa et al. 2018).

Despite the potential of aeglids as bioindicators of aquatic environmental quality, little is known about this topic. However, studies (Faria et al. 2018) provide integrative analyses, correlating phylogeny, environmental parameters, and antioxidants, demonstrating that the aeglids are niche specific, although there is some interspecific plasticity. According to these authors, the dosage of metals and analyses involving the antioxidant defense system (ADS) in these decapods can be an alternative for monitoring freshwater environments. The knowledge of such information would have practical applications and would be relevant for the mapping of preservation areas for aeglids as well as the environments habitats they occupy.

However, even if precarious, records about the occurrence and the extinction risk category allow the identification of some priority regions for the conservation of aeglids. The northeast region of the State of Rio Grande do Sul and the watercourses that make up the Guaíba River Basin (Figure 9.8) stand out.

9.6 CONSERVATION OF AEGLIDS

The conservation of aeglids is closely associated with the conservation of their habitats, primarily because the ecological niche is extremely restricted in these endemic species. Therefore, it is fundamental to know more about their distribution, the threats that impact the species, and their origin for the assessment of the extinction risk. Only a joint analysis of the threats versus the distribution of the species will permit the elaboration of public policies that seek to eliminate or mitigate the main threats to aeglids.

The calculation of the extent of occurrence (EOO) of aeglids, as well as for other freshwater species is based on the number of localities where the species has been recorded and represents the area where the main threats to the species operate. Otherwise, there will be only a superficial and generic indication of the threats,

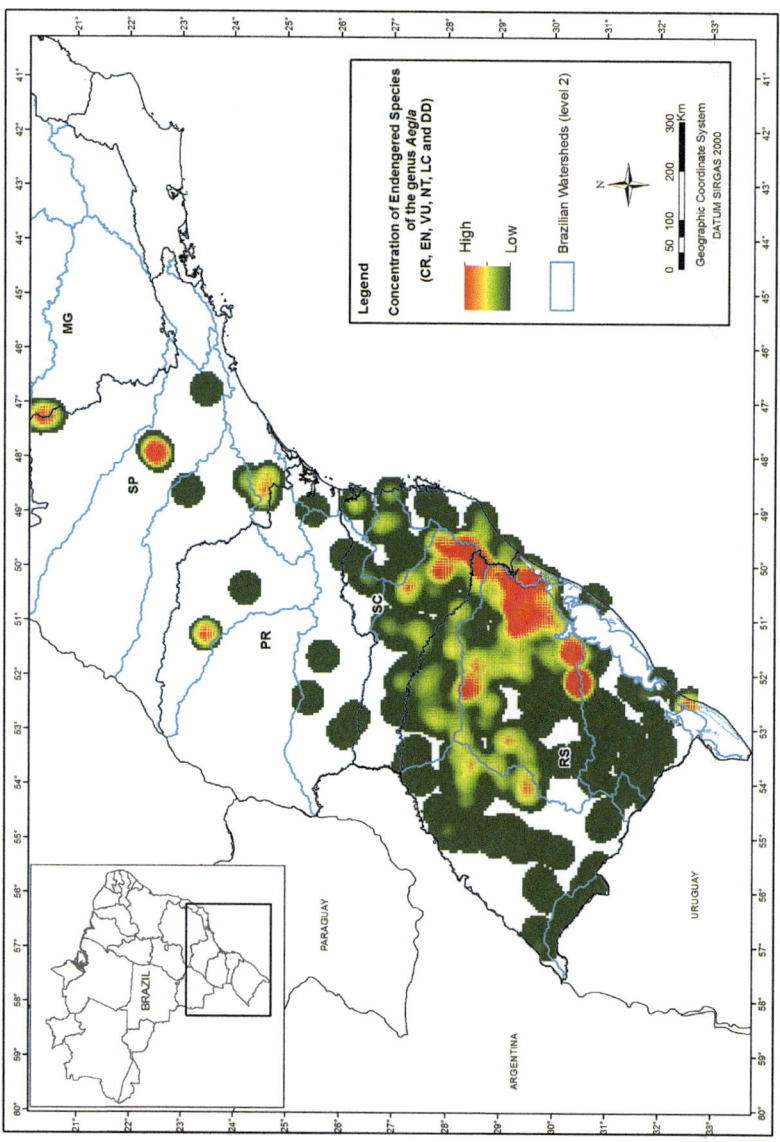

Figure 9.8 Concentration of threatened aeglid species in Brazil, resulting from Kernel density estimation tool by ArcGIS 10.5. The contours indicate the locations where the threatened species (CR, EN, and VU) are more concentrated.

hindering both the identification of cause and effect relationships and the elaboration of effective strategies to revert the extinction risk.

On the other hand, conservation measures should incorporate technological advances, especially genetic evidence. This approach has full adherence to the most recent conservation actions. In this sense, the prioritization of freshwater ecoregions (Abell et al. 2008), which encompasses a greater genetic diversity and number of aeglid species (Pérez-Losada et al. 2009), could help in the decision making for the allocation of resources for the conservation of these animals, which in general is very limited.

There are still no conservation measures for the anomuran freshwater crabs in South America (Magalhães et al. 2016). However, in Brazil, the "National Action Plan for the Conservation of Atlantic Forest Fish and Aeglids" (in prep.) aims to protect 26 endangered species of aeglids (Figure 9.9). Action plans identify and guide priority actions in order to save species at risk of extinction, and the implementation of these plans involve different sectors of society, such as governmental organizations and organized civil society (Brazil 2012).

Environmental education is a crucial part of these priority actions. Following the classic "must-know-to-conserve" idea, conservation initiatives for aeglids will only be supported when these fascinating animals are known not only by researchers but also by most of the society. In this sense, the "Biodiversity of Campos de Cima da Serra: popularizing knowledge" project coordinated by Bond-Buckup (2008a, b) is an example of a valuable initiative. The elaborated books describe educational activities for teachers and students of the cities located in the "Campos de Cima da Serra," within the Atlantic Forest biome in southern Brazil.

The conservation of aeglids represents an important challenge for people and institutions to reverse the extinction risk of these animals. The unique biology and restricted distribution of many aeglid species are a challenge for their conservation, which will only be met if researchers, environmentalists, and decision makers pool their efforts to reverse the trend towards increasing threats for these species.

ACKNOWLEDGMENTS

The authors thank our institutions ICMBio/CEPSUL (Brazilian Ministry of Environment) and UNESP IB/CLP, and the reviewers and editors for their valuable contributions. MAAP also thanks CNPq for the research fellowship grant awarded (Grants # 303286/2016-4).

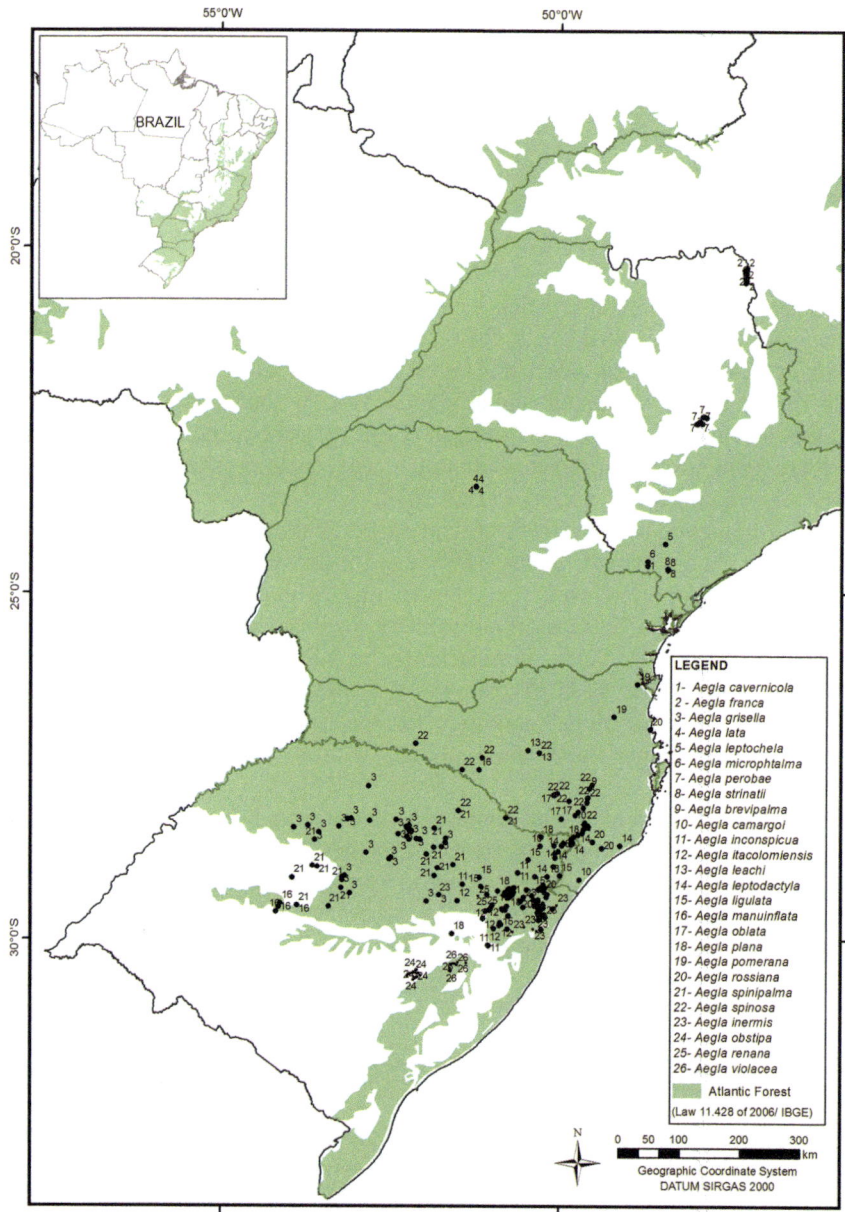

Figure 9.9 Species of *Aegla* that will be included in the "National Action Plan for the Conservation of Atlantic Forest Fish and Aeglids." Source: Bond-Buckup and Buckup (1994), Boos et al. (2012), Bueno et al. (2016), and ICMBio (2018).

REFERENCES

Abell, R., M. L. Thieme, C. Revenga, et al. 2008. Freshwater ecoregions of the world: a new map of biogeographic units for freshwater biodiversity conservation. *BioScience* 58(5):403–14.

Arantes, C. C., K. O. Winemiller, M. Petrere, L. Castello, L. L. Hess, and C. E. C. Freitas. 2018. Relationships between forest cover and fish diversity in the Amazon River floodplain. *Journal of Applied Ecology* 55:386–95.

Bahamonde, N., and M. T. López. 1963. Decápodos de aguas continentales en Chile. *Investigaciones Zoologicas Chilenas* 10:123–49.

Bertrand, L., M. V. Monferrán, C. Mouneyrac, and M. V. Amé. 2018. Native crustacean species as a bioindicator of freshwater ecosystem pollution: A multivariate and integrative study of multi-biomarker response in active river monitoring. *Chemosphere* 206:265–77.

Bond-Buckup, G. 2008a. *Biodiversidade dos campos de Cima da Serra*. Porto Alegre: Libretos.

Bond-Buckup, G. 2008b. *Biodiversidade dos campos de Cima da Serra – Livro de Atividades*. Porto Alegre: Libretos.

Bond-Buckup, G., and L. Buckup. 1994. A família Aeglidae (Crustacea, Decapoda, Anomura). *Arquivos de Zoologia* 32:159–346.

Bond-Buckup, G., A. Fransozo, A. V. Barreto, et al. 2009. Crustacea. In *Estado da arte e perspectivas para a Zoologia no Brasil*, eds. R. M. Rocha, and W. A. P. Boeger, pp. 101–30. Curitiba: Sociedade Brasileira de Zoologia – Editora UFPR.

Bond-Buckup, G., C. G. Jara, M. Pérez-Losada, L. Buckup, and K. A. Crandall. 2008. Global diversity of crabs (Aeglidae: Anomura: Decapoda) in freshwater. *Hydrobiologia* 595:267–73.

Boos, H., G. Bond-Buckup, L. Buckup, et al. 2012. Checklist of the Crustacea from the state of Santa Catarina, Brazil. *Check List* 8(6):1020–46.

Boos, H., M. A. A. Pinheiro, and R. A. Magris. 2016. O processo de avaliação do risco de extinção dos crustáceos no Brasil: 2010–2014. In *Livro vermelho dos crustáceos do Brasil: Avaliação 2010–2014*, eds. M. Pinheiro, and H. Boos, Chapter 1, pp. 28–34. Porto Alegre: Sociedade Brasileira de Carcinologia.

Bueno, S. L. S., A. L. Camargo, and J. C. B. Moraes. 2017. A new species of stygobitic aeglid from lentic subterranean waters in southeastern Brazil, with an unusual morphological trait: short pleopods in adult males. *Nauplius* 25:e2017021.

Bueno, S. L. S., A. L. Camargo, B. F. Takano, and F. P. A. Cohen. 2010. Crustáceos eglídeos (*Aegla* sp.): Uma história única na América do Sul. *O Carste* 22(1):8–11.

Bueno, S. L. S., F. L. M. Mantelatto, S. S. Rocha, and E. C. Mossolin. 2018. *Aegla franca* Schmitt, 1942. In *Livro vermelho da fauna Brasileira ameaçada de extinção*, ed. ICMBio – Instituto Chico Mendes de Conservação da Biodiversidade, vol. VII, 376–79. Brasília: Instituto Chico Mendes de Conservação da Biodiversidade.

Bueno, S. L. S., S. Santos, S. S. Rocha, K. M. Gomes, E. C. Mossolin, and F. L. Mantelatto. 2016. Avaliação dos Eglídeos (Decapoda: Aeglidae). In *Livro vermelho dos crustáceos do Brasil: Avaliação 2010–2014*, eds. M. Pinheiro, and H. Boos, Chapter 2, pp. 35–63. Porto Alegre: Sociedade Brasileira de Carcinologia.

Brazil. 1968. Portaria IBDF nº 303. Lista oficial brasileira das espécies de animais e plantas ameaçadas de extinção. Available at: http://www.mma.gov.br/estruturas/179/_arquivos/179_05122008034305.pdf (accessed June 01, 2018).

Brazil. 2004. Instrução Normativa MMA n° 5. Reconhece espécies de invertebrados aquáticos e peixes ameaçados de extinção e espécies sobreexplotadas ou ameaçadas de sobreexplotação. Available at: http://www.mma.gov.br/estruturas/179/_arquivos/179_05122008033927.pdf (accessed June 01, 2018).

Brazil. 2008. Decreto n° 6514. Dispõe sobre as infrações e sanções administrativas ao meio ambiente, estabelece o processo administrativo federal para apuração destas infrações, e dá outras providências. Available at: http://www.planalto.gov.br/ccivil_03/_ato2007-2010/2008/decreto/d6514.htm (accessed June 01, 2018).

Brazil. 2012. Instrução Normativa ICMBio n° 25. Procedimentos para a elaboração, aprovação, publicação, implementação, monitoria, avaliação e revisão de planos de ação nacionais para conservação de espécies ameaçadas de extinção ou do patrimônio espeleológico. Available at: http://www.icmbio.gov.br/portal/images/stories/biodiversidade/fauna-brasileira/normativas/IN_PLANO_DE_ACAO_25-2012.pdf (accessed June 01, 2018).

Brazil. 2014. Portaria MMA n°. 445. Reconhece espécies de peixes e invertebrados aquáticos da fauna brasileira ameaçadas de extinção constantes da "Lista Nacional Oficial de Espécies da Fauna Ameaçadas de Extinção – Peixes e Invertebrados Aquáticos". Available at: http://www.icmbio.gov.br/cepsul/images/stories/legislacao/Portaria/2014/p_mma_445_2014_lista_peixes_amea%C3%A7ados_extin%C3%A7%C3%A3o.pdf (accessed June 01, 2018).

Chile. 2014. Decreto Supremo MMA n° 52. Aprueba y oficializa clasificación de especies según su estado de conservación. Available at: http://www.mma.gob.cl/clasificacionespecies/Anexo_decimo_proceso/DS%2052_2014_DiariOfcial_OficializaDecimoProceso.pdf (accessed June 01, 2018).

Chiquetto-Machado, P. I., L. C. M. Vieira, R. M. Shimizu, and S. L. S.Bueno. 2016. Life cycle of the freshwater anomuran *Aegla schmitti* Hobbs, 1978 (Decapoda: Anomura: Aeglidae) from southeastern Brazil. *Journal of Crustacean Biology* 36:39–45.

Di Dario, F., C. B. M Alves, H. Boos, et al. 2015. A better way forward for Brazil's fisheries. *Science* 347:1079.

de Sosa, L. L., H. C. Glanville, M. R. Marshall, A. P. Williams, and D. L. Jones. 2018. Quantifying the contribution of riparian soils to the provision of ecosystem services. *Science of the Total Environment* 624:807–19.

Devictor, V., R. Julliard, and F. Jiguet. 2008. Distribution of specialist and generalist species along spatial gradients of habitat disturbance and fragmentation. *Oikos* 117:507–14.

Duarte, L. F. A., C. A. Souza, C. R. Nobre, C. D. S. Pereira, and M. A. A. Pinheiro. 2016. Multi-level biological responses in *Ucides cordatus* (Linnaeus, 1763) (Brachyura, Ucididae) as indicators of conservation status in mangrove areas from the western Atlantic. *Ecotoxicology and Environmental Safety* 133:176–87.

Duarte, L. F. A., C. A. Souza, C. D. S. Pereira, and M. A. A. Pinheiro. 2017. Metal toxicity assessment by sentinel species of mangroves: In situ case study integrating chemical and biomarkers analyses. *Ecotoxicology and Environmental Safety* 145:367–76.

EMBRAPA – Empresa Brasileira de Pesquisa Agropecuária. 2017. *Cadastro vitícola do Rio Grande do Sul: 2013 a 2015.* Brasília: EMBRAPA.

Faria, S. C., R. D. Klein, P. G. Costa, et al. 2018. Phylogenetic and environmental components of inter-specific variability in the antioxidant defense system in freshwater anomurans *Aegla* (Crustacea, Decapoda). *Scientific Reports* 8:2850.

Flores, V. G. B. 2010. Identificación de factores que afectan la estructura poblacional del cangrejo de río (*Aegla septentrionalis*) en dos vertientes del Municipio de Tupiza (Dpto. Potosí). Graduation thesis. Facultad de Agronomia, Universidad Mayor de San Andrés, La Paz, Bolívia.

Galloway, J. N., and E. B. Cowling. 1978. The effects of precipitation on aquatic and ter-restrial ecosystems: a proposed precipitation chemistry network. *Journal of the Air Pollution Control Association* 28:229–35.

GLOBO. 2018. Restrição a agrotóxico usado no milho e na soja começa a valer nesta quinta (22/03/2018). Available at: https://g1.globo.com/ciencia-e-saude/noticia/restricao-a-agrotoxico-usado-no-milho-e-na-soja-comeca-a-valer-nesta-quinta.ghtml (accessed June 01, 2018).

Gregory, S. V., F. J. Swanson, W. A. McKee, and K. W. Cummins. 1991. An ecosystem per-spective of riparian zones. *BioScience* 41:540–51.

IBGE – Instituto Brasileiro de Geografia e Estatística. 2010. *Atlas de saneamento 2001.* Rio de Janeiro: IBGE.

ICMBio – Instituto Chico Mendes de Conservação da Biodiversidade. 2018. *Livro vermelho da fauna Brasileira ameaçada de extinção: Volume VII – Invertebrados.* Brasília: ICMBio/MMA.

IUCN – International Union for Conservation of Nature. 2001. *IUCN Red List Categories and Criteria: Version 3.1.* IUCN Species Survival Commission. Gland and Cambridge: IUCN.

IUCN – International Union for Conservation of Nature. 2017a. *Guidelines for Using the IUCN Red List Categories and Criteria. Version 13.* Standards and Petitions Subcommittee. Available at: http://www.iucnredlist.org/documents/RedListGuidelines.pdf (accessed June 01, 2018).

IUCN – International Union for Conservation of Nature. 2017b. *Assessment Methods in Freshwater.* Available at: http://www.iucnredlist.org/initiatives/freshwater/process/methods (accessed June 01, 2018).

Larramendi, M. L. 2017. *Ecotoxicology and Genotoxicology: Non-traditional Aquatic Models. Issues in Toxicology*, vol. 33. London: Royal Society of Chemistry.

Liyanage, C., and K. Yamada. 2017. Impact of population growth on the water quality of natural water bodies. *Sustainability* 9(8):1405.

Mace, G. M. 2014. Whose conservation? *Science* 345:1558–60.

Mace, G. M., N. J. Collar, K. J. Gaston, et al. 2008. Quantification of extinction risk: IUCN's system for classifying threatened species. *Conservation Biology* 22:1424–42.

Magalhães, C., M. R. Campos, P. A. Collins, and F. L. Mantelatto. 2016. Diversity, distribu-tion and conservation of freshwater crabs and shrimps in South America. In *A Global Overview of the Conservation of Freshwater Decapod Crustaceans*, eds. T. Kawai, and N. Cumberlidge, pp. 303–22. Cham: Springer International Publishing.

Magris, R. A., G. Bond-Buckup, C. Magalhães, et al. 2010. Quantification of extinction risk for crustacean species: an overview of the National Red Listing process in Brazil. *Nauplius* 18:129–35.

Malcolm, J. R., C. Liu, R. P. Neilson, L. Hansen, and L. Hannah. 2006. Global warming and extinctions of endemic species from biodiversity hotspots. *Conservation Biology* 20:538–48.

McClain, M. E., E. W. Boyer, C. L. Dent, et al. 2003. Biogeochemical hot spots and hot moments at the interface of terrestrial and aquatic ecosystems. *Ecosystems* 6:301–12.

Mittermeyer, R. A., P. R. Gil, M. Hoffmann, et al. 2004. *Hotspots. Biodiverisad amenazada II.* Mexico: CEMEX.

Moraes, J. C. B., M. Tavares, and S. L. S. Bueno. 2017. Taxonomic review of *Aegla marginata* Bond-Buckup and Buckup, 1994 (Decapoda, Anomura, Aeglidae) with description of a new species. *Zootaxa* 4323(4):519–33.

Moraes, J. C. B., M. Terossi, R. C. Buranelli, M. Tavares, F. L. M. Mantelatto, and S. L. S. Bueno. 2016. Morphological and molecular data reveal the cryptic diversity among populations of *Aegla paulensis* (Decapoda, Anomura, Aeglidae), with descriptions of four new species and comments on dispersal routes and conservation status. *Zootaxa* 4193(1):1–48.

Myers, N., R. A. Mittermeier, C. G. Mittermeier, G. A. B. Fonseca, and J. Kent. 2000. Biodiversity hotspots for conservation priorities. *Nature* 403:853–58.

Oliveira, U., B. S. Soares-Filho, A. P. Paglia, et al. 2017. Biodiversity conservation gaps in the Brazilian protected areas. *Scientific Reports* 7:9141.

Pacca, H. M., T. C. G. Sebrian, and E. Trajano. 2007. Conservação. In *Sistemas areias – 100 anos de estudo. São Paulo*, ed. E. Trajano, pp. 113–19. São Paulo: Redespeleo Brasil.

Pandit, S. N., J. Kolasa, and K. Cottenie. 2009. Contrasts between habitat generalists and specialists: an empirical extension to the basic metacommunity framework. *Ecology* 90:2253–62.

Pérez-Losada, M., G. Bond-Buckup, C. G. Jara, and K. A. Crandall. 2009. Conservation assessment of southern South American freshwater ecoregions on the basis of the distribution and genetic diversity of crabs from the genus *Aegla*. *Conservation Biology* 3(23):692–702.

Pérez-Losada, M., C. G. Jara, G. Bond-Buckup, and K. A. Crandall. 2002. Conservation phylogenetics of Chilean freshwater crabs *Aegla* (Anomura, Aeglidae): assigning priorities for aquatic habitat protection. *Biological Conservation* 105(3):345–53.

Pignati, W. A., F. A. N. Souza e Lima, S. S. de Lara, et al. 2017. Spatial distribution of pesticide use in Brazil: a strategy for Health Surveillance. *Ciência and Saúde Coletiva* 22(10):3281–93.

Pinheiro, A. P., and W. Santana. 2016. A new and endangered species of *Kingsleya* Ortmann, 1897 (Crustacea: Decapoda: Brachyura: Pseudothelphusidae) from Ceará, northeastern Brazil. *Zootaxa* 4171(2):365–72.

Pinheiro, M. A. A., C. B. M. Alves, H. Boos, et al. 2015. Conservar a fauna aquática para garantir a produção pesqueira. *Ciência e Cultura* 67(3):56–59.

Pinheiro, M. A. A., C. A. Souza, F. P. Zanotto, R. A. Torres, and C. D. S. Pereira. 2017. The crab *Ucides cordatus* (Malacostraca, Decapoda, Brachyura) and other related taxa as environmental sentinels for assessment and monitoring of tropical mangroves from South America. In *Ecotoxicology and Genotoxicology: Non-traditional Aquatic Models*, ed. M. L. Larramendi, pp. 212–41. London: Royal Society of Chemistry, Issues in Toxicology n° 33.

Ribeiro, F. B., L. Buckup, K. M. Gomes, and P. B. Araujo. 2016. Two new species of South American freshwater crayfish genus *Parastacus* Huxley, 1879 (Crustacea: Decapoda: Parastacidae). *Zootaxa* 4158(3):301–24.

Ribeiro, F. B., A. F. Huber, C. D. Schubart, and P. B. Araujo. 2017. A new species of *Parastacus* Huxley, 1879 (Crustacea, Decapoda, Parastacidae) from a swamp forest in southern Brazil. *Nauplius* 25:e2017008.

Santos, S., G. Bond-Buckup, L. Buckup, M. Pérez-Losada, M. Finley, and K. A. Crandall. 2012. Three new species of *Aegla* (Anomura) freshwater crabs from the upper Uruguay River hydrographic basin in Brazil. *Journal of Crustacean Biology* 32(4):529–40.

Santos, S., G. Bond-Buckup, A. S. Gonçalves, M. L. Bartholomei-Santos, L. Buckup, and C. G. Jara. 2017. Diversity and conservation status of *Aegla* spp. (Anomura, Aeglidae): an update. *Nauplius* 25:e2017011.

Sartori, F., and E. Vidrio. 2018. Environmental fate and ecotoxicology of paraquat: a California perspective. *Toxicological and Environmental Chemistry.* doi:10.1080/027 72248.2018.1460369

Schepis, W. R., T. V. Medeiros, D. M. S. Abessa, and S. A. Silva. 2016. Toxicidade aguda e contaminação por metais em sedimentos do Rio dos Bugres, Ilha de São Vicente, SP. *Brazilian Journal of Aquatic Science and Technology* 20(1):42–53.

Trajano, E. 2000. Cave faunas in the Atlantic tropical rain forest: Composition, ecology, and conservation. *Biotropica* 32(4b):882–93.

Veríssimo, G., A. Bast, and A. R. Weseler. 2017. Paraquat disrupts the anti-inflammatory action of cortisol in human macrophages in vitro: therapeutic implications for paraquat intoxications. *Toxicology Research* 6:232–41.

Wen, Y., G. Schoups, and N. van de Giesen. 2017. Organic pollution of rivers: Combined threats of urbanization, livestock farming and global climate change. *Scientific Reports* 7:43289.

Zimmermann, B. L., C. S. Dambros, and S. Santos. 2016. Association of microhabitat variables with the abundance and distribution of two neotropical freshwater decapods (Anomura: Brachyura). *Journal of Crustacean Biology* 36:198–204.

CHAPTER 10

Sampling and Data Analysis for Population Studies on the Life History of *Aegla* spp.

Roberto Munehisa Shimizu and Sergio Luiz de Siqueira Bueno

CONTENTS

10.1 INTRODUCTION

Methods for collecting and manipulating specimens and for analyzing the obtained data for investigations on the life history of *Aegla* species are reviewed in this chapter. Advantages and disadvantages of these procedures are discussed and recommendations for proper use are proposed.

Since many *Aegla* species are currently threatened with extinction, the section on sampling and observations encourages maintaining live individuals for examinations and returning them alive to the environment after the procedures are completed. The section on data analysis proposes an integrated approach for studies on different aspects of the life history of the species aimed at biologically sound interpretations of the gathered information.

10.2 FIELD TECHNIQUES

10.2.1 Sampling

For samples to be duly representative of the studied population, animals should be collected in a portion of the water body in which all demographic categories of interest (juveniles, adult males, and ovigerous as well as non-ovigerous females) are present. Data obtained from a preliminary survey of different sectors of the studied site will provide the necessary information for a delimitation of the collection area.

The sampling techniques that are often employed in studies on the life history of aeglids can be categorized as active or passive regarding the collector. In the former approach, a collector captures animals by dislodging specimens from their hiding place (Bueno and Bond-Buckup 2000; Noro and Buckup 2002; Colpo et al. 2005; Gonçalves et al. 2006; Silva-Gonçalves et al. 2009). This is done by removing pebbles and small rocks from the streambed while a hand net or a Surber sampler is positioned immediately downstream to catch the dislodged animals (Figure 10.1A). Dragging fine-mesh bag-shaped nets along a previously standardized area of the bottom of a river/stream has also been reported (Jara et al. 1995). Small juveniles can be actively captured by using hand nets or sieves in low-speed water areas where these individuals tend to concentrate under small rocks and accumulated plant debris (Rocha et al. 2010; Takano et al. 2016) (Figure 10.1B,C).

In the passive approach, baited traps are employed (Figure 10.1D,E). Since aeglids from epigean habitats are more active during night hours (Ayres-Peres et al. 2011), traps are typically set late in the afternoon and checked for captured animals the following morning (Bueno et al. 2007, 2014; Rocha et al. 2010; Cohen et al. 2011; Dalosto et al. 2014; Trevisan and Santos 2014; Copatti et al. 2016a; Takano et al. 2016). A variant of this technique does not involve the use of traps but baits only (e.g., pieces of meat), which are randomly laid under the water between pebbles in areas of calm water to attract aeglid specimens, which are then captured manually (Rodrigues and Hebling 1978). Whenever abundance estimations are required, both active and passive procedures can be standardized by units of effort (e.g., numbers of nets per unit of time or traps per unit of area).

Collections employing nets and traps differ in selectiveness in relation to the size of the captured individuals. The active technique provides samples composed of juveniles (including smaller individuals that may show no clear development of primary sexual traits, thus hindering any possible sex identification whatsoever), young adults and, in lower proportions, large adults (Dalosto et al. 2014; Trevisan and Santos 2014; Copatti et al. 2016b). On the other hand, traps are more efficient in collecting larger juveniles and adults, since smaller juveniles are much less frequently attracted and captured. In spite of this higher selectiveness, animals captured

Figure 10.1   Procedures for collecting and handling *Aegla* specimens during fieldwork. (A) Active sampling technique: one collector removes rocks and pebbles to dislodge animals into the water current to be captured by a hand net held by another collector in position immediately downstream. (B) Active sampling technique: use a sieve to scoop up plant debris accumulated on the streambed. (C) Three aeglid juveniles (*Aegla jaragua*) collected with the sieve. (D) Passive sampling technique: a plastic trap set in the stream. Large rocks are placed on top of the trap lid to keep the trap in position on the streambed. (E) Sample of adult aeglids captured by a plastic trap (trap lid has been removed and canister containing bait is not shown). (F) Three creels (arrows) kept partially immersed in water containing live aeglids, which have been captured by traps. (G) A collection of live aeglids (*Aegla schmitti*) kept in a tray filled with stream water and containing scattered tree leaves to provide shelter for the animals. (H) An ovigerous female of *Aegla jaragua* prepared for egg counting: the specimen is rendered immobile ventral-side up in a large excavated glass slide, with the pleon unfolded and held still by a rubber band and completely immersed in freshwater. Bar: 3 mm.

by traps are less susceptible to injuries since they are not forced out of their shelters nor trod on accidentally by the collector. Considering the complementary nature of data obtained with active and passive techniques, Copatti et al. (2016a) proposed the use of both techniques in population samplings. It should be noted, however, that studies based on frequency distribution in size classes in which the relative number of individuals in the classes is relevant, data obtained by each sampling technique should be analyzed separately, since the procedures are not standardized. In such a situation, pooling the data might provide biased information on the actual relative proportion among demographic categories of interest.

Traps also tend to capture more males than females. Overall sex ratios skewed towards males occur whenever this sampling technique is employed, including cases in which nets are used in combination (Cohen et al. 2011; Trevisan and Santos 2014; Copatti et al. 2016a). Conversely, 1:1 ratios have been observed when samples were collected actively [see data compilation in Table 1 in Cohen et al. (2011); Trevisan and Santos (2011); Oliveira and Santos (2011); Bueno et al. (2014); Copatti et al. (2015); Chiquetto-Machado et al. (2016); Takano et al. (2016); Marçal et al. (2018) for updates]. This has been demonstrated in studies where both techniques were employed and the results were compared (Trevisan and Santos 2014; Copatti et al. 2016a). While the results obtained with traps might be biased in terms of descriptions of population structure, they provide indirect evidence of differences in behavioral patterns between males and females. Overall sex ratios skewed towards males have been attributed to lower female activity during the period of the year when they are incubating eggs and are, thus, less likely enter in traps (see discussion in Cohen et al. 2011). Differential intensity of locomotory behavior between sexes during reproductive and non-reproductive periods can be detected by calculating the sex ratio for each sampling date and testing for significant departures from the expected 1:1 sex ratio each time and analyzing the results as a temporal sequence (Cohen et al. 2011). An alternate interpretation for male predominance in samples collected by traps is that females (and juveniles) move less frequently into traps in which larger males are present to avoid agonistic interactions (Trevisan and Santos 2014; Copatti et al. 2016a).

10.2.2 Handling of Individuals and Data Collection

Since more than half of the currently described species of *Aegla* are regarded as being in danger of extinction (Bueno et al. 2016a, b; Santos et al. 2017), it is highly recommended that field data should be obtained from live individuals as quickly as possible after collection and that the animals should be returned to the environment immediately after all observations and measurements are completed. Even though aeglids are tough animals, some care is required to minimize harm during manipulation. Live specimens should be kept underwater and inside a holding container, such as plastic mesh creels (Figure 10.1F) to prevent unnecessary stress until they are required for examination. When preparing the animals for observations, they should be kept in trays containing water from their environment and placed in a shady area. Several fallen tree leaves readily available in the working area should be added to the trays to provide shelter for the animals (Figure 10.1G). They actively hide underneath the leaves and tend to remain in resting position, minimizing excessive walking activity and chances of agonistic confrontations.

Measurements of body parts should be done as quickly as possible to minimize the time animals are exposed to air. Although their branchial chamber can retain moisture for a long period of time, the areas of attachment of their walking legs and chelipeds to the body become increasingly fragile, making appendages likely to come off quite easily with time due to desiccation.

As in the case of most decapod groups, carapace length is the standard measurement adopted as representative of body size in studies on *Aegla* species. How this measurement is taken, however, varies among studies depending on the criteria of including or excluding the rostrum (from the tip of rostrum or from the orbital sinus to the mid-posterior border, respectively). Although the latter criterion avoids biased data due to an injured rostrum (or loss of information in cases in which individuals bearing an injured rostrum are simply not measured), the former criterion has been used more frequently. To make the results comparable, a reasonable procedure is to take both measures and provide a linear regression equation to allow conversion from one variable to the other (Bueno and Shimizu 2008, 2009; Rocha et al. 2010; Cohen et al. 2011; Chiquetto-Machado et al. 2016; Takano et al. 2016).

Observations of eggs attached to the pleopods of females in order to determine the stage of embryonic development and counting of eggs for fecundity analysis are carried out under dissecting scopes and take longer than measurements. An example of this procedure is provided by Bueno and Shimizu (2008), in which the movements of an ovigerous female maintained ventral-side up are gently restrained on a large excavated glass slide, with the aid of a double-twisted rubber band pressing down the cephalothoracic region and the tail fan to forcibly keep the pleon extended for observation (Figure 10.1H). The female is then placed in a smaller tray and maintained completely immersed in water taken from the stream and observed under the dissecting scope throughout all procedures.

Since each aeglid female produces few large eggs (Bahamonde and López 1961; Jara 1980; Noro and Buckup 2002; Lizardo-Daudt and Bond-Buckup 2003; Gonçalves et al. 2006; Bueno and Shimizu 2008; da Silva et al. 2016), counting them without detaching them from the pleopods is recommended to minimize the loss of offspring. This is carried out with females prepared as described in the previous paragraph by gently raising each pleopod at a time with the aid of a blunt stylet and counting the set of attached eggs thereto, preferably more than once and considering the rounded-up average of the counts (Bueno and Shimizu 2008). Whenever possible, the examined aeglids should be returned to the stream, preferably in the same area from where they were collected. When releasing the animals, the container should be slowly tilted near the water surface to transfer them back to the stream as gently as possible.

10.3 DATA ANALYSIS

10.3.1 Temporal Size Structure

Obtaining a temporal sequence of frequency distribution in size classes and building the corresponding histograms is a simple but important procedure that helps to attain an overview of the life history of the studied species. This is often the first step in analyzing and describing the life history pattern of a population.

Polymodal distributions of data collected at one-to-three-month intervals afford recognition of different cohorts (age groups) as groups of frequencies associated to each of the various modes of the distributions (Figure 10.2B). Cohorts, in turn, provide clues on the occurrence of recruitment (frequency increase in the smaller size classes), individual growth (temporal progression of the modes or means), and longevity (disappearance of the largest size classes from samples). Each of these aspects can be assessed in detail when analyzed jointly with information on the reproductive cycle and ontogenetic development of the species.

For a clear and biologically adequate illustration of the size structure of the analyzed samples, it is important to devote some time to adjusting the size-class interval to the trends shown by the data set. Excessively narrow classes emphasize random frequency variations making recognition of age groups difficult or, in extreme cases, leading to a separation of more cohorts than there actually are (Figure 10.2A). Conversely, classes that are too wide hinder the correct separation of cohorts and obscure temporal shifts of the mode/mean corresponding to each age group. As a rule, the slower the growth of the individuals, the narrower the class interval should be (Figure 10.2C,D).

Adoption of different size-class intervals for males and females of a single species is sometimes required. In the case of *Aegla perobae* and *A. schmitti* (Bueno et al. 2014; Chiquetto-Machado et al. 2016) the frequencies of carapace length of females were distributed in classes of 0.5 mm intervals because they grew much slower than males (frequencies distributed in 1.0 mm class intervals). Although somewhat time consuming, the most appropriate size-class interval for each case should be found by trial and error, changing the class interval, for example, by 0.5 mm each time. Data on the reproductive cycle of the population, particularly the records of one or more breeding peaks during the year, are helpful in determining the proper number of cohorts present in a yearly sequence of samples and, thus, the adequate width of size-class intervals. Since a peak of ovigerous females precedes an increase in the frequencies of classes corresponding to recruited juveniles in virtually all *Aegla* species studied so far, the temporal progression of this group of frequencies serves as a reference for the recognition of the number of cohorts in each distribution in the sequence (Figure 10.3).

Pooling data obtained in two or three consecutive samplings is often required when sample sizes are small. Although this is a common practice, some care should be taken when analyzing data grouped this way, since temporal variation in size structure might be obscured, especially in fast-growing individuals that show considerable size variation from one sampling date to the next. When growth is slower, the effect of pooling becomes less evident.

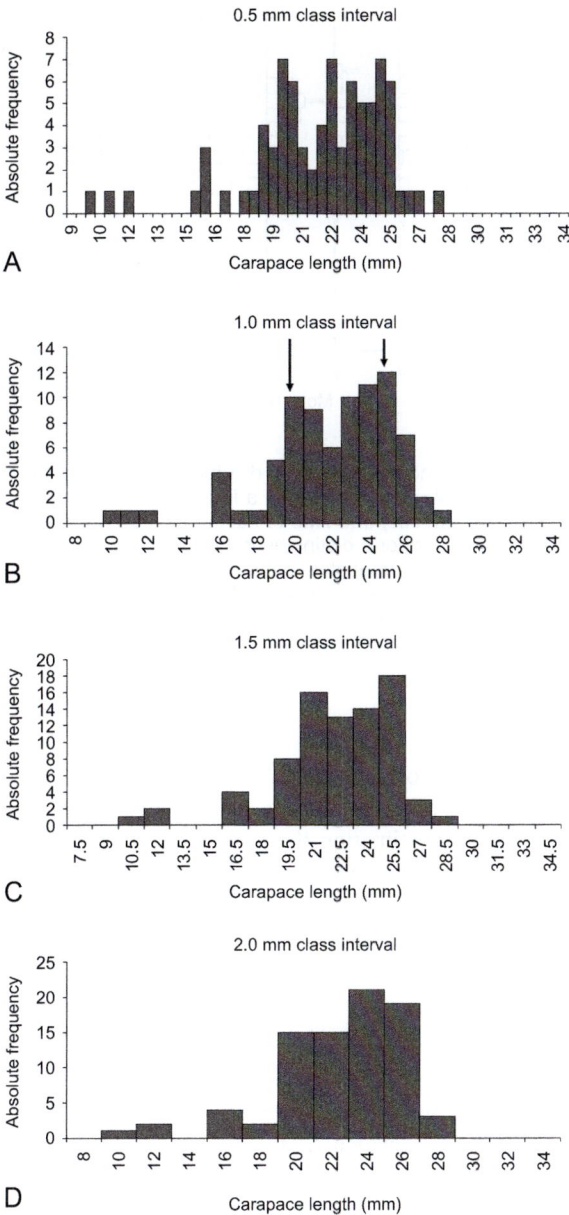

Figure 10.2 *Aegla schmitti*. Frequency distribution in classes of carapace length with different intervals of data collected in February 2011. According to the reproductive period of the species and the pattern shown by the temporal sequence of histograms for the entire study period, there should be two adult cohorts in this sample. An incipient cohort of juveniles (10–12 mm of carapace length) generated in the previous reproductive period is also discernible. The clearest distinction of these cohorts (arrows indicating modes) was obtained when the 1.0 mm class interval was employed (see Chiquetto-Machado et al. 2016 for details).

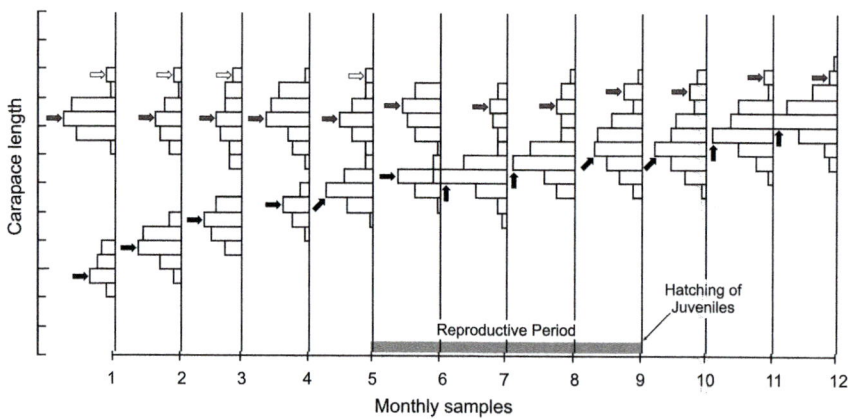

Figure 10.3 Temporal sequence of frequency distribution in classes of carapace length of a hypothetical *Aegla* species with a four-month reproductive period (presence of ovigerous females) and time of hatching of juveniles (thin arrow) annotated. Modes of three cohorts distinguished visually are indicated with thick arrows. According to the reproductive pattern, juveniles of the youngest cohort (black arrow) hatched from eggs produced in the reproductive period four months before sample 1. The intermediate-age (grey arrow) and the oldest cohort (white arrow) were 16 and 28 months old, respectively, in sample 1.

10.3.2 Individual Growth

After identification of cohorts, describing the growth of individuals with an equation relating body size to age is the next analytical procedure. The obtained growth curve is useful for estimating the age at which individuals undergo major biological events related to ontogenetic development or reproduction, or for estimating average longevity.

The most widely accepted model used for this purpose is the von Bertalanffy growth equation (von Bertalanffy 1938):

$$L_t = L_\infty \left(1 - e^{-k(t-t_0)} \right) \tag{10.1}$$

where L_t = body size at time t, L_∞ = asymptotic (theoretical maximum) body size, k = growth constant (= rate at which L_∞ is attained), and t_0 = time when body size equals zero.

The Ford–Walford method (Sparre and Venema 1998) and non-linear curve fitting have been employed for estimating the parameters of growth curves in studies with *Aegla* species. In the Ford–Walford method, the parameters L_∞ and k are calculated from the slope and the intercept of the linear function obtained by regression on [X = mean or mode of cohort size at time t, Y = mean or mode of cohort size at time $t+1$] data points (see Sparre and Venema 1998 for details). Non-linear curve fitting is based on [X = mean or mode of cohort size, Y = cohort age] data points and is performed with statistical programs that include user-defined function options in the curve-fitting routine or the Solver tool of Microsoft® Excel (for comprehensive description of procedures for this latter option, see Harris 1998; Brown 2001).

Since non-linear curve fitting is performed by iteration, it requires seed values of the parameters to start the computations. Therefore, it is desirable that these initial values are close to the converged results to save processing time or to avoid inaccurate estimates. Seed values that can be used for this purpose are the following:

- L_∞: the largest size contained in the overall data set divided by 0.95 (Bray and Pauly 1986)
- k: the mean of the values estimated for other species from the same geographic region as that of the population under study
- t_0: the length of period in which females incubate the attached eggs

While the estimation of the parameters of the growth curve is less laborious with the Ford–Walford method, non-linear curve fitting has the advantage of providing statistics that allow one to assess the adequacy of the fitted curve (r^2, standard error, and confidence interval). In addition, when both methods are applied on the same data, one can note that the Ford–Walford method tends to overestimate the growth parameter k resulting in curves with less adequate fitting than those obtained by non-linear (Figure 10.4). Thus, caution is required when comparing growth between studies that employed different methods of the estimation of curve parameters.

Computer programs that contain routines for splitting polymodal frequency distributions into unimodal components, such as the Bhattacharya routine of the FiSAT II software (Gayanillo et al. 2005) and the Mixture analysis routine of the PAST software (Hammer et al. 2001) are employed to separate cohorts and obtain corresponding means. Ages are attributed to each cohort in each sample based on the time of juvenile hatching (age zero) estimated by reproductive cycle analysis (see Section 10.3.3 and Figure 10.3 for illustration).

For a biologically appropriate cohort separation, a preliminary visual assessment of the temporal sequence of the frequency distributions (histograms) is recommended to identify the cohorts present in each sample according to a coherent temporal sequence of frequency groups based on the time of juvenile hatching. Cohorts identified this way and noted on graphs [Figure 10.3; additional examples are available in Figure 5 of Bueno et al. (2014) and in Figure 4 in Chiquetto-Machado et al. (2016)] offer a valuable reference when it comes to splitting polymodal frequency distributions into unimodal components (cohorts). In this procedure, the number of unimodal components is defined by the user and, since this is done sample by sample, having the temporal sequence of histograms with cohort annotations at hand will help to ensure that the separated distributions and the corresponding estimated statistics correspond to the pattern seen first.

Once the growth curve equation is obtained and the sizes at which individuals undergo relevant biological events are available (e.g., onset of morphometric and

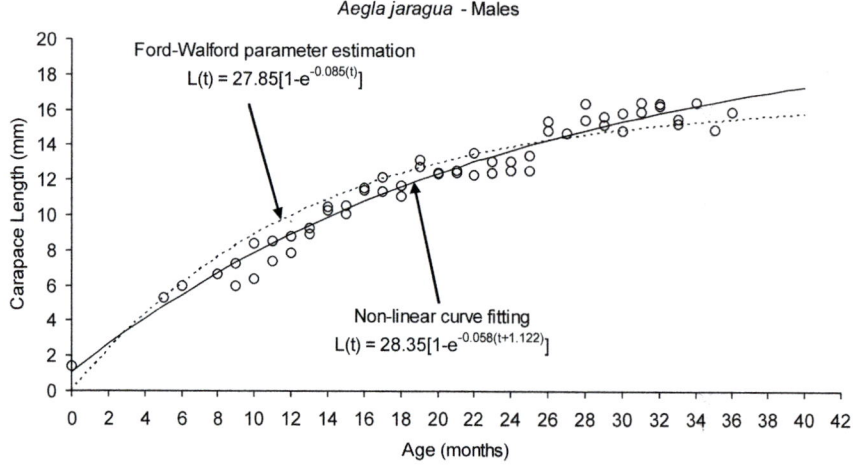

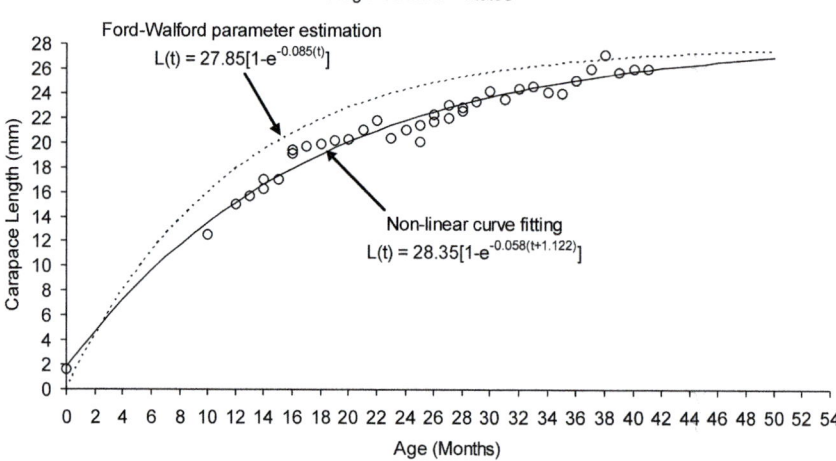

Figure 10.4   Two examples of comparison of growth curves with parameters estimated with the Ford–Walford method and with non-linear curve fitting. In both cases the growth parameter *k* was overestimated when the Ford–Walford method was employed. Data of *Aegla jaragua* (as *A. paulensis* sensu lato) are from Cohen et al. (2011) and those of *A. schmitti* from Chiquetto-Machado et al. (2016).

functional maturities, reproductive period; see Sections 10.3.3 and 10.3.4), the corresponding ages can be estimated with the inversed form of the von Bertalanffy growth equation:

$$t(L) = t_0 - \frac{1}{k}\ln\left(1 - \frac{L_t}{L_\infty}\right) \tag{10.2}$$

where the size is entered as L_t, and $t(L)$ is the estimated age. For estimation of the average longevity of individuals, the 95% percentile of the overall size data set can be used as L_t. This value is preferable to the size of the largest individual in the sample which tends to vary among study cases due to the random effect of sampling.

10.3.3 Reproductive Cycle, Fecundity, and Average Size at the Onset of Functional Maturity of Females

The description of the reproductive cycle of a given population allows an estimation of the number and duration of breeding events occurring in a year, and of the starting time of an age group. The obtained information, coupled with data on temporal size structure, is useful as a reference point for establishing the body size vs. age relationship discussed in Section 10.3.2.

While most studies on reproductive cycles of *Aegla* species are based on the presence and on the temporal variation of proportions of ovigerous females, a broader view of the breeding process is achieved when proportions of females exhibiting late ovarian development is included in the analysis (Bueno and Shimizu 2008, Figure 3; Rocha et al. 2010, Figure 2; Chiquetto-Machado et al. 2016; Takano et al. 2016, Figure 3). The late ovarian development condition can be recognized macroscopically when the posterior lobes of the bright red colored gonads, observed through the translucent exoskeleton of the ventral side of the pleon, extend to the second pleomere or further posteriorly (Bueno and Shimizu 2008). The relevance of this approach is illustrated by the interpretation of the occurrence of two adult male morphotypes, which differed in relative chela dimensions, as observed in *A. franca* and *A. castro* (Bueno and Shimizu 2009; Takano et al. 2016). Morphotype II (having larger chela than mophotype I male of similar size) was considered to be related to reproduction based on the close correlation detected between their increased abundance proportion and those of females with developed ovaries in the population as mating period draws to a close (see Section 10.3.4).

If females are sufficiently abundant in the samples (>20 individuals in each), the analysis of the reproductive cycle can be further detailed by discriminating the proportions of the stages of both ovarian development and embryonic development of eggs in each temporal sample (descriptions of stages of both ovaries and embryonic stages of the eggs are available in Bueno and Shimizu 2008; see also Chapter 6 of this book for illustrations of the embryonic stages of the eggs). With this procedure, it is possible to follow the successive changes in gonad development and in the embryonic stages of the eggs, which, in turn, provide relatively precise estimates of impending egg extrusion as well as the imminent hatching of juveniles, respectively. This latter aspect is particularly useful as a reference for attributing ages to periodically sampled cohorts in individual growth analyses (Section 10.3.2). The time of juvenile hatching is estimated as the month in which the proportion of females carrying eggs in the late embryonic development stage decreases to half (or less) of the preceding peak (Bueno et al. 2016b). When data on the embryonic development stage of eggs are not available, the same criterion is adopted for the overall proportion of ovigerous females (Cohen et al. 2011; Chiquetto-Machado et al. 2016).

The proportion of reproductive females should be calculated in relation to the number of adult females rather than the total number of females. While this might seem obvious, the latter procedure is not uncommon in the literature concerning reproduction of decapod crustaceans. Proportions of reproductive females obtained in relation to the total number of females are often distorted results because the number of immature females is included in the calculations. Immature females do not take part in the breeding processes (which alone makes the exclusion of their numbers from computations a sensible decision) and vary greatly in number during the year, with marked increases when a new cohort begins to be sampled, and decreases when they attain morphometric maturity and are computed as adults. When included in the total number of females, these quantities lead to underestimations of the proportions of reproductive females and an alteration of the temporal pattern of breeding intensity (Figure 10.5). Therefore, data on females smaller than the size at the onset of morpho-metric maturity should be excluded from calculations (see Section 10.3.4).

Fecundity is usually assessed by the number of eggs carried by females. Since it is a size-dependent variable, the reported values expressed as means or ranges should be accompanied by the corresponding body sizes (carapace lengths) of the ovigerous females. Whenever sufficient data are available, it is also appropriate to report the number of eggs vs. body size relationships described as a power (Noro and Buckup 2002; Gonçalves et al. 2006; Bueno and Shimizu 2008) or linear (Bueno et al. 2014) function equation. These procedures afford a proper comparison of results, minimiz-ing biases due to the effect of size.

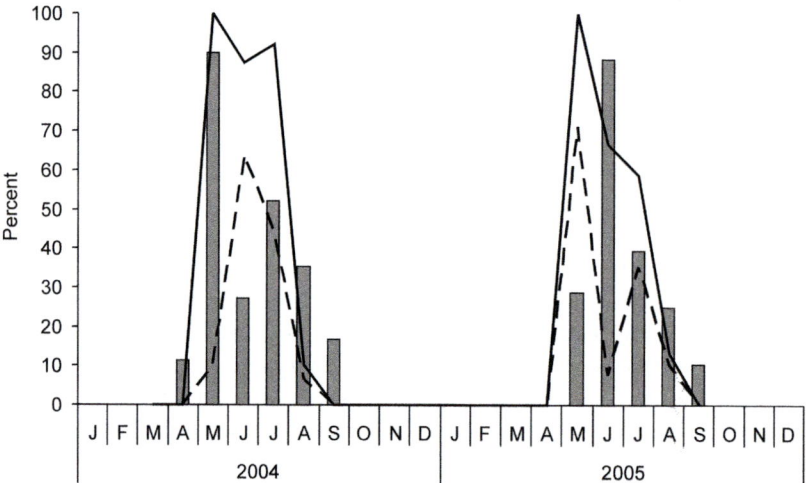

Figure 10.5 *Aegla franca.* Temporal variation of the proportion of immature females (grey bars) and ovigerous females, calculated in relation to the number of adult females only (solid line) and in relation to the total number of immature plus adult females (dashed line). Note that the inclusion of immature female data in the calculations causes underestimation of the proportions of ovigerous females and alters the temporal pattern of the reproductive periods. [Data from Bueno and Shimizu (2008)].

When analyzing fecundity, data should be preferably restricted to those obtained from eggs at the initial embryonic development stage to avoid underestimated results due to egg losses that occur with time during the egg incubation period. If the data on these eggs are insufficient for a reliable analysis, the following procedure can be adopted (see Bueno and Shimizu 2008):

- Build a scatter plot with the number of eggs vs. body size. If the number of eggs varies according to a power function as the body size increases, transform both variables with natural logarithm (ln) to make the relationship suitable for description using a linear function. If the relationship is best described by the linear function, no data transformation is needed.
- Apply linear regression on the [X=body size, Y=number of eggs] or [X=ln(body size), Y=ln(number of eggs)] data points of each embryonic development stage separately.
- Compare the resulting equations with an analysis of covariance and the post-hoc comparison test if a difference of slope or elevation is detected. Only the data that do not differ from those of eggs in the initial embryonic development stage are suitable to be added to the analysis.

Functional maturity refers to females that are capable of mating and carrying eggs (López-Greco and Rodríguez 1999). The estimation of the average size at the onset of this condition in females is also based on the recognition of the reproductive traits of the individuals. Two criteria have been adopted for this purpose in studies of *Aegla* species: (1) the size of the smallest ovigerous female in the sample (Viau et al. 2006; Trevisan and Santos 2014), and (2) the size at which 50% of females show definitive reproductive traits (ovaries in late development stages or eggs attached to the pleopods). When this latter criterion has been adopted, the average size at the onset of functional maturity has been computed by interpolation of a logistic function obtained by curve fitting on a proportion of reproductive females vs. carapace length class plot (Bueno and Shimizu 2008; Grabowski et al. 2013) or by logistic regression in which the response variable is the reproductive condition of each individual (0 = not showing reproductive traits; 1 = showing reproductive traits) and the grouping variable is body size (carapace length) (Rocha et al. 2010; Takano et al. 2016).

From the conceptual point of view, the smallest ovigerous female criterion is biologically appropriate, pointing to the minimum size at which females are involved in reproduction. This value, however, might vary among study cases due to the randomness imposed by the sampling procedures. Thus, adoption of this criterion should be restricted to cases where large samples are available. Conversely, the estimation of average size at the onset of functional maturity as the size at which 50% of females are capable to reproduce successfully provides standardized results that are readily comparable among studies. The two methods based on logistic functions produced similar results for *A. franca*: with the logistic function fit for the proportion of reproductive females vs. carapace length class, Bueno and Shimizu (2008) obtained the average size at the onset of functional maturity of 12.75 mm; performing the logistic regression on the same data set, the estimated value was 12.17 mm.

One important point to be considered when estimating the average size at the onset of functional maturity by employing methods based on logistic functions is to restrict the data included in the analysis to those obtained from samples where the proportion of reproductive females is higher than 20%. In seasonally breeding *Aegla* species, this corresponds to samples collected from the beginning of the reproductive period through the time prior to a marked decrease in the proportion of ovigerous females. This restriction minimizes the risk of obtaining overestimated sizes caused by the inclusion of large post-ovigerous females that are no longer showing reproductive traits in the analysis.

10.3.4 Allometric Growth and Average Size at the Onset of Morphometric and Functional Maturity

Allometric or relative growth analysis is based on the relationship between a body part dimension and the carapace length or width and has been extensively employed in studies on decapod populations aimed at estimating the size at the onset of morphometric maturity. A significant alteration in the relationship between these dimensions indicates the occurrence of the puberty molt during the juveniles-to-adult transition (Hartnoll 1978). Thus, the studied dimensions are mainly from body parts that are related to reproductive activity, such as chelipeds (Colpo et al. 2005), chelae (Viau et al. 2006; Bueno and Shimizu 2009; Oliveira and Santos 2011; Trevisan and Santos 2012; Copatti et al. 2016b; Takano et al. 2016; Marçal et al. 2018), and the pleon (Colpo et al. 2005; Viau et al. 2006; Oliveira and Santos 2011; Trevisan and Santos 2012; Copatti et al. 2016b; Marçal et al. 2018), the latter body dimension being adopted exclusively in analyses of females.

The body part size (analyzed variable Y) vs. body size (reference variable X) relationship is described by the allometric (power function) equation (Hartnoll 1978):

$$Y = aX^b \tag{10.3}$$

where b is the slope (= allometric growth coefficient) and a is the intercept on the Y axis. For further analyses that employ procedures based on a linear function, the straight line version of the equation

$$\ln Y = \ln a + bX \tag{10.4}$$

is employed (Hartnoll 1978). The primary procedure of allometric growth analysis is the separation of (X,Y) data points corresponding to juveniles and adults. While computer programs specifically developed for this purpose such as Mature I (Somerton 1980a), Mature II (Somerton 1980b), and Regrans (Pezutto 1993) have been employed by several authors (Colpo et al. 2005; Viau et al. 2006; Oliveira and Santos 2011; Trevisan and Santos 2012; Copatti et al. 2016b), data point separation has also been successfully performed with the non-hierarchical k-means clustering method available in several standard statistical software (Bueno and Shimizu 2009;

Takano et al. 2016; Marçal et al. 2018). This iterative method splits the total data point set into groups (number defined by the user) maximizing between-group variability and minimizing within-group variability.

This single juvenile/adult separation has been performed for males and females of several *Aegla* species. Actually, for females of the species analyzed to date, this procedure suffices, since the allometric growth associated with the adult phase in females is expressed by one single straight line on the ln$X$ vs. ln$Y$ scatter plots. However, in the case of males of species from lower latitudes, two adult morphotypes have been detected based on an increase in the variability of ln$Y$ values corresponding to high values of ln$X$ (Bueno and Shimizu 2009; Takano et al. 2016; Marçal et al. 2018). In these cases, the variability of ln$Y$ increases in larger animals of the sample, enabling a visual distinction of two groups of points on the scatter plot (Figure 10.6A,C, for lnCarapace length > 2.7). The group of points with higher ln$Y$ values corresponds to morphotype-II adults and should be separated prior to the juvenile/adult separation, with the following procedure (according to Bueno and Shimizu 2009):

- Separate all data points with ln$X$ equal to or higher than that corresponding to the beginning of the increase in variability of ln$Y$.
- Run linear regression on the separated points and obtain standardized residuals of each ln$Y$ ($= Z_{\ln Y}$) with the formula:

$$Z_{\ln Y} = \frac{R \ln Y}{S.D.\left[R \ln Y\right]} \qquad (10.5)$$

where R ln$Y$ is the residual of the ln$Y$ value (observed ln$Y$ minus the corresponding ln$Y$ estimated by the regression equation) and S.D.[Rln$Y$] is the standard deviation of residuals (square root of the residual mean square provided by the analysis of variance of the regression). The options for automatic calculations of $Z_{\ln Y}$ values are commonly available in the linear regression routines of statistical software or of electronic spreadsheet software.

- Run k-means clustering analysis on the set of (ln$X$, $Z_{\ln Y}$) data points, informing the program that the set of points is to be separated in two groups. This procedure splits the cloud of points in groups corresponding to the two morphotypes. The (ln$X$, ln$Y$) data points corresponding to the separated group with the highest $Z_{\ln Y}$ values are those of the morphotype-II adults. The remaining (ln$X$, ln$Y$) points corresponding to the group of points with the lower $Z_{\ln Y}$ values are pooled with those not included in the above-described procedure and then the k-means clustering analysis is performed again, this time on the (ln$X$, ln$Y$) data points for separating juveniles from morphotype-I adults.

Each of the separated groups of data points (two for females; two or three for males) is submitted to linear regression analysis. Significant differences in slope or elevation among the resultant straight-line equations which demonstrate alterations in the body part dimension vs. body size relationships is verified either with specific t-tests (see Zar 1996 or more recent editions for description of these procedures) or with analysis of covariance.

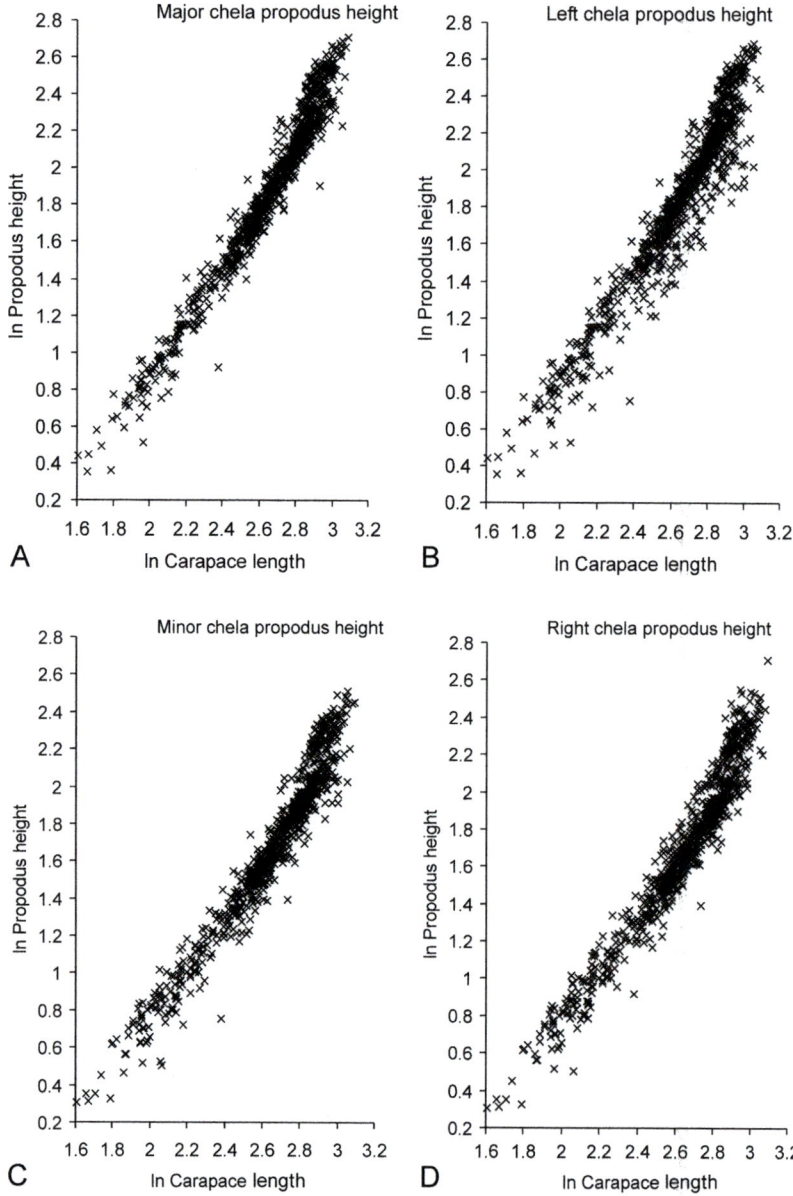

Figure 10.6 *Aegla franca*. Length (A, B) and height (C, D) of chelae plotted against carapace length (all variables transformed with natural logarithm). Scatter plots A and C were constructed with data points classified by chela size (major vs. minor) while B and D were based on points classified by the side of the chela (left vs. right). Note how the relationships are less defined and groups of points corresponding to different life history stages are less distinguishable on graphs B and C. This effect was caused by 109 out of 821 (13.3%) individuals that exhibited a major chela on the right side.

The two adult male morphotypes are sequential life history stages as shown by constructing monthly or bimonthly $\ln X$ vs. $\ln Y$ scatter plots from which gradual transition from morphotype I to morphotype II are visualized (see Bueno and Shimizu 2009, Figure 4; and Takano et al. 2016, Figure 4). It has also been shown that monthly or bimonthly values of the proportion of morphotype-II adults (calculated in relation to the total number of morphotype-I adults) are closely correlated to those of the proportion of females with ovaries in late development stages (see Bueno and Shimizu 2009, Figure 7; and Takano et al. 2016, Figure 3). This strongly indicates that morphotype II corresponds to the adult males involved in mating while the predominance of the non-reproductive morphotype I in the population is linked to periods when females show few, if any, late ovarian development.

The estimation of the average size at the onset of morphometric maturity of males and females is performed through logistic regression, considering the individual conditions of juveniles = 0 and adults (morphotype I in the case of males) = 1 against body size (carapace length). With the same procedure, the size at the onset of functional maturity in males is estimated considering morphotype I (non-reproductive) = 0 and morphotype II (reproductive) = 1 against body size, based on the distinction of the adult morphotypes as discussed above. The set of data points included in the analysis should be restricted to those obtained from samples in which the proportion of morphotype II is higher than 20% to avoid overestimated results. In cases in which several body part dimensions are employed to estimate the average size at the onset of morphometric maturity of males and females, and of functional maturity of males, the mean of the results obtained with each dimension is considered as the final estimate (Bueno and Shimizu 2009; Takano et al. 2016).

Recommendations for adequately performing the analyses described in this section are as follows:

- Classify chelae dimensions by size (as major and minor) and not by side (left and right) of chelae. Even though *Aegla* species exhibit strong handedness, with the major chela occurring predominantly on the left side (Bueno and Shimizu 2009; Trevisan and Santos 2012), a significant percentage of individuals do have a larger right claw. Classifying data points by side produces sets containing those of the predominant chela type plus some proportion of those pertaining to the other chela type. This proportion is not negligible, blurring the trends of points in scatter plots and making recognition of the groups corresponding to different life history stages difficult (Figure 10.6 B,D). Sets of data points defined this way will certainly affect separation of the different life history stages and their description by regression lines. Sets of data points classified by side (which is easier under fieldwork conditions) can be easily transformed to sets separated by major and minor chelae with the aid of the maximum and minimum functions available in electronic spreadsheet programs.
- Exclude from analyses any point corresponding to individuals with a chela showing signs of injury or regeneration, or obtained from specimens with one of the chelae missing.
- When building scatter plots, the scale intervals of both variables should be the same. Also, format the graph so that the intervals of both variables have the same width.
- When visually analyzing the ln(body part dimension) vs. ln(body size) scatter plots, identify and exclude points placed too far from the cloud of points (obvious outliers).

- Build a scatter plot corresponding to each step of recognizing and separating different groups of data points to visually monitor the adequacy of the obtained results.
- When running linear regression on each of these group of data points, it is important to verify whether outliers are present in the analyzed group of data points. This is done by obtaining standardized residuals of each $\ln Y$ $(= Z_{\ln Y})$ with formula 10.5 or selecting the option for automatic calculations available in the software. If points corresponding to $Z_{\ln Y}$ values lower than -2.57 or higher than 2.57 $(p<0.01)$ are found, they are considered outliers and should be excluded from the analysis. After this initial removal of outliers, re-run regression and inspect for outliers among the $Z_{\ln Y}$ values recalculated based on the new regression equation. If outliers are found, exclude them from the data set and repeat this procedure until only $-2.57 < Z_{\ln Y} < 2.57$ values are obtained.

10.4 DURATION OF THE STUDY PERIOD

A final consideration on methods for studies on the natural history aspects of *Aegla* species concerns the extent of time of data collection. Since the currently available information indicates that aeglids live approximately two years or more (see data compilation in Cohen et al. 2011, Table 3; and Trevisan and Santos 2011; Bueno et al. 2014; Chiquetto-Machado et al. 2016; Copatti et al. 2016b for updates), data collection for a period of more than one year (ideally, two years or more) is recommended. Information gathered with such a sampling schedule enables accurate identification and follow-up of age groups (e.g., Cohen et al. 2011; Bueno et al. 2014), the evaluation of consistency of periodical biological events that are important for the recognition of life history patterns (e.g., reproductive period) (Bueno and Shimizu 2008; Rocha et al. 2010; Bueno et al. 2014), and also the detection of changes in population features due to marked alterations in the environmental conditions (Bueno et al. 2014).

ACKNOWLEDGMENTS

We are very grateful to Angela Christine Charity for reviewing the English text and to Antonio Leão Castilho for critically reviewing the manuscript. Special thanks to Carolina Moraes Martins de Barros and to Jeniffer Kim for their kind assistance in demonstrating sampling techniques and fieldwork procedures (Figure 10.1A,B,C,F).

REFERENCES

Ayres-Peres, L., C. Coutinho, J. S. Baumart, A. S. Gonçalves, P. B. Araujo, and S. Santos. 2011. Radio-telemetry techniques in the study of displacement of freshwater anomuran. *Nauplius* 19(1):41–54.

Bahamonde, N., and M. T. López. 1961. Estudios biologicos en la población de *Aegla laevis laevis* (Latreille) de el Monte (Crustacea, Decapoda, Anomura). *Investigaciones Zoológicas Chilenas* 7:19–58.

Bray, T., and D. Pauly. 1986. Electronic length frequency analysis – a revised and expanded user's guide to ELEFAN 0, 1 and 2. *Berichte aus dem Institut für Meereskunde an der Christian-Albrechts-Universität Kiel* 149:1–76 (ICLARM Contribution, number 261).

Brown, A. M. 2001. A step-by-step guide to non-linear regression analysis of experimental data using a Microsoft Excel spreadsheet. *Computer Methods and Programs in Biomedicine* 65:191–200.

Bueno, A. A. P., and G. Bond-Buckup. 2000. Dinâmica populacional de *Aegla platensis* Schmitt (Crustacea, Decapoda, Aeglidae). *Revista brasileira de Zoologia* 17(1):43–49.

Bueno, S. L. S., S. Santos, S. S. Rocha, K. M. Gomes, E. C. Mossolin, and F. L. Mantelatto. 2016a. Avaliação dos Eglídeos (Decapoda: Aeglidae). In *Livro vermelho dos crustáceos do Brasil: Avaliação 2010–2014*, ed. M. Pinheiro, and H. Boos, pp. 35–63. Porto Alegre: Sociedade Brasileira de Carcinologia.

Bueno, S. L. S., and R. M. Shimizu. 2008. Reproductive biology and functional maturity in females of *Aegla franca* (Decapoda: Anomura: Aeglidae). *Journal of Crustacean Biology* 28:652–62.

Bueno, S. L. S., and R. M. Shimizu. 2009. Allometric growth, sexual maturity, and adult male chelae dimorphism in *Aegla franca* (Decapoda: Anomura: Aeglidae). *Journal of Crustacean Biology* 29:317–28.

Bueno, S. L. S., R. M. Shimizu, and J. C. B. Moraes. 2016b. A remarkable anomuran: The taxon *Aegla* Leach, 1820. Taxonomic remarks, distribution, biology, diversity and conservation. In *A Global Overview of the Conservation of Freshwater Decapod Crustaceans*, ed. T. Kawai, and N. Cumberlidge, pp. 23–64. Cham: Springer International Publishing.

Bueno, S. L. S., R. M. Shimizu, and S. S. Rocha. 2007. Estimating the population size of *Aegla franca* (Decapoda: Anomura: Aeglidae) by mark-recapture technique from an isolated section of Barro Preto stream, County of Claraval, state of Minas Gerais, southeastern Brazil. *Journal of Crustacean Biology* 27(4):553–59.

Bueno, S. L. S., B. F. Takano, F. P. A. Cohen, et al. 2014. Fluctuations in the population size of the highly endemic *Aegla perobae* (Decapoda: Anomura: Aeglidae) caused by a disturbance event. *Journal of Crustacean Biology* 34:165–73.

Chiquetto-Machado, P. I., L. C. M. Vieira, R. M. Shimizu, and S. L. S. Bueno. 2016. Life cycle of the freshwater anomuran *Aegla schmitti* Hobbs, 1978 (Decapoda: Anomura: Aeglidae) from southeastern Brazil. *Journal of Crustacean Biology* 36:39–45.

Cohen, F. P., B. F. Takano, R. M. Shimizu, and S. L. S. Bueno. 2011. Life cycle and population structure of *Aegla paulensis* (Decapoda: Anomura:Aeglidae). *Journal of Crustacean Biology* 31:389–95.

Colpo, K. D., L. O. Ribeiro, and S. Santos. 2005. Population biology of the freshwater anomuran *Aegla longirostri* (Aeglidae) from south Brazilian streams. *Journal of Crustacean Biology* 25:495–99.

Copatti, C. E., J. V. V. de Machado, and A. Trevisan. 2015. Morphological variation in the sexual maturity of three sympatric aeglids in a river in southern Brazil. *Journal of Crustacean Biology* 35(1):59–67.

Copatti, C. E., R. P. Legramanti, A. Trevisan, and S. Santos. 2016a. Method of capture and population structure of *Aegla georginae* Santos and Jara, 2013 (Decapoda: Anomura: Aeglidae) in a tributary of the Ibicuí River in southern Brazil. *Brazilian Journal of Biology* 7(4):1035–42.

Copatti, C. E., R. P. Legramanti, A. Trevisan, and S. Santos. 2016b. Growth, sexual maturity and sexual dimorphism of *Aegla georginae* (Decapoda: Anomura: Aeglidae) in a tributary of the Ibicuí River in southern Brazil. *Zoologia* 33(3): e20160010.

Dalosto, M. M., A. V. Palaoro, D. Oliveira, E. Samuelsson, and S. Santos. 2014. Population biology of *Aegla platensis* (Decapoda: Anomura: Aeglidae) in a tributary of the Uruguay River, state of Rio Grande do Sul, Brazil. *Zoologia* 31(3):215–22.

da Silva, A. R., M. R. Wolf, and A. L. Castilho. 2016. Reproduction, growth and longevity of the endemic South American crab *Aegla marginata* (Decapoda: Anomura: Aeglidae). *Invertebrate Reproduction & Development* 60(1):59–72.

Gayanillo, F. C., P. Sparre Jr., and D. Pauly. 2005. *FISAT II (FAO ICLARM Stock Assessment Tools)*. Food and Agriculture Organization of the United Nations (version 1.2.0), Rome.

Gonçalves, R. S., D. S. Castiglioni, and G. Bond-Buckup. 2006. Ecologia populacional de *Aegla franciscana* (Crustacea, Decapoda, Anomura) em São Francisco de Paula, RS, Brasil. *Iheringia, Série Zoologia* 96:109–14.

Grabowski, R. C., S. Santos, and A. L. Castilho. 2013. Reproductive ecology and size of sexual maturity in the anomuran crab *Aegla parana* (Decapoda: Aeglidae). *Journal of Crustacean Biology* 33:332–38.

Hammer, Ø., D. A. T. Harper, and P. D. Ryan. 2001. PAST: paleontological statistics software package for education and data analysis. *Palaeontologia Electronica* 4:9.

Harris, D. C. 1998. Nonlinear least-squares curve fitting with Microsoft Excel Solver. *Journal of Chemical Education* 75(1):119–21.

Hartnoll, R. G. 1978. The determination of relative growth in Crustacea. *Crustaceana* 34:281–93.

Jara, C. 1980. Dos nuevas especies de *Aegla* Leach (Crustacea, Decapoda, Anomura) del sistema hidrográfico del rio Valdivia. *Anales del Museo de Historia Natural* 13:255–66.

Jara, C. G., M. Cerda, and A. Palma. 1995. Distribuición geográfica de *Aegla papudo* Schmitt, 1942 (Crustacea: Decapoda: Anomura: Aeglidae) y estado de conservación de sus poblaciones. *Gayana Zoologia* 59(1):13–22.

Lizardo-Daudt, H. M., and G. Bond-Buckup. 2003. Morphological aspects of the embryonic development of *Aegla platensis* (Decapoda, Aeglidae). *Crustaceana* 76(1):13–25.

López-Greco, L. S., and E. M. Rodríguez. 1999. Size at the onset of sexual maturity in *Chasmagnathus granulatus* Dana, 1851 (Grapsidae: Sesarminae): a critical overall view about the usual criteria for its determination. In *Crustaceans and the biodiversity crisis. Proceedings of the Fourth International Crustacean Congress. Vol. 1. Amsterdam, The Netherlands*, eds. F. R. Schram, and J. C. von Vaupel Klein, pp. 675–89. Leiden: Brill.

Marçal, I. C., L. M. Ioshimura, J. J. S. Rosa, and G. M. Teixeira. 2018. Population structure and sexual maturity of *Aegla castro* (Decapoda, Anomura), an endemic freshwater crab from Brazil. *Invertebrate Reproduction and Development* 62(1):35–42.

Noro, C. K., and L. Buckup. 2002. Biologia reprodutiva e ecologia de *Aegla leptodactyla* Buckup and Rossi (Crustacea, Anomura, Aeglidae). *Revista Brasileira de Zoologia* 19:1063–74.

Oliveira, D., and S. Santos. 2011. Maturidade sexual morfológica de *Aegla platensis* (Crustacea, Decapoda, Anomura) no Lajeado Bonito, norte do estado do Rio Grande do Sul, Brasil. *Iheringia, Série Zoologia* 101(1–2):127–30.

Pezzuto, P. R. 1993. REGRANS: a "basic" program for an extensive analysis of relative growth. *Atlântica* 15:91–105.

Rocha, S. S., R. M. Shimizu, and S. L. S. Bueno. 2010. Reproductive biology in females of *Aegla strinatii* (Decapoda: Anomura: Aeglidae). *Journal of Crustacean Biology* 30:589–96.

Rodrigues, W., and N. J. Hebling. 1978. Estudos biológicos em *Aegla perobae* Hebling and Rodrigues, 1977 (Decapoda, Anomura). *Revista Brasileira de Biologia* 38:383–90.

Santos, S., G. Bond-Buckup, A. S. Gonçalves, M. L. Bartholomei-Santos, L. Buckup, and C. G. Jara. 2017. Diversity and conservation status of *Aegla* spp. (Anomura, Aeglidae): an update. *Nauplius* 25:e2017011.

Silva-Gonçalves, R., G. Bond-Buckup, and L. Buckup. 2009. Crescimento de *Aegla itacolomiensis* (Crustacea, Decapoda) em um arroio da Mata Atlântica no sul do Brasil. *Iheringia, Série Zoologia* 99(4):397–402.

Somerton, D. A. 1980a. Fitting straight lines to Hiatt growth diagrams: a re-evaluation. *Journal du Conseil international pour l'Exploration de la Mer* 39:15–19.

Somerton, D. A. 1980b. A computer technique for estimating the size of sexual maturity in crabs. *Canadian Journal of Fishery and Aquatic Sciences* 37:1488–94.

Sparre, P., and S. C. Venema. 1998. *Introduction to Tropical Fish Stock Assessment Part l: Manual*. Rome: Food and Agriculture Organization of the United Nations.

Takano, B. F., F. P. Cohen, A. Fransozo, R. M. Shimizu, and S. L. S. Bueno. 2016. Allometric growth, sexual maturity and reproductive cycle of *Aegla castro* (Decapoda: Anomura: Aeglidae) from Itatinga, state of São Paulo, southeastern Brazil. *Nauplius* 24:e2016010.

Trevisan, A., and S. Santos. 2011. Crescimento de *Aegla manuinflata* (Decapoda, Anomura, Aeglidae) em ambiente natural. *Iheringia, Série Zoologia* 101(4):336–42.

Trevisan, A., and S. Santos. 2012. Morphological sexual maturity, sexual dimorphism and heterochely in *Aegla manuinflata* (Anomura). *Journal of Crustacean Biology* 32:519–27.

Trevisan, A., and S. Santos. 2014, Population dynamics of *Aegla manuinflata* Bond-Buckup and Santos 2009 (Decapoda: Aeglidae), a threatened species. *Acta Limnologica Brasiliensia* 26(2):154–62.

Viau, V. E., L. S. López-Greco, G. Bond-Buckup, and E. M. Rodríguez. 2006. Size at the onset of sexual maturity in the anomuran crab, *Aegla uruguayana* (Aeglidae). *Acta Zoologica* 87:253–64.

von Bertalanffy, L. 1938. A quantitative theory of organic growth (Inquiries on Growth Laws. II). *Human Biology* 10:181–213.

Zar, J. H. 1996. *Biostatistical Analysis*, 3rd edn. New Jersey: Prentice Hall.

Index

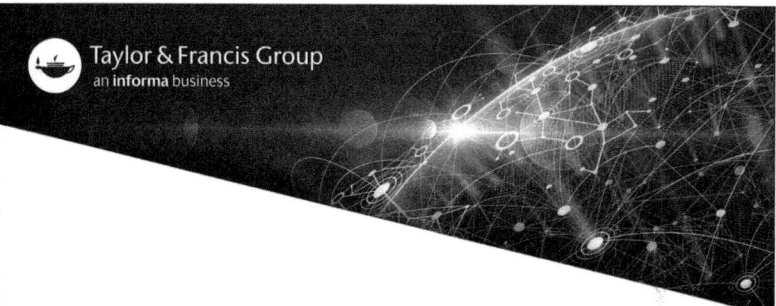